LINCOLN CHRISTIAN COLLEGE AND SEMINARY

Doubts
about
Darwin

Doubts
about
Darwin

A History of Intelligent Design

Thomas Woodward

Baker Books

A Division of Baker Book House Co
Grand Rapids, Michigan 49516

© 2003 by Thomas Woodward

Published by Baker Books
a division of Baker Book House Company
P.O. Box 6287, Grand Rapids, MI 49516-6287
www.bakerbooks.com

Second printing, December 2003

Printed in the United States of America

All rights reserved. No part of this publication may be reproduced, stored in a retrieval system, or transmitted in any form or by any means—for example, electronic, photocopy, recording—without the prior written permission of the publisher. The only exception is brief quotations in printed reviews.

Library of Congress Cataloging-in-Publication Data
Woodward, Thomas, 1950–
    Doubts about Darwin : a history of intelligent design / Thomas Woodward.
        p.   cm.
    Originally presented as the author's thesis (doctoral)—University of South Florida, 2001. Includes bibliographical references and index.
        ISBN 0-8010-6443-0 (pbk.)
    1. Evolution. 2. Darwin, Charles, 1809–1882. 3. Intelligent design (Teleology)
I. Title.
B818.W76 2003
576.8′2—dc21   2003004172

Contents

108150

Foreword

The meaning of the term "rhetoric" has been distorted in our time by some unfortunate associations. When we dismiss some politician's speech as "mere rhetoric," we mean that it contains nothing but bombast. People today need to be reminded that rhetoric is actually a noble art which has been the subject of serious study since the time of Aristotle and before. Put simply, rhetoric is the art of framing an argument so that it can be appreciated by an audience, even one which is relatively uneducated in the subject or predisposed not to appreciate it. My favorite example is that of a lawyer who has to persuade a biased jury to take seriously some evidence or argument that they do not at first understand or would much rather dismiss on the basis of prejudice. I faced a classic rhetorical problem involving the perennial elements of *ethos, pathos,* and *logos* when, in an incident described in these pages, I debated the famous Darwinian authority Stephen Jay Gould before a select academic audience while I was yet a relative novice in the subject.

It was just as difficult to persuade reluctant Christian audiences that they needed to make the effort to understand the case I made in *Darwin on Trial,* when it would have been so much easier for them to relax and continue relying upon the officially sanctioned experts. Later, when such colleagues as the biochemist Michael Behe and the mathematician William Dembski came forward to take leading roles in the Intelligent Design Movement, our problem was how to persuade our audiences and the media that we were taking a legitimate scientific position, so we could avoid being buried under the stigmatizing label "religion."

Tom Woodward knows as much about the history of Intelligent Design as any person could, and here he combines that history with a scholar's appreciation of rhetorical theory. The happy result of this combination is a fine book that tells an exciting story accurately. I am sure readers will enjoy the story and finish it with an appreciation of the difficulties we in that movement have faced and how we have overcome them.

Phillip E. Johnson

Preface

Most Americans felt anxiety after the shocks of the 9-11 terrorist assault against the United States, but I experienced a different kind of apprehension during those days. Eight days after the attacks, I was scheduled to appear before my doctoral committee on the campus of the University of South Florida in Tampa, Florida, to sweat through one of those timeless rituals of academia—a dissertation defense. It had been three years since I started writing, and now the committee members had in front of them my 345-page history of a controversial new movement in science called "Intelligent Design" (sometimes called "ID" or "Design" for short). My chosen title, "Aroused from Dogmatic Slumber," referred to an astonishing pattern of scientific conversion stories that I had woven together. As I prepared for my doctoral Day of Reckoning, I was reassured by a simple fact: The four professors on my committee had already read and commented extensively on earlier drafts. I had dutifully incorporated their corrections and responded to their questions.

Two professors played a special role during this three-year grind. At the outset my committee chair David Payne had challenged me to tackle the broad sweep of a "movement study." My original idea was to focus simply on the rhetoric of ID theorist Michael Behe, but Payne pointed out that the work of Behe was so interwoven with the situational strands of other Design figures such as Phillip Johnson and Michael Denton, it would be an excellent idea to survey the *rhetorical history* of the entire movement. I agreed wholeheartedly and set to work under Professor Payne's watchful eye.

After seemingly endless months of reading, interviewing, writing, and polishing, I finally had a manuscript to defend. Along the way, Gerald Robinson, professor of biology at the University of South Florida, played a crucial role, carefully reading each draft and verifying my scientific accuracy. His numerous suggestions enhanced the flow and clarity of my narrative. By the time I appeared at my defense in September 2001, two other professors, Carol

Jablonski and Mark Neumann, also had helped me strengthen the manuscript. I am greatly indebted to all four for their pivotal assistance.

During the two-hour defense in a sweltering classroom, all went well and I found myself actually *enjoying* the session. As expected, the committee asked tough, searching questions. My answers were apparently satisfactory, because the committee emerged from their deliberations with an approval, on the condition that I edit a flabby footnote.

A few months later, as I rewrote "Aroused from Dogmatic Slumber" for Baker Book House, I rephrased the more abstruse technical jargon and strove to explain communication ideas and theories clearly. Circumlocutions were snipped; the flow was smoothed; repetitions were deleted. Where possible, I incorporated the latest developments in the unfolding drama. In short, everything was done to make the book "reader friendly."

Launching into *Doubts about Darwin*, the reader eventually will sense that I have a close relationship with Intelligent Design. This will become especially clear for those who peruse the endnotes, where additional background information is laid out. As a friendly participant in some parts of the story, I was able to attend a few early meetings of the proto-ID group called the "Ad Hoc Origins Committee." I have known Phillip Johnson since 1989, and Michael Behe and William Dembski since 1992 when we met at the Darwinism Symposium at Southern Methodist University in Dallas. Because of these connections, skeptical questions naturally arise: How can this history be a balanced, objective account? Can it possibly attempt to be *above the fray?* Is it not inevitable that this book itself will function (or at least be seen) as another piece of advocacy, another rhetorical artifact, on behalf of Design? These are good questions and were raised by my committee during the defense.

Let me make two points. First, it is inevitable that any writer of history will come to some point of view about the incidents he or she is describing. As a theist, it is understandable that I am sympathetic to the theories and thought of ID. I would like to point out, however, that this connection between theism and sympathy for ID should not be taken as a foregone conclusion. I know many evangelical Christians who are strong theistic evolutionists and are largely opposed to the research perspective of Design. In any case, the point of telling history is not to avoid taking a position but to make a conscious effort to maintain as much objectivity and balance as possible—*and to substantiate one's viewpoint clearly, with compelling evidence and analysis.* I'm well aware that any "straw man" I might erect eventually will be knocked flat. Thus, I do not pretend to be without a point of view, and my hope is that, far from nullifying my work, this fact will imbue the narrative with a certain intellectual energy.

My second point is that to a certain degree the story of Design tells itself. By this I mean that in the course of hearing how key Design advocates came to their current view, it becomes clear that their entry into the movement stemmed from intellectual or scientific—not religious—reasons. Thus, my own bias becomes somewhat irrelevant. As I point out throughout the book and sum up in my concluding chapter, these personal stories are factually *recalcitrant:* they are unyieldingly "true." This leads naturally to the conclusion that a pro-ID perspective is, to some extent at least, rationally justified. In the last few years, a small stream of anti-ID books have appeared, penned by Niles Eldredge, Robert Pennock, Kenneth Miller, Michael Ruse, and others. At times, especially in the hands of Miller, Pennock, and Eldredge, these attacks have portrayed Design in a way that points to "religious motives" and even "fear" as crucial motivations. In my view, this picture is a severe distortion, and my book is designed to controvert and correct this caricature.

In fact, the attacks on ID are now becoming so numerous and vehement that probably an entire second book could be written just to cover this intense phase of the Design story. Do I sense a sequel in the wings? For this book I chose to concentrate on the founders and current leaders of Design, bringing in criticisms that were closely tied to the main books of each figure.

I want to acknowledge many wonderful people without whose help I never could have brought this book to completion. First, the patient encouragement and support of my wife, Normandy, were the foundation of all my work. Chad Allen, my editor at Baker Books, played the leading role in the final reshaping of the manuscript; I am indebted to him for unflagging patience and wise guidance at every turn. My brilliant cousin, Nancy McCabe, was a sharp-eyed editor of the entire manuscript, and two leaders of Trinity College, Drs. Bill Lanpher and James Lanpher, read each chapter and gave helpful feedback. Without the teaching and inspiration of my many professors in the fields of rhetoric and narrative theory, I could not have even contemplated this project. Among these superlative mentors were Eric Eisenberg, Art Bochner, Carolyn Ellis, Marsha Vanderford, Elizabeth Bell, and John Angus Campbell. And finally, a special thanks goes to those who helped me relive the past as I interviewed them and corresponded with them: Phillip Johnson, David Raup, Charles Thaxton, Murray Eden, Mike Woodruff, Charles Haynes, Michael Smith, William Dembski, and Michael Behe. I am also grateful to the dozens of Design activists who responded to my surveys and the countless others who helped with the accuracy of the manuscript.

Finally, I wish to dedicate the book to Dr. and Mrs. Richard Williams of Tampa, Florida, without whose unflagging encouragement this entire project never would have been completed.

1

Aroused from Dogmatic Slumber

An Introduction to Intelligent Design and a Paradigm Crisis

Although teaching evolution in public schools has been controversial, it is rare for the controversy to surface in presidential politics. One of those rare occasions was 11 August 1999, when the Kansas Board of Education voted to deemphasize the teaching of biological "macroevolution," or large-scale morphological change over time, in the public schools throughout the state. The board's decision, which narrowly passed (six to four), avoided any hint of a ban on the teaching of Darwin's view of origins. Rather, the decision was to leave it to local school boards to decide how to structure their biology curriculum and also whether and how much each district would teach on macroevolution. The board mandated the continued teaching of the more limited process of "microevolution"—the development of sister species within a given type. Still, the news story was clear and shocking: the Darwinian notion of common ancestry was deemed sufficiently problematic in terms of evidentiary support to merit its demotion from "mandated inclusion" in the science curriculum of Kansas high schools.[1]

All three Republican candidates who commented on the decision referred to it as defensible, perhaps even positive, although Elizabeth Dole essentially ducked the issue: "I am a person of strong faith. . . . I'm going to leave that to the states." Even Vice President Gore released a statement that initially signaled support for local control on how evolution was to be taught in public schools, though he modified his stand on the issue the next day to reemphasize the legal strictures that determine how creation may be mentioned in public schools. Only Bill Bradley forthrightly criticized the Kansas decision.

Reactions to the Kansas Board vote within editorial pages throughout the United States and many foreign countries were opposite those of the politicians. Recoiling with horror, the writers poured forth a cascade of condemnatory editorials and guest columns. Some attacks on the Kansas Board were shrill and frantic. John Rennie, editor in chief of *Scientific American,* went so far as to propose that college admissions officers write the Kansas Board, warning them that high school applicants from Kansas will need to have their qualifications scrutinized with extra care from now on "in light of the newly lowered education standards."[2]

In one of the more restrained attacks, *Boston Globe* columnist Ellen Goodman wrote, "Removing evolution from a required science curriculum is a bit like removing verbs from the English curriculum." She wove into her column a brief refresher in biology, asserting that there is "no serious scientific dispute about the fact of evolution. It's supported by anatomy, fossils, carbon-dating, genetic evidence, the ages of rocks if not the rock of ages." Referring more than once in her article to the fundamentalist agendas of Genesis literalists, Goodman invoked images of the Scopes "Monkey Trial."[3]

However, in the midst of editorial denunciation, two important pieces of dissent appeared in prominent newspapers: "The Church of Darwin" by Berkeley law professor Phillip E. Johnson in the *Wall Street Journal* and "Teach Evolution—and Ask Hard Questions" by Lehigh biochemist Michael Behe in the *New York Times.* These opinion pieces argued that there are legitimate concerns about the dogmatic teaching of evolution that motivated the Kansas Board. Johnson and Behe proposed a rhetorically clever but scientifically heretical solution: *Instead of teaching less evolution, schools should teach "far more about evolution."* That is, schools should continue teaching about macroevolution, but they should add a crucial new segment to such teaching—the legitimate scientific controversy over Darwinism.

In his *Wall Street Journal* piece, Johnson wrote, "So one reason the science educators panic at the first sign of public rebellion is that they fear exposure of the implicit religious content in what they are teaching. An even more compelling reason for keeping the lid on public discussion is that the official neo-Darwinian theory is having serious trouble with the evidence." Johnson proceeded to cite a litany of evidentiary problems for the theory.[4]

Likewise, Behe's opinion column advised Kansas to "teach Darwin's elegant theory" but not stop there. He added, "Discuss where it also has real problems accounting for the data, where data are severely limited, where scientists might be engaged in wishful thinking, and where alternative—even 'heretical'—explanations are possible."[5]

These closing words of Behe's column serve as the symbolic quintessence of the deepening rhetorical challenge which evolutionists now face. When presidential candidates sided with the Kansas Board, critics could easily attribute their words to ignorance, a failure of nerve, or careful posturing for political advantage. Yet it was a much more difficult task to neutralize the complex skepticism of macroevolution expressed by Behe, a biochemist in good standing at a major research university.

This was not the first time he had diffused such "heresy." Three years earlier Behe began to establish his own journalistic rhetoric of anti-Darwinism when he argued in the *New York Times* that something called "intelligent design theory" does a far better job of explaining the origin of complex biochemical machines within the cell than does Darwinian theory ("Darwin Under the Microscope," 29 October 1996). Indeed, the August 1996 publication of Behe's book on molecular complexity, *Darwin's Black Box,* with its coinage of the now famous phrase "irreducible complexity," was the first time sophisticated skepticism of naturalistic evolution was brought to center stage in American society.[6]

Behe's doubts about Darwin, which differ in many ways from traditional creationism, have energized an entire network of dissidents in academia who share his skepticism. As a result, a flank of professors, known after 1996 as the Intelligent Design Movement, have organized themselves under the leadership of Behe, Johnson, and others for the purpose of scientific persuasion in the universities.[7] Their stream of persuasion spilled over into new channels of political action, including efforts to modify the policies and curricula of classroom instruction in evolution. An excellent example of their spillover effect is Alabama's mandated inclusion of a statewide "disclaimer" on evolution that, starting in 1996, has been pasted in the front of every biology textbook in the state's public high schools. The adoption of the disclaimer was led by Norris Anderson, an educator who worked closely with Johnson and received advice and assistance from many of Johnson's scientific colleagues.

These "sophisticated skeptics" of Darwinian evolution have not gone unnoticed by the theory's defenders. Dr. Eugenie Scott, a leading consultant and spokesperson for the evolutionist side in educational issues, began devoting her energies in 1987 to tracking all types of creationist or antievolutionist movements. As director of the National Center for Science Education (NCSE), a proevolution watchdog group in Berkeley, California, she has led the effort to repel attempts to dilute the teaching of evolution (or insert creationist teachings) in U.S. schools. Just months before the Kansas Board vote, she conveyed in her NCSE fundraising letter a sense of growing frustration:

> [T]he "pincer movement" I spoke of is squeezing us not just from the grassroots direction, but from the "top down" as well—thanks to the rise of university-based antievolutionism in the form of "intelligent design theory" and other well-camouflaged varieties of creationism. University of California lawyer and neo-creationist guru Phillip E. Johnson boasts of his "wedge" strategy. . . .
>
> But when a topic has religious implications, such as the Big Bang, the origin of life, or the origin of human beings, they don't want us to look for natural

explanations. They claim these phenomena are "too complex"; they claim that they couldn't have happened "by chance" . . . ; they claim that "intelligence" must have been involved. "Intelligence," of course, means divine creation, a subject outside of science.

The new rhetorical situation, described by Scott as a "pincer movement," is no brief anomaly. It is now being generally acknowledged as a major and fundamental shift in the landscape of the creation-evolution debate.

In her letter Scott notes the new terms of the old debate, especially "intelligence" taking the place of "divine creation" as the causal agency. She implies that the "inference to design"—a connection between intelligence and extreme complexity—is an argument that is spreading and winning converts in American society. Of course, Scott denies the validity of this inference, and in her rhetorical analysis of the key term "intelligence" she raises an important question: What is the relationship between this alleged "intelligence" and "divine cre-

The Alabama Textbook Disclaimer

A Message from the Alabama State Board of Education:

This textbook discusses evolution, a controversial theory some scientists present as a scientific explanation for the origin of living things, such as plants, animals, and humans.

No one was present when life first appeared on earth. Therefore, any statement about life's origins should be considered as theory, not fact.

The word "evolution" may refer to many types of change. Evolution describes changes that occur within a species. (White moths, for example, may "evolve" into grey moths.) This process is microevolution, which can be observed and described as fact. Evolution may also refer to the change of one living thing to another, such as reptiles into birds. This process, called macroevolution, has never been observed and should be considered a theory. Evolution also refers to the unproven belief that random, undirected forces produced a world of living things.

There are many unanswered questions about the origin of life which are not mentioned in your textbook, including:

Why did the major groups of animals suddenly appear in the fossil record (known as the "Cambrian Explosion")?

Why have no new major groups of living things appeared in the fossil record for a long time?

Why do major groups of plants and animals have no transitional forms in the fossil record?

How did you and all living things come to possess such a complete and complex set of "Instructions" for building a living body?

Study hard and keep an open mind. Some day you may contribute to the theory of how living things appeared on earth.

ation"? Implied here is Scott's central rhetorical question: If divine creation is what the skeptics of Darwinism have in view when they speak of "intelligence," how can this theory be called scientific at all, since divine creation is a "subject outside of science"? And of course, being a key rhetorical question, it invites its own rhetorical as well as philosophical analysis. For example: What are the defining traits that mark off science from other endeavors, and how are they known? Who decides what questions may be posed and what explanations may be considered within the scientific arena?

As the world begins to tune in to this public conversation, it is learning about a unique turn in the road that has occurred during the last two decades. The perennial cultural struggle over Darwinism—tugged and shaped through a web of social, ideological, political, and educational forces in the United States—has produced a "rhetorical crisis" for the evolutionary worldview and the scientific establishment that embraces it and to some extent depends upon it for its quotidian work.

A Controversy That Won't Go Away

One front of the debate—the Kansas School Board—has fallen quiet, as conservative Republican members of the board who led the move to change policy were voted out in the August 2000 primary election. As expected, in February 2001 the new board reinstated the mandate to teach macroevolution.[8] Nevertheless, neither side of the intelligent design debate views the rhetorical battle over Darwinism as receding. Rather, it seems to be steadily mounting, as seen in many recent developments. I'll cite two.

In early November 2000 Eugenie Scott delivered a guest lecture in an undergraduate course on intelligent design at the University of California at San Diego. In her introduction she held up a copy of the newly released *The Icons of Evolution* by Jonathan Wells, a Berkeley-trained biochemist and key member of the Intelligent Design Movement.[9] *Icons* selects ten of the most popular "visual proofs" of evolution, displayed in virtually every textbook used in high school biology classes. Some of these (like the peppered moth) combine classic photographs with a famous story; others are sketches or diagrams (like the "tree of common ancestry") that proclaim an evolutionary narrative. Wells scrutinizes the explanatory prose surrounding each of these ten icons and subjects it to withering criticism, pointing out significant omissions and inaccuracies, some of which border on fraud. Scott's warning to the class about *Icons* was blunt: "This book will be a royal pain in the fanny" for those who are dutifully teaching Darwinian evolution in high schools.[10] In fact, reviews of Wells by prominent evolutionists have betrayed similar signs of foreboding about rising public hostility toward the teaching of macroevolution as "proven fact" and about the potential of Wells's book to aggravate that situation.[11]

A second recognition of the growing momentum of Intelligent Design's challenge to Darwinism was a pair of front-page stories that appeared on the new movement in March and April 2001 in the *Los Angeles Times* and the *New York Times,* respectively. They constitute the first important media coverage of the Design Movement. Both articles were evenhanded, presenting the movement in a generally positive light, while quoting some strong criticism from evolutionists. The articles enhanced the scientific legitimacy of Intelligent Design by contrasting it with the older genre of biblical creationism. A good example of this comes from *New York Times* science writer James Glanz:

> This time, though, the evolutionists find themselves arrayed not against traditional creationism, with its roots in biblical literalism, but against a more sophisticated idea: the intelligent design theory. Proponents of this theory, led by a group of academics and intellectuals and including some biblical creationists, accept that the earth is billions of years old, not the thousands of years suggested by a literal reading of the Bible. But they dispute the idea that natural selection, the force Darwin suggested drove evolution, is enough to explain the complexity of the earth's plants and animals. That complexity, they say, must be the work of an intelligent designer.[12]

Glanz then reinforced this "separateness from biblical creationism" in several ways. First, he said that while God is one candidate for the role of the designer, other ideas are discussed as possibilities within the movement. Also, his article showcased the work of Behe and mathematical theorist William Dembski. Embedded within the article was a visual legitimizer—a large picture of Behe, bearded, bespectacled, and dressed in his trademark lumberjack shirt, leaning against a counter filled with chemical flasks in his biochemical laboratory at Lehigh University. Glanz implied that evolutionists sense "seductive danger" lurking within the theory of intelligent design, thereby adding an air of tension and even a hint of melodrama: "Supporters of Darwin see intelligent design as *more insidious* than creationism, especially given that many of its advocates have mainstream scientific credentials, which creationists often lack. 'The most striking thing about the intelligent design folks is their *potential to really make anti-evolutionism intellectually respectable,*' said Dr. Eugenie Scott" (emphasis added).[13]

Design proponents viewed these two articles as a significant milestone—marking the "end of the beginning," according to Phillip Johnson. Suddenly, at the turn of the century, a platoon of several hundred skeptics of Darwinian evolution were striving as a cohesive scientific-rhetorical movement toward the goal of radical transformation in the way biological origins is taught in public schools and universities. This initiative constitutes a rare attempt at revolutionary defiance in the world of science. The movement's own rhetorical vision, which is based on a symbolic "projection" of what can be accomplished through concerted action, amounts to an ambitious campaign of paradigm "sabotage."[14]

Yet, these are no ordinary saboteurs. As alluded to in Glanz's article, many in the movement are university professors[15] and relatively few of them adhere to the traditional cluster of beliefs that define classic creationism.[16]

Glanz pointed out that the leader of the Intelligent Design Movement is Phillip E. Johnson, emeritus professor of law at the University of California at Berkeley. Johnson has been the tireless organizer of this movement since 1990 when he joined forces with a preexisting group called the "Ad Hoc Origins Committee." Even more than Behe, he has been the skeptics' most visible media presence. ABC's *Nightline* interview with him on 27 July 2000 was typical of his media appearances. During the interview, which was prompted by network coverage of the primary election of members to the Kansas Board of Education,[17] Johnson defended the controversial 1999 actions of the state board. He said that a "public movement is beginning to question the dominant religious philosophy of our time—effectively the established religion of our culture—which is scientific naturalism."[18]

The character of Intelligent Design as a public movement rather than a movement only among the educational elite has tended to blur the distinction between it and the older "scientific creationism." Both movements have large and somewhat overlapping public constituencies. While the literature on traditional creationism is voluminous,[19] relatively few have written about this newer movement, and very few have made more than a cursory attempt to trace its history step-by-step.[20]

Many who have written on the recent outbreaks of public activism against evolution, even when such outbreaks have involved Intelligent Design scholars in university venues, have stereotyped the activists as biblical literalists. Evolutionists who write about the Intelligent Design Movement typically characterize its theorists as dangerous ideologues, driven by impulses that are ultimately religious or emotional, not empirical and rational.

Such a picture is at best a poor caricature and at worst a stark inversion of reality. I shall argue in this book that all who ply this stereotype are ignoring crucial characteristics that distinguish Intelligent Design as a social and scientific movement. Often overlooked or distorted are its diverse social composition, its guiding assumptions and values, its vision and goals, its conceptual and critical content, and its alliances and strategies. In fact, the Intelligent Design Movement (to which I will also refer as the Design Movement, Intelligent Design, or simply Design) displays an entire network of contrasts from traditional creationist activism.

Most importantly, opponents of Design commonly ignore (or purposely avoid) the *larger historical and rhetorical frame of the movement,* which gives insight into why and how opposition to macroevolution has gained such energetic recruits from the ranks of university professors. Rather than carefully trace the personal accounts of how Design scholars began to doubt Darwin, critics often have rooted such doubters in the soil of biblical creationism and at times have hinted that they have covert plans to insert Genesis into the science class.

Such criticism can be understood as part of a stigmatizing "anti-Design rhetoric" or as part of Darwinism's "fantasy theme," employed to stir up its own ranks, but it does not count as responsible scholarship. Indeed, such criticism paints an astonishingly skewed picture of Design, one in which the social reality is seriously misreported and from which a huge part of the story is absent.

The purpose of this book is to probe and trace this missing historical frame—the story of the rapid emergence of the Intelligent Design Movement as a unique and snowballing rhetorical phenomenon in American society. My investigation will emphasize the *historical dimension,* tracing the key events, personalities, sociocultural forces, and other factors that shaped this movement and coaxed it into the paths it took. I will also stress the *rhetorical dimension* by focusing on the processes of Design's persuasion. Thus, I shall take advantage of the perspectives, insights, and analytical tools that come from the accumulated scholarship about how informative and persuasive discourse achieves its goals within a given rhetorical situation. I work in a rapidly growing field known as the "rhetoric of science," whose labor is to explore the question, *How do scientists—and the public at large—come to be persuaded* that they are in the possession of solid scientific knowledge on a given topic?[21]

As a rhetorical historian of Design, I view the field of narrative drama as revolving around two poles representing two kinds of scientific drama. The first pole is the *drama of the rise of the movement's leaders.* Observers of Intelligent Design often overlook the pivotal role played by this experiential reality. A growing fund of stories detail how many of the movement's leaders, to borrow Kant's phrase, were "aroused from dogmatic slumber." Several of the founders frequently relate a vivid tale of how they previously had assumed the validity of Darwinian scenarios and were later shocked to discover major weaknesses in the case for Darwinism. Typically this intellectual epiphany leads to further reading and research, which cements the new radical doubt about the theory's plausibility.[22]

When Design theorists tell such stories of intellectual conversion, they invariably stress their "empirical persuasion," pointing to the key evidentiary points as opposed to religious factors that led them into skepticism. Typically, lively interactions with evolutionists with whom they raised the issue only confirm their doubts. Such stories form a significant part of the bedrock of the persuasive power of Design. I will begin to flesh out many of these microdramas to reveal their rhetorical importance and function.

Once led into doubt about Darwin, these leaders begin to speak out, both orally and in print. Thus, the second historical pole, which I shall explore as well, is the *drama of the design discourses themselves.* Some discourse stories are inherently factual, such as the story of *how a discourse was developed and produced,* the embedded *scientific narratives within design discourses,* and the stories of the *discourse responses* (reviews, reactions, affirmations, or rebuttals of Design's books, articles, and speeches). On the other hand, some discourse stories, while they are linked to facts, employ clearly imaginative themes. I prefer to call these stories "projection themes," although rhetoricians refer to them as "fantasy themes." Such projections, which I will explain in more detail in the next section of this chapter, play a crucial role. They gradually link together and accumulate, creating the substance and shape of Design's rhetorical vision.

The Pervasive Power of Narrative Rationality

Stephen Jay Gould acknowledged that humans are essentially "storytelling creatures" and pointed out "the centrality of narrative style in any human discourse (though scientists like to deny the importance of such rhetorical devices—while using them all the time—and prefer to believe that persuasion depends upon fact and logic alone)."[23] Walter Fisher, a leading communication theorist, would agree with Gould. In contrast to the "rational-world paradigm," Fisher has identified and elaborated in recent decades a "narrative paradigm" of communication—an alternative and much more pervasive structure of reasoning and persuasion. Fisher argues that human societies use two basic modes of persuasion. One is the "rational-world" type of persuasion, which relies on formal argumentative skill. People arguing in this mode use *cogent logic* and *compelling data* within well-defined structures of reason; they speak from a certain level of expert knowledge. The second mode, says Fisher, is "narrative reasoning," which is common to all human beings. It is part of our normal socialization and thus is virtually a universal and often more effective mode of discourse and persuasion.

While distinguishing between the ground rules and function of these two kinds of rationality, Fisher resists the notion that one must choose one or the other. Rather, he implies, to use my own crude analogy, the "narrative pudding" in any discourse or interactive setting will naturally contain some "argumentative raisins." Or, to quote Fisher, speaking about narrative rationality's perspective on all kinds of social interactions, "Discourse often contains structures of reason that can be identified as specific forms of argument and assessed as such. Narrative rationality incorporates this fact but goes beyond it to claim that reason occurs in human communication in other than traditional argumentative structures."[24]

Fisher's concept of a narrative paradigm of communication provides a fruitful perspective for my research on the history of Intelligent Design. Although the more formalized types of argumentative persuasion are clearly detectable in both Darwinist and Design discourses, the overall rhetorical situation is so sprawling and eclectic that it virtually guarantees the predominance and supreme relevance of the narrative paradigm. From this perspective a valuable picture emerges of a vast, societywide debate that is building in suspense from year to year and is culturally central to all people on planet earth. Yet, at the same time, this debate is centered and focused on a smaller, more "elite" context. I'm referring here to the rhetorical sparks now flying in the university world, as described in the previously mentioned *New York Times* article by Glanz. It is in this center circle where the "Design versus Darwin" debate is elevated to its peak of intensity, yielding the most exciting clashes and drawing interest from the watching public. Both sides, Darwinism and Design, are locked in this inner narrative struggle; both are vigorously deploying their own stories and busily attacking the stories of their opponents.

"Factuality" in Scientific Narrative

Earlier, when speaking of discourse stories, I distinguished between two different levels of "factuality" in scientific narratives. It is important to understand these two levels. In fact, both Design advocates and Darwinists are constantly deploying stories at both levels. One level can be called "factual-empirical narrative."[25] This level embraces any kind of story or even a proposed plausible scenario that seeks to align itself carefully with the best available data. (Of course, skillful art and purposive selectivity are always involved in such factual storytelling.) Any purely factual stories are intended to be taken as such, and they may even take place in the distant past, such as the story of human evolution from early hominids. Alternatively, a story may pertain to the recent past, like the account of Darwin's development of his theory. Some factual stories, such as the current debate over Behe's critique of Darwinism, are literally in today's science news. In a later chapter I will deploy a three-part toolkit—a taxonomy of such factual stories—to help in our story sleuthing. For now, I shall note that most of the narrative action takes place at this factual level, and the plausibility of such stories, as told by those on both sides of the debate, is constantly being evaluated by others for empirical and evaluative credibility.

The second level of narrative is "semi-imaginative." Mentioned earlier under my term "projection," it is literally *a mixture of fact and faith.* I call such stories "projection themes," although the standard term for such stories among rhetoricians is "fantasy themes." (I don't care for the term "fantasy" because of the natural implication to the lay reader that such stories are entirely invented—and "fantastic" or "implausible" at that.)[26] Regardless of one's terminology, these stories are striking and recognizable. From the facts at hand they "project" into the realm of imagination *what is really happening now, what can be accomplished in the future, or what surely must have happened in the past.* Such projection themes play a leading role in the formation of the rhetorical visions of both Darwinism and Design.[27]

A school of rhetorical analysis called Symbolic Convergence Theory (SCT), whose origins go back to 1972, specializes in the study of fantasy themes and their role in all kinds of discourse and social interaction. According to SCT, these imaginative stories are not hard to spot. They are typically populated with heroes and villains and bear recognizable plot lines. As effective fantasy themes are generated, they strike an enthusiastic response within a social grouping or movement, and then they are reproduced or retold by others in that group with modifications and variations. In other words, they "chain out" through the social movement and beyond. This chaining process builds up and coalesces into (or contributes to) a *rhetorical vision.* SCT pioneer Ernest Bormann at the University of Minnesota described this process: "A rhetorical vision is constructed from fantasy themes that chain out in face-to-face interacting groups, in speaker-audience transactions, in viewers of television broadcasts, in listeners to radio programs, and in all the diverse settings for public and intimate communication in a given society. Once such a rhetorical vision emerges it contains *dramatis personae* [characters in a play] and typical plot lines that can be alluded to in all communication contexts and spark a response reminiscent of the original emotional chain."[28]

One such theme spun by Phillip E. Johnson was a vivid allegory sketched at the close of an epilogue in the revised edition of his *Darwin on Trial.*[29] Darwinism was pictured as a seemingly impregnable battleship which, though armored and equipped with huge guns, had sprung a metaphysical leak and was now mustering "high-tech" escape craft. In 1996, in Johnson's concluding speech at the Mere Creation conference in Los Angeles (sometimes identified as the public birthplace of Design), Johnson repeated this projection theme in his conclusion (i.e., he chained it out), adding there would likely be "academic wine and cheese parties on the deck" even as the ship of Darwinism began to sink. The editor of *Mere Creation,* the 1998 compendium of the eighteen papers read at that conference, also did his part in the chaining, entitling Johnson's contribution "Afterword: How to Sink a Battleship." Others since have carried on this chaining process.[30]

SCT theorists claim that this approach can powerfully account for the formation of a movement's shared consciousness. Donald Shields explains this perspective:

> SCT provides the account of how consciousness is created, raised and sustained and thereby effects human action. . . . SCT does so by demonstrating that the existence of the communicative force of fantasy, a force that continuously effects consciousness in individuals, groups, and large publics, is due to our intrinsic nature as fantasizers and our ontological inquisitiveness. . . . In other words, as Bormann . . . put it, "SCT posits the communicative force of fantasy (fantasy-sharing and -chaining)—interlinked with ontological need to provide explanation—as the causative component that ignites the creation, raising and sustenance of consciousness."[31]

Clearly, the insights of SCT apply to the unique rhetorical-historical struggle between the Intelligent Design Movement and its detractors. Let me recap my overview of the two levels of factuality of narrative:

Factual Scientific Narrative—relatively stable, nonfiction "constructions
of reality"

Semi-imaginative Narrative—projection themes (SCT's fantasy themes),
which are retellings of or extrapolations from "perceived reality"

Together, the factual narratives and projections feed into and shape Design's
"rhetorical vision." The same is true of Darwinism's rhetorical vision.

I shall now posit my central thesis about the Intelligent Design Movement: *The narrative of the movement itself functions as the central integrating
and motivating factor of all the rhetorical projects that the movement is pursuing.*
The movement narrative of Design, which encompasses both the past and
the present, is continually under dramatic tension as it seeks a convergence
in its trajectory with its own ultimate rhetorical vision of a paradigm shift. As
a result, the Design Movement story is not only the movement's own *central
integrating narrative,* it is also an underreported phenomenon, badly in need
of fleshing out and clarification. The coming chapters are structured around
the major stages of this story.

The Leaders of the Intelligent Design Movement

Four principal spokespersons represent the Design Movement. Proceeding
in the order that their work in this area was published, the first is Michael Denton, whose *Evolution: A Theory in Crisis* inspired the second and third figures,
Phillip E. Johnson and Michael Behe. The fourth figure, William Dembski,
is both the leading intellectual theorist of Design and a symbol of the rising
generation of young scholars who are joining the movement. My rhetorical
history will devote two chapters to Denton and the setting in which he wrote,
three to Johnson, one to the "aftermath" of Johnson's *Darwin on Trial* and the
rapid growth of Design in the 1990s, and one each to the labors of Behe and
Dembski.

Because I will be focusing on the key texts, I will begin with the story
of Michael Denton's *Evolution: A Theory in Crisis.* Denton, a self-described
agnostic, argued that Darwinian microevolution is quite plausible but that
the macroevolutionary thesis suffers a chronic weakness of empirical support.
After surveying several fields of scientific evidence, he concluded his book on
a biting, sarcastic note:

One might have expected that a theory of such cardinal importance, a theory that
literally changed the world, would have been something more than metaphysics,
something more than a myth.

Ultimately the Darwinian theory of evolution is no more nor less than the
great cosmogenic myth of the twentieth century. . . . In the final analysis we
still know very little about how new forms of life arise. The "mystery of myster-

ies"—the origin of new beings on earth—is still largely as enigmatic as when Darwin set sail on the *Beagle*.[32]

In 1987 Johnson was drawn to the topic of evolution when he read Denton alongside *The Blind Watchmaker*, a vigorous defense of Darwinism from Richard Dawkins, the Oxford champion of evolution.[33] He was impressed by Dawkins's rhetorical brilliance, but it was Denton who persuaded him that macroevolution by natural selection was more mythological than empirical. Johnson was intrigued with the rhetorical techniques and strategies that were discernable in the literature of Darwinism, and he dedicated his sabbatical year in London (1987–88) to researching the topic. He returned to California with a lengthy research paper on Darwinism, and in September 1988 he defended this paper at a Berkeley faculty seminar. Encouraged by the forthright interaction, he did further research and rewriting on the manuscript until *Darwin on Trial* was published in 1991. The 1993 edition added an epilogue, which replied to the published criticisms of Stephen Jay Gould and many others. During the nineties and beyond, Johnson spoke widely in universities and expanded his criticism in five sequels: *Reason in the Balance* (1994), *Testing Darwinism* (1997), *Objections Sustained* (1998), *The Wedge of Truth* (2000), and *The Right Questions* (2002).

In all six books Johnson argues that every area of relevant scientific evidence tends to falsify Darwinism rather than confirm it. How can this be, if the texts declare Darwinism a "fact"? Johnson says that key philosophical assumptions, especially "metaphysical naturalism," buttress evolutionary biology and protect it from questioning.[34] He defined naturalism as the belief that the universe is a "closed system" of material causes and effects that cannot be influenced by any "outside entity" like God. Johnson said his central purpose was to legitimize well-informed dissent that asks the key question, Is Darwinism true? Not being a literalist himself, he asserted that his critique had nothing to do with bringing a literal reading of Genesis into the biology class.[35]

The books of Denton and Johnson have functioned as the early intellectual manifestos of Design. Yet it was Michael Behe's *Darwin's Black Box*, discussed briefly already, which propelled Design into the spotlight of media attention in 1996 and firmly lodged the "design inference" as a *plausible scientific notion* in the American consciousness. Several reasons account for this. Whereas Johnson was often dismissed as a curmudgeonly lawyer, meddling with matters outside his own expertise, Behe wrote as a tenured professor of biology. Also, Behe's attack on Darwinism was highly focused. His entire case for the scientific consideration of design was built out of recent discoveries in his own field of biochemistry.

The fourth leader, William Dembski, who completed a doctorate in mathematics and another in the philosophy of science, is a research professor at Baylor University.[36] In 1996 he formally proposed a procedure for detecting design called the "explanatory filter," and it quickly began to play a central role in the movement. The filter is a step-by-step matrix of statistical and logical criteria whereby Dembski claimed an investigator can reliably detect which phenomena or objects in the universe are designed and which are not. His ideas, first published in *Mere Creation* (1998), were presented in a highly technical form in *The Design Inference* (1998) and at a more accessible level in *Intelligent Design* (1999) and *No Free Lunch* (2002).

Although Dembski's explanatory filter is a fundamental "plank" in the Design platform, *it is in the work of Behe* that critics of the Darwinian paradigm detect firm ground for administrating a challenge against macroevolution. Here amid Behe's spinning rotary motors and molecular cascades, the skeptics sense that they have been granted the weaponry to foment an unmistakable paradigm crisis and ultimately a paradigm shift within the scientific community, hence throughout the world. As I have hinted, however, the business of this challenge is not (and cannot) be confined to the arguments, beliefs, and practices of science, because implicated in this particular scientific challenge—or rather *exposed* in this controversy—is the important issue of the existence of a creator of biological diversity. At this stage it is sufficient to say that the core argument of Design, which claims to supply evidence for some sort of creator, is Behe's concept of irreducible complexity.

Behe's development of irreducible complexity starts with Charles Darwin. Behe seizes upon a quote from Darwin's *The Origin of Species:* "If it could be demonstrated that any complex organ existed which could not possibly have been formed by numerous, successive, slight modifications, *my theory would absolutely break down*" (emphasis added).[37] Behe then argues that in molecular biology, Darwin's "wager" can at last be put to the test. Scientists have identified and researched many subcellular "machines" that are extremely complex. Scientists have no idea how these systems could have evolved step-by-Darwinian-step. Thus, says Behe, a compelling case for design can be drawn from irreducibly complex molecular systems, such as blood clotting and the flagellum. As a result, Behe argues, evolutionary theory has, in Darwin's own words, absolutely broken down.

What is an irreducibly complex machine? Behe says it is a system with several working parts, wherein the removal of any one part would prevent the machine from functioning. Behe's famous illustration for irreducible complexity is the mousetrap. It has five parts, yet it does not catch a single mouse until all parts are present and properly fitted together. It could not plausibly evolve, step-by-step, and is therefore irreducibly complex. Likewise, says Behe, such systems as the cilium and the biochemical vision-cascade were apparently designed, since they appear to have no plausible evolutionary pathway (none has ever been published in scientific literature). From here he introduces a controversial proposal. Behe argues that biologists should begin applying basic tests to see which cellular systems are clearly designed and which are plausibly evolved from earlier systems. Yet, he cautions that science alone cannot determine "who" or "what" this designing intelligence is. He suggests that scientists engage in dialogue with scholars in other relevant fields on this crucial question.[38]

Behe's ideas and proposals have not evoked widespread agreement among professional biologists. In the *New York Times,* James Shreeve spoke respectfully of this idea but rejected it as premature, asking, "Shouldn't we leave something for our children and grandchildren to puzzle out besides which systems in the cell are intelligently designed and which are not? Because something is beyond our understanding today does not mean it will be beyond theirs." A number of Behe's opponents attacked his basic idea of bringing *design*—a "religious concept"—into science as an explanation. They derided this as "thinly veiled creationism."[39]

On this point, Behe and his colleagues have faced their greatest rhetorical problem. They have struggled simply to gain a hearing.[40] The movement has tried to separate itself from traditional biblical creationism and thereby escape the "Inherit the Wind" (i.e., Bible versus science) stereotype. This is one reason the renegades began to identify themselves in the mid-1990s as the Intelligent Design Movement.[41]

Several defining traits distinguish design theory from traditional creationism. Some of these were mentioned above, in conjunction with the *New York Times* coverage, such as the fact that the vast majority of design advocates are open to a universe that is billions of years old. Another mark is the avoidance of any discussion of the Noahic flood, the creation of Adam and Eve, or any other details that come from a literal reading of Genesis. Echoing Behe, members of the Design Movement say that science can only indicate that some "intelligent cause" is the agent of biological complexity. *Science, by its very nature, is ill-equipped to identify God as the creator.*[42]

However, such differences do not prevent Darwinian critics from linking this movement with its fundamentalist cousins. Opponents of Design often refer to members of the movement as "neo-creationists" or "Intelligent Design Creationists."[43] Joining the two groups of creationists is one way that defenders of Darwinism have acted symbolically to prevent the dawning in the public mind that macroevolution has indeed entered a crisis. And that brings us to the notion of "crisis" as a key conceptual focus.

The Rhetorical Crisis of Darwinism

In his landmark book, *Evolution: A Theory in Crisis,* Denton uses "crisis" in the sense coined by Thomas Kuhn in *The Structure of Scientific Revolutions.* Kuhn's general theory that scientific paradigms change over time is well known and widely accepted among scientists, despite criticism of some of Kuhn's descriptions and details. In Kuhn's view, a crisis emerges when a scientific paradigm has accumulated too many anomalies to ignore and is approaching the brink of a scientific revolution.[44] Ever since 1962, when Kuhn's ideas began to spread and penetrate deeply in academia, it has not been uncommon for scientists in a given field to speculate about the next paradigm or to measure ideas and events in terms of changing paradigms.

Denton deemed evolution a theory currently in (or ready to undergo) such a crisis, as though it were on the verge of dismissal, but he doubtlessly intended to be more provocative than descriptive. By whatever standards one might use to define a total paradigm revolution in historical-biological science, Darwinian evolution seems secure at the moment. This is best symbolized by the fact that not one of the mainstream biological or other scientific journals has published an article favorable to the detection of true design in nature.[45] Yet Denton's rhetoric no doubt persuaded Behe, Johnson, and many others that the current scientific narrative—that our biosphere was continuously constructed through Darwinian forces—was now empirically bankrupt. Moreover, this "awakening from dogmatic slumber" has acted as the core of a larger historical drama, an implicit paradigm crisis, something destined to culminate in what Darwin's new skeptics "Kuhnistically know" will be the next great paradigm revolution—the overthrow of Darwinism.[46] This historical framework is a key to understanding the rhetorical vision of the Design Movement and the dramatic scene in which Behe and Johnson are acting.

Johnson views the movement as the "wedge" and himself as the "leading edge." The purpose of this wedge is to pry open a crack in the Darwinian paradigm, to dislodge the stubborn log of Darwinian domination in the biological sciences and in universities generally.[47] The long-range goal is both clear and radical—the replacement of the current evolutionary paradigm with one that is open to both natural and intelligent causes in the history of biological origins.

One of the most exciting features of this clash of paradigms, as illustrated in the writing and speaking of those such as Behe and Johnson, is that it demonstrates a scientific crisis is, and must be, a *rhetorical* crisis as well. That is, the "science" of Darwinism or Design will stand or fall in relationship to the *arguments* that are made on behalf of these two movements. While Darwinism has been variously challenged and assimilated in relationship to religious beliefs from the outset (see Russett's *Darwin in America,* 1972), nevertheless the persuasive case for macroevolution appears to be eroding in the minds of many university-educated citizens—the same group of people whose beliefs and practices will constitute what is and what is not scientifically valid, or indeed what is or is not science itself.

At least part of the erosion of certainty about the truth of Darwinism is the product of a *rhetorical onslaught*—the persuasive case making of skeptics, critics, and motivated rhetors like Behe and Johnson. Johnson has even used the projection theme of "vietnamization" to describe this spreading "grass roots guerrilla campaign" that is difficult for the Darwinian establishment to deal with.[48] Whether or not sufficient dissent exists to put Darwinism in what we might call a "scientific crisis," organized doubting such as Design represents has served to put Darwinism in an incipient rhetorical crisis, one in which the "priests of scientific orthodoxy" are called upon to answer the challenge from a heretical minority.

One important piece of evidence that demonstrates we have reached this point surfaced in April 2002 when, against the protests of some senior scientists, the American Museum of Natural History featured contributions from Design theorists in its magazine, *Natural History*. Moreover, in conjunction with the magazine coverage, *Natural History* sponsored a public debate at the museum. Behe and Dembski were permitted to give brief addresses, and two prominent Darwinists subjected them to intense questioning.[49]

Readily apparent in the opening remarks of the debate were the nervousness and internal strain that have arisen among evolutionary biologists when confronting the enigma of Intelligent Design. Darwinists seem divided in their strategy for confronting this threat, but they are united in their puzzlement. In fact, they are asking the same questions I am posing: How has a relatively tiny band of academicians managed to spread a remarkably virulent new strain of dissent, with its vision of "scientific crisis," and the triumphant "paradigm shift" predicted to follow? Of equal interest to me is how the new infection has prompted the production of Darwinian "rhetorical antibodies," counter-arguments and counter-projections that constitute a counter-rhetorical vision whose purpose is to defeat the invaders.

Dramatic Supremacy and Cultural Centrality

Design's rhetorical vision—the arrival of a "Darwinian paradigm crisis"—possesses two unique characteristics that pump into the vision a unique quality of energy and boost the general perception of its cultural importance. First, Design's vision of paradigm crisis possesses a relatively rare quality that I call "dramatic supremacy."[50] By this I mean that any measure of success in overturning or drastically modifying the Darwinian paradigm would dwarf nearly all previous revolutions in the history of science. This is so because of the ramifying intellectual and social effects that the installation of a Design paradigm would have on the academic world and on modern culture worldwide. *The cultural stakes of this science drama could not be more elevated.* Any such "supremely dramatic" process will have great interest for the rhetorician, the historian, and even the anthropologist who maps and studies the quirks and patterns of human nature.[51]

"Dramatic supremacy" is a powerful, energizing force for Design, even as it strives for the interim goal, a sort of halfway house on the road to revolution, which is the collapse of neo-Darwinism's monopoly in public education. The rhetorical challenge in reaching even this interim goal is quite daunting, not just because neo-Darwinism is securely entrenched and well funded, but because the proposed replacement—design theory—is viewed by its opponents with

Following is part of the transcript of the opening remarks of the debate over intelligent design sponsored by *Natural History.* Richard Milner, who is on the board of editors for *Natural History,* delivered the words of welcome.

Welcome to what promises to be an unusually interesting evening—our forum on the Intelligent Design or ID controversy. This program has been organized jointly by Matt Johnson of the Education Department of the museum and by *Natural History Magazine,* which has published a printed version in our April issue.

Ms. Goldensohn deserves special thanks for giving us the green light to go ahead with this venture, which was a courageous editorial decision and a controversial one. Rarely at *Natural History Magazine* have we received so many impassioned letters and comments on a single feature. The sole exception I believe was when the astrophysicist, Neil Tyson, wrote a column decrying mathematical illiteracy and inadvertently made a minor math mistake, which an inordinate number of our readers took a great delight in pointing out.

Now, before I introduce Dr. Eugenie Scott, our moderator for tonight's forum, I want to give you just a little bit of behind-the-scenes background. Several prominent scientists emphatically disagree with our plan to sponsor this forum. "We should ignore Intelligent Design proponents," they urged, "and offer them no credibility by giving them a platform in the magazine or at the museum." This institution, after all, is a bastion of evolutionary biology. It has been so for almost a century and a half.

In their published articles and books, some Intelligent Design proponents have characterized Darwinian evolutionists as status quo ideologues, defenders of high orthodoxy, parrotters of the dominant paradigm, dogmatic priests of Darwinism—which some of them view as a secular religion bolstered by uncritical faith and demonstrably bogus icons. [To Dembski—"Did I get that right?"—Laughter.]

On the other hand, in the view of the vast majority of life scientists, geologists and paleontologists, Intelligent Design is sometimes characterized as stealth creationism, anti-evolutionism (even anti-science) and neo-Paleyism. The Reverend William Paley, you may recall, was the eighteenth-century cleric who said if he found a watch in a field, he would have to conclude that somewhere there was a watchmaker. There! Now we've got that all out in the open.

Whether or not Intelligent Design proposes a serious threat or challenge to Darwinian biology, it cannot be ignored as a sociopolitical phenomenon at least. And I know this from firsthand experience—in my travels around the country, [presenting] my own Darwin program, I've often been asked about Intelligent Design. So, *Natural History* has decided not to ignore the dissidents but instead to turn a spotlight upon the controversy. We welcome you and we welcome this panel to Evolutionville. We have tried and will try tonight to keep the focus on scholarly issues, to keep *ad hominem* arguments to a minimum and attempt to proceed in an atmosphere of mutual respect and truth seeking.

deep hostility and suspicion. The intensity of Design's opposition adds high drama to the rhetorical ambience of "dramatic supremacy." And yet Design is gaining friends in the media; its voice is being heard.[52] This friendly recognition in universities and the media adds still another nuance of excitement to the drama and continues to reshape and boost the rhetorical dynamics of the debate.

Second, the Darwinism-Design debate that is building to a fever pitch in American society is not peripheral but rather *culturally central,* even though it may not dominate the news at this point.[53] It is central because questions posed in any debate on origins touch the deepest level of our personal and societal notions of what it means to be human. The cultural stakes of the Darwinism-Design debate are high. The debaters are contending over the fundamental cultural story of humankind, and those who succeed at crafting and telling the most convincing story of origins hold in their hands supreme cultural authority. If any group, religious or scientific, gains the authority to present its own story as uniquely true and to label other stories as mythological, this group functions as the high priesthood of our time.

The current priesthood, that of Darwinian science, like the theological hierarchy it parallels (and in one sense replaces), carries an aura of scientific objectivity and infallibility. Thomas Kuhn showed that in general any prevailing scientific metanarrative, or master story, appears to possess infallibility in its field. However, this infallibility remains only during the tenure of a paradigm. It is shattered in the course of a scientific revolution, when the new paradigm is embraced and old stories may completely disappear as textbooks are rewritten.[54]

Kuhn's notions have virtually become academic clichés and the subject of various criticisms. Nevertheless, Design spokespersons who are picking up the conceptual battle implements scattered across the post-Kuhnian landscape have predicted the impending collapse of the Darwinian paradigm.[55] If such a paradigm collapse took place in the public universities, one could expect it to trigger incalculable and highly unpredictable shock waves through the secularized bedrock upon which so much of our civilization is built.

A Final Word: Design's Motivation

A central purpose of this book is to understand more fully the development of Design's rhetorical vision. One of the most important components of this vision is its own perceived set of "noble motives," which both legitimize the existence of the movement and bestow a peculiar kind of urgency upon it. By following the stories of each of the major figures, I can highlight their own recollections and expressions, which reveal the deep "reformist impulse" that drives the movement. Along the way I will clearly note key additions to the *moral core* of that impulse. The primary target of Design's attack is the perceived suppression of free speech—a pervasive misrepresentation about "what science really knows" when it comes to biological origins.[56] Thus, Design theorists see themselves as spearheading an intellectual reformation within science, seeking to restore academic integrity and to expose and dissolve networks of self-deception as well as public deception.

Many Design participants believe that if the scientific establishment recognized that life and humankind arose from an intelligent designer, a huge spiritual payoff would result. But for many the main motive is not the foreseen spiritual implications but simply the denuding of what they believe is a contemptible flow of misinformation. Those driven by this motivation see their task as simply telling the truth. We detect this most vividly in the recollections of Behe made during my 1997 videotaped interview ("Opening Darwin's Black Box"). Twice he recalled his profound anger upon reading and hearing things in Darwinian presentations that were blatantly false and *known to be false by experts in the field*.[57] He said such statements make one want to "fight" for change. In a similar vein Johnson has said that academia views macroevolution as a "given." It is something that only religious fundamentalists could doubt, and it would be *irrational* for clear-thinking citizens to question it on scientific grounds.[58] As far back as 1988, in a seminar he organized with twenty faculty members at Berkeley, Johnson described this "factual status" of Darwinism as a cognitive "illusion" and expressed his intention to expose such.[59]

Both Johnson and Behe were converted to this state of robust skepticism by reading the work of Michael Denton. It was Denton, more than anyone else, who triggered the birth of Design. We now turn to his story.

2

Murmurs of Dissent

The Prelude to Michael Denton

The Darwin Centennial

The 1959 Centennial Celebration in Chicago was Darwinism's finest hour. One hundred years after the publication of Darwin's *The Origin of Species,* several hundred scholars converged on the campus of the University of Chicago to pay homage to perhaps the greatest scientific revolution of all time. One of the most honored speakers on this occasion was Sir Julian Huxley, grandson of Darwin's "bulldog," T. H. Huxley. Julian Huxley's speech was a glittering oration on the majestic grandeur of Darwin's achievement, coupled with a vision of its totalizing implications for the future.

> Future historians will perhaps take this Centennial Week as epitomizing an important critical period in the history of this earth of ours—the period when the process of evolution, in the person of inquiring man, began to be truly conscious of itself. . . . This is one of the first public occasions on which it has been frankly faced that all aspects of reality are subject to evolution, from atoms and stars to fish and flowers, from fish and flowers to human societies and values—indeed, that all reality is a single process of evolution. . . .
>
> In the evolutionary pattern of thought there is no longer either need or room for the supernatural. The earth was not created; it evolved. So did all the animals and plants that inhabit it, including our human selves, mind and soul as well as brain and body. So did religion. . . . Finally, the evolutionary vision is enabling

us to discern, however incompletely, the lineaments of the new religion that we
can be sure will arise to serve the needs of the coming era.[1]

Huxley's aspirations for evolution, though expansive, were not surprising.
After all, Darwin and his successors had elaborated a biological theory that
apparently explained the evolution of all the diversity of life from one or a few
simple progenitors, without the direction or participation of any designing
intelligence. *This was no longer viewed as a speculative notion.* Huxley again:
"The first point to make about Darwin's theory is that it is no longer a theory
but a fact . . . Darwinianism has come of age so to speak. We are no longer
having to bother about establishing the fact of evolution."[2]

What's more, the Centennial participants basked in the afterglow of
recent discoveries bearing on our chemical origins. Just six years earlier in
1953, through a key experiment that took place in a nearby laboratory at
the University of Chicago, the theoretical reach of biological evolution had
been extended backward to its molecular genesis. Chemist Harold Urey
and his graduate student Stanley Miller had captured the imagination of
evolutionary biologists and the scientifically literate public by synthesiz-
ing the building blocks of proteins—amino acids—from a simple mixture
of gases swirling through a loop-shaped glass apparatus that contained a
"spark chamber." Through this spark chamber flowed a hot fog of water
vapor, methane, ammonia, and hydrogen gas. Periodically, a spark jumped
from one electrode to another, simulating lightning strikes in the primordial
atmosphere of the earth.

Miller's identification of several amino acids, collected in a tiny trap in his
loop, was not just an exciting breakthrough. It was a gigantic inspiration to
that generation of evolutionary biologists and many others. The Miller-Urey
experiment promised to extend Darwin's cosmological narrative back into the
prebiotic evolution of the first cell from lifeless chemicals. The theological
implications were obvious to anyone familiar with *The Origin of Species,* since
it was at the point of the very beginning of life where Darwin found it rhetorically useful to inject the creator into his narrative.[3]

Other researchers—astronomers, physicists, and planetary geologists—were
rapidly developing their own cosmological theories to explain the origin and
development of the entire universe with its vast numbers and immense variety
of heavenly bodies. As Huxley said, it promised to be just a matter of time and
effort to fill in the details of the evolution of all human reality as well: consciousness, reason, and the entire range of social and cultural phenomena—including morality and religion. If religion is nothing more than a fundamental
psychosocial impulse of human nature, then why not take charge of that part
of humanity and reshape it (along with everything else) in light of the reality
of universal evolution?

Evolution had ascended as the new universal metanarrative, and the scientific
mindset that had given birth to it seemed to possess the key to unlock any

question humanity could ever pose about the observed universe. In fact, in 1959 a halo of overwhelming triumph encircled this metanarrative. Unlike the superstitious or mystical stories that had undergirded Western cultures previously, the Darwinian narrative was now rhetorically secured on the foundation of empiricism. Scientists assured the public that their account of our origins was based on scientific objectivity, not on subjective belief. Darwinism's triumph, along with its promise of unlimited explanatory potential, functioned as a powerful "projection theme." This was also an excellent example of a special narrative situation that I earlier called *dramatic supremacy*. This attribute of "unprecedented historical importance" (described in chapter 1 in relation to the rhetorical program of Design) not only boosted the energy and intellectual dynamics of this evolutionary enterprise, it also acted as a motive with vast emotional power and global reach. Julian Huxley's remarks above bear this out.

Another part of the excitement of this historical moment was that in a young but crucial branch of biology—molecular biology—a new odyssey had just begun. In 1953, the same year that the world was told that a spark apparatus could reenact earth's creation of life, James Watson and Francis Crick finally solved a massive, lingering puzzle—the structure of DNA. The discovery of the double-helix architecture seemed to open a new scientific world, especially to the biologist. Now at last, the precise nature of genetic mutation and inheritance could be explored at the molecular level. This was pictured as a vital cog in the Darwinian apparatus. In the preceding decades, the neo-Darwinian synthesis had identified *mutations* (instead of Darwin's "variation") as the true raw material of evolution.

There is little doubt that the Centennial guests saw the coming generation of molecular biology as one of steady extension, solidification, and application of the Darwinian paradigm. These same biologists would have laughed at any prediction that within thirty-five years the molecular biological revolution would generate strange discoveries that would be shaped into arguments by which scientific dissenters (e.g., Michael Denton and Michael Behe) would cast radical doubts on Darwinism itself.

The modern phenomenon known as "scientific creationism" did not even exist in 1959. It was not until 1962 that theologian John Whitcomb and hydraulic engineer Henry Morris gave birth to that movement when they published their *Genesis Flood*.[4] Not until 1967 was the Institute for Creation Research (ICR) founded by Morris and his debating partner, biochemist Duane Gish. Even then, this new creationism was more of an oddity, a nagging annoyance, and was never any substantive threat. No hint of dissent emerged in any refereed scientific journal or in any mainstream biological or geological textbooks.[5] Questioning of the fundamental tenets of Darwinism by scientists at private colleges or public universities was virtually unheard of. All was quiet on the Darwinian front in the early 1960s.

Denton's Dawning of Doubt and the Early Dissent

In this season of the triumph of Darwinian orthodoxy, a bright high school student named Michael Denton was learning the basics of evolutionary biology in Gateshead, situated near Newcastle in County Durham in the far northeast of England. Born in 1943, Denton completed high school in 1961 and then left for Bristol University. He spent eight years at Bristol, earning an undergraduate degree with honors in physiology in 1964 and continuing on for a medical degree in 1968. After a short time in medical practice, Denton entered the doctoral program in biochemistry at King's College, London University, receiving his Ph.D. in 1974. Later he served in various teaching and chemical pathology posts at hospitals and universities in England, Canada, and Australia. During these years he coauthored two dozen published papers on topics in biochemistry and genetics.[6]

While completing his doctorate in biochemistry, Denton began to think deeply about problems with evolution. Just as Darwin had experienced his own conceptual epiphany on the HMS *Beagle* and its famous survey of the Galapagos Islands in the 1830s, Denton's voyage of awakening unfolded in the lecture halls, laboratories, and libraries of King's College in London. As he followed the discoveries in his field of molecular biology, such as the molecular clock hypothesis and the structure of the MS2 gene, he intuitively sensed a poor fit with Darwin's mechanism. As Denton later recalled, "Here [in the MS2] was the nesting of two functions in the same sequence. . . . Here was an information bearer that had physical functions—a very sophisticated form of technology." He realized that at this level "nature is astonishingly complex." Denton said that in particular, "The multifunctionality of things . . . struck me as an extraordinary thing to experience, and this level of complexity was not easily reducible to a simple, continuous, random process." The supposed ability of such a Darwinian process to achieve these ends, as he pondered it, seemed to violate "one's common sense. One would not accept this without really good supporting evidence—the reality was so astonishing."[7]

Creationism was not an alternative for Denton. Although he had grown up in a conservative Christian home, during his university years he had become an agnostic. Denton felt that creationist notions entailed the breaking of natural law. Even if God did exist, he reasoned, it was absurd to think that the one who created the laws of the universe would then break those same laws and intervene miraculously to make life. Thus, he was convinced, contrary to popular belief, that with the failure of Darwinism as a credible explanation of the biosphere, there must be another cogent *natural* explanation. His astonishment and curiosity grew year by year. The sense of a poor fit between Darwinian theory and the empirical evidence of biology settled upon Denton's mind and grew into a voracious intellectual hobby.

Denton quickly sensed that these problems were not merely routine puzzles that should be expected to yield to scientific analysis but rather were *highly recalcitrant patterns*—deeply rooted anomalous structures that indicated to him a fundamental implausibility of Darwinian theory. The birth of his interest in evolution coincided with the first odd murmurs of doubt that had begun to arise here and there in the mid-1960s. The earliest indications of public dissent began with the Wistar Symposium in 1966, which was a response to the findings of Murray Eden and his colleagues.

By 1965 Murray Eden, a professor of electrical engineering at MIT, along with the French mathematician Marcel Schutzenberger and others had begun to model natural selection of random mutations using probability theory. After numerous attempts to model the Darwinian mechanism, Eden's group was struck with their consistently negative results. Staying in close communication, they experimented with new algorithms and grew increasingly skeptical of the mutation-selection mechanism. Finally, their skepticism became known to evolutionary biologists, and within a matter of months a meeting was organized that attracted several well-known Darwinian scientists to discuss the problem with Eden's group.[8]

The meeting took place in July 1966 at the Wistar Institute, a scientific research center on the campus of the University of Pennsylvania in Philadelphia. The focus of the skeptics' attack was on the notion of "randomness" in mutations, the raw material of evolution. For example, mathematician D. S. Ulam argued that it was highly improbable that the eye could have evolved by the accumulation of small mutations because the number of mutations would have to be so large and the time available was not nearly long enough for them to appear. Nobel laureate Sir Peter Medawar replied that the mathematicians were thinking backwards in their scientific appraisal. Clearly, he pointed out, the eye had evolved. He implied that this notion was simply not in doubt. Therefore, the plausibility problem must have arisen due to errors or oversights in the mathematicians' equations. Harvard biologist Ernst Mayr, a leading evolutionary theorist, said, "Somehow or other by adjusting these figures we will come out all right. We are comforted by the fact that evolution has occurred."[9]

The wrangling at Wistar resulted in a rhetorical stalemate, but it produced a highly readable transcript of the conference (albeit with the technically forbidding title, *Mathematical Challenges to the Neo-Darwinian Interpretation of Evolution*). The volume recorded not only the lectures but also the lively discussions. From the point of view of scientific communication and criticism, these interactions symbolize the tensions inherent in the rhetorically barren landscape into which both sides were venturing. The Wistar discussions were literally unprecedented in terms of mathematical and empirical objections to Darwinian mechanisms. Misunderstanding and a certain level of mutual incomprehension were predictable, and they were indeed common in the symposium.

Wistar thus reveals important and complex psychosocial factors that complicated discussions of Darwinism's alleged basic flaws. In particular, both sides were extremely sensitive to the connection with, and perception of, "creationism" in their debate. At one point, skeptic Schutzenberger announced, "There is a considerable gap in the neo-Darwinian theory of evolution, and we believe this gap to be of such a nature that it cannot be bridged within the current conception of biology." Darwinist C. H. Waddington sensed a move toward religion and replied, "Your argument is simply that life must have come about by special creation." Schutzenberger, joined by others in the audience, shouted, "No!"[10] In the end, the Darwinists were not significantly moved by the problems raised by the mathematicians, and the latter were not at all mollified by the responses from the evolutionists.

Mathematicians were not the only skeptics who raised such questions in the 1960s. Three years after Wistar, British journalist-philosopher Arthur Koestler organized the 1969 Alpbach Symposium entitled "Beyond Reductionism" for the "express purpose of bringing together biologists critical of orthodox Darwinism."[11] Invitations to the conference, said Koestler, "were confined to personalities in academic life with undisputed authority in their respective fields, who nevertheless share that holy discontent."[12] Paul Weiss and several other biologists participated, and from the symposium came Koestler's provocative work of the same title as the symposium, *Beyond Reductionism.*

The Wistar and Alpbach Symposia were of rhetorical significance in more ways than one. They signaled the birth of an articulate skepticism of evolution that was not connected to biblical objections. Skeptics Eden and Schutzenberger constituted just two nodes of a growing "underground network," with whose help Denton began to do research for his book.

These conversations and discourses also became a testing opportunity, a laboratory and proving ground for incipient skeptics of Darwinian orthodoxy to hone their arguments and test various rhetorical strategies. As their arguments were raised and discussed, and as Darwinian replies were received, assessed, and replied to, an embryonic rhetorical tradition was taking shape.

At the same time a new narrative framework was born. These early skeptics were beginning to sketch a radical alternative to the standard tale of how the truth of biological origins had been decisively revealed. Eden and his allies may not have realized it, but they themselves were contributing to that new narrative—a "skeptics' tale" of empirically grounded dissent. It is also significant that the Wistar Symposium was repeatedly cited in all early Design publications. This controversial and highly revealing encounter became a rhetorically powerful micronarrative that in the following decades became what rhetoricians called a "commonplace"—an influential resource that could be used over and over, in this case, by early Design advocates.[13]

Pierre Grassé and Thomas Kuhn

The skeptical rhetoric that began in the 1960s gradually took shape as a proto-genre and was tagged with several labels, one of the most common being "nonliteralist antievolutionism." Critiques that were written in this new genre did not always question macroevolution itself, but they almost always attacked the Darwinian mechanism of mutation and selection. Perhaps the most prominent example of this genre is the book *L'Evolution du Vivant* (1973) by the renowned French zoologist Pierre Grassé, which was translated into English in 1977.[14] Grassé, reviewing the state of knowledge about evolution, launched a barrage of vigorous and at times bitter and sarcastic attacks on the Darwinian mechanism of natural selection. His attitude was that of a frustrated but insistent schoolmaster, repeatedly attacking and dismissing the confident assertions of leading Darwinists. He reminded the reader that fossils, and *only fossils,* can shed ultimate light on the story of evolution. Having no detailed replacement to propose for Darwin's mechanism, Grassé suggested only that "mysterious internal factors" in organisms enabled them somehow to evolve toward complexity and diversity. In his conclusion, after rejecting neo-Darwinism, Grassé suggested, "It is possible that in this domain, biology—impotent—yields the floor to metaphysics."[15]

Grassé's bombshell was immediately reviewed by Columbia biologist Theodosius Dobzhansky, the father of modern genetics. Dobzhansky was respectful but resistant: "The postulate that evolution is 'oriented' by some unknown force explains nothing. . . . But to reject what is known and to appeal to some wonderful future discovery which may explain it all, is contrary to sound scientific method."

Several parts of Dobzhansky's review later took their place as classical rhetorical commonplaces of the new antievolutionism. A few of these proved to be especially influential in the movement's early rhetorical development. One of the most oft-quoted was Dobzhansky's summary of Grassé: "The book of Grassé is a frontal attack on all kinds of Darwinism. Its purpose is 'to destroy the myth of evolution, as a simple, understood, and explained phenomenon,' and to show that evolution is a mystery about which little is, and perhaps can be, known."[16]

A second popular excerpt from Dobzhansky's review upholds Grassé's positive "ethos," something rhetoricians emphasize as vitally important—the perceived *good character and reputation of the speaker,* which influences the reception of the discourse: "Now one can disagree with Grassé but not ignore him. He is the most distinguished of French zoologists, the editor of the 28 volumes of *Traite de Zoologie,* author of numerous original investigations, and ex-president of the Academies des Sciences. *His knowledge of the living world is encyclopedic"* (emphasis added).[17] This certification of Grassé's familiarity with the evidence was extremely useful to all who opposed Darwinism. Grassé's "encyclopedic knowledge" of biology gives much greater weight to his strong doubts about natural selection's role in macroevolution.

How important was Grassé? Even though Denton does not mention Grassé (a puzzling omission), later Design advocates, most notably Phillip Johnson, emphasize his critique strongly. They were undoubtedly motivated to do so for several reasons: First, the 1977 English translation of Grassé, though somewhat technical, is clear, feisty, and eloquent when attacking the mutation-selection mechanism. Second, given Dobzhansky's historical eminence and his certification of Grassé's ethos, his review placed an enduring aura of mysterious quirkiness around the book.[18] Thus, both Dobzhansky (inadvertently) and Grassé (directly) played a part in the rhetorical process of Design's genre development. This story became grafted alongside the Wistar story as part of an emerging history-of-science narrative.

During this same period (the late 1960s, early 1970s) Kuhn's *Structure of Scientific Revolutions* (1962) was being read, quoted, discussed, debated, and widely applied across the university world by historians, philosophers, and practitioners of science. Some readers were hesitant or critical, but many began to imbibe deeply his iconoclastic shattering of the traditional view that science was stable, gradually progressive, and strictly objective. The philosophical and cross-disciplinary fallout of Kuhn's work occurred about the same time as the first sophisticated attacks on Darwinism. Kuhn's ideas were clearly part of a synergy of different scientific criticisms and modes of questioning that made what had been unthinkable a reality: an image of the Darwinian paradigm as a prolonged but passing phase, hobbled with its own Kuhnian phenomenon of "blindness to anomalies." This was in direct contrast to the way in which Darwinists would interpret Kuhn in relation to their own work. In their own Kuhnian projection theme (glimpsed earler in Julian Huxley's address), the Darwinian paradigm was the "final paradigm," which could be extended, filled in, and fine-tuned, but *never superseded.*

Missing Fossils and the Birth of Punctuated Equilibria

Grassé argued insistently that the fossil evidence reigns supreme in showing what really has happened in evolution. If his principle is correct, then the most important part of the changing rhetorical situation before Denton can be traced to a pair of American paleontologists. Niles Eldredge, Curator of Invertebrates at the American Museum of Natural History, rose to the pinnacle of evolutionary stardom along with Stephen Jay Gould in the 1970s as together they developed a new model of evolutionary change that they called "punctuated equilibria."

What nagging rhetorical situation in fossils impelled Eldredge and Gould to develop their new theory? Primarily, the problem was the persistent absence of "gradualism" in fossil series, which had been noticed by paleontologists for decades. In Gould's famous and oft-quoted words, "The extreme rarity of transitional forms in the fossil record persists as the trade secret of paleontology. The evolutionary trees that adorn our textbooks have data only at the tips and nodes of their branches; the rest is inference, however reasonable, not the evidence of fossils. . . . I wish in no way to impugn the potential validity of gradualism. I wish only to point out that it was never 'seen' in the rocks."[19]

Predictably, Gould's admission, though it was made in the context of his new evolutionary explanation of the phenomenon, was soon pressed into service as a weapon against macroevolutionary claims. This threat to evolution from fossils was not a new situation. Since Darwin's day, paleontologists had experienced puzzlement and chagrin when speaking of the overall patterns of fossil evidence. Occasionally this pattern had been described to the outside world, usually in the context of scholarly appraisals of the fossil record.[20] But this pattern was not widely known to the educated lay public. What pattern? Gould says that the

> history of most fossil species includes two features particularly inconsistent with gradualism: (1) *Stasis*. Most species exhibit no directional change during their tenure on earth. They appear in the fossil record looking much the same as when they disappear; morphological change is usually limited and directionless. (2) *Sudden Appearance*. In any local area, a species does not arise gradually by the steady transformation of its ancestors; it appears all at once and "fully formed."[21]

In tackling this problem, Eldredge and Gould proposed a solution that was both revolutionary and conservative. It was moderately revolutionary in that it argued (contra Darwin) that a significant portion of evolutionary change is not taking place in a gradual transformation of large, central populations. Rather, Eldredge and Gould argued evolution commonly occurs in small isolated populations, and it happens rapidly, in evolutionary spurts, taking merely a few thousand years to evolve a new species, instead of millions of years. So new species typically come about not by steady, slow accumulation of favorable mutations over millions of years but by sudden and rapid bursts of change in small, isolated groups at the peripheries of populations. Gould and his colleagues suggested that these changes are sparked by lucky "macromutations" that are able to spread quickly through the tiny, isolated group.[22] After coming into being, the new species then settles into a lengthy period of stability or "stasis" during which no directional change takes place. This "stasis" is the "equilibria" part of the process.

The Gould-Eldredge theory of *punctuated equilibrium,* as it later became known, with its exposure of the absence of fossil transitions, turned out to be a vast rhetorical bonanza for the skeptics of Darwin. The new theory drew attention to the absence of gradualism in the fossil record and made it much easier to build a cogent case against macroevolution, even though this is the last thing that Gould or Eldredge wanted to encourage.[23] This exposure was a major boon to nonliteralist anti-Darwinists like Denton.

Gould referred to the "*embarrassment* of a record that seems to show so little of evolution directly."[24] The word *embarrassment* is sufficiently vivid and tinged with drama to suggest to onlookers that a significant *anomaly* had finally been acknowledged in evolutionary biology. As we have seen, those anomalies which resist explanation, despite arduous efforts to account for them, act as conceptual triggers of a rhetorically construed Kuhnian crisis. The explosive feature of the new fossil discussions was that they drew attention to the fact that these uncomfortable fossil patterns had not been previously publicized to any significant degree. They were not quite "covered up," but it did seem that the public had been misled to think fossils generally support Darwinian continuity—that they manifest a pattern of steady, gradual change. All such misinformation, once revealed, added to the skeptics' reservoir of indignation against blithe assertions of "macroevolution as fact." Thus, the controversy over the "sudden appearance" and "stasis" of fossils became an enduring motivation for the diverse membership of the Design Movement.

This fossil controversy also contributed immensely to Denton's critique, especially since his attack was aimed at a Darwinian foundational doctrine—the "continuity" of biological life forms through time. If Darwinism is true, said Denton, there had to have existed a seamless "tree of evolution" with various branches that trace modification. Tracing the process of development of all life forms would show they have blended from one form into another across eons of time. According to this view, the divisions between species that appear today are a result of the extinction or inferior viability of the formerly living transitional intermediates. Evolutionists retain the perennial hope of finding more and more transitional fossils that bear out this "phylogenic continuity."

On the contrary, sudden appearance, coupled with stasis, speaks of gaps that are stubbornly persistent. Such a pattern does not readily coincide with Darwinism. Gould and Eldredge did their best to confront and solve this problem with their idea of punctuated equilibrium. Yet one of the most important results of their introduction of this idea was that they opened this entire area for public discussion. These discussions yielded useful rhetorical ammunition for arguing for the *discontinuity of biological forms.*

Several other cultural factors made the missing transitional fossils a major rhetorical resource for Design. First, fossils are a direct link with the ancient past and thus are inherently interesting and appealing to the general public. Rhetorically they have broad appeal in terms of *sheer interest.* Second, the phenomenon of "missing fossil links" is understandable without great mental effort or extensive background knowledge, making it well suited for both public and scholarly consumption. And consume they did, because Gould's felicitous and brilliant articulations of the problem provided a mother lode of rich material, which the antievolutionists repeatedly mined and quoted to their advantage.

Hoyle, Patterson, and the Enigma of Life's Origin

Several other outbreaks of anti-Darwinian skepticism occurred before 1985, helping to shape the rhetorical landscape of unanswered (and sometimes unasked) questions and providing Denton with additional motivation and resources. Space does not permit a discussion of all of these, but I shall mention three, beginning with the controversial critiques of astrophysicist Sir Fred Hoyle and his colleague Chandra Wickramasinghe. They argued that chance processes could not have formed the biochemical machinery of the cell, especially the enzymes. In their *Evolution from Space,* they estimated the probability of forming a single enzyme or protein at random, in a rich ocean of amino acids, was no more than one in 10 to the 20th power. They then calculated the likelihood of forming by chance all of the more than two thousand enzymes used in the life forms of earth. This probability is calculated at one in 10 to the 40,000th power.[25] They wrote, "The theory that life was assembled by an intelligence has, we believe, a probability vastly higher than one part in 10 to the 40,000th power of being the correct explanation of the many curious facts discussed in preceding chapters. . . . The speculations of *The Origin of Species* turned out to be wrong. . . . It is ironic that the scientific facts throw Darwin out but leave William Paley, a figure of fun to the scientific world for more than a century, still in the tournament with a chance of being the ultimate winner."[26]

A vivid analogy from Hoyle became perhaps the most prominent commonplace—virtually a creationist cliché—used also by some Design rhetors: Belief in the chemical evolution of the first cell from lifeless chemicals is equivalent to believing that a tornado could sweep through a junkyard and form a Boeing 747.[27]

Their analysis was a fairly sober assessment of the difficulties in the origin-of-life field, but their specific hypothesis of "panspermia" seemed very far-fetched. They said the first life forms and subsequent jumps in complexity, which show up as gaps in the fossil record, were probably due to living things' incorporation of genetic material that rained down from space. This "genetic fallout" was produced, they suggested, not by God but by some intelligent life elsewhere in the universe, and the genes were sent here by means of the drifting currents of solar wind. This "panspermia" idea of Hoyle and Wickramasinghe was rhetorically contaminated, and Hoyle's credibility as a theoretician was damaged by his weird speculation that the alien intelligence may have actually taken up a covert residence on earth in the form of the numerous insect species![28]

A more cautious line of skepticism came from Colin Patterson at the British Museum. By 1981 Patterson had developed a reputation as something of a heretical freethinker in his field of cladistics (taxonomy of species and other groupings).[29] As a disgruntled "evolutionary agnostic," he was traveling from conference to conference in 1981 asking everywhere the same embarrassing question, "Can you tell me one thing about evolution that is true—any one thing at all?"[30] Denton highlights Patterson in his rhetorical strategy, as does Johnson later, as an authority who considered evolution as virtually irrelevant to his work as a taxonomist.

The materialistic theory of the origin of life became yet another problem area by the early 1980s. After the heady days of the Miller-Urey experiment in 1953, an entire interdisciplinary field coalesced to investigate prebiotic evolution. Chemists, biologists, physicists, astronomers, geologists, and geochemists worked together to unravel the pathways by which nature brought about the production of life's building blocks (nucleotides and amino acids) and their subsequent linking as polymer-chains (proteins, DNA, and RNA), leading to larger and more complex structures called "protocells." Most biology texts featured Miller's circulating spark apparatus and conveyed the impression that the mystery of life's origin was virtually solved. Books on the topic exuded confidence and displayed the latest photos of bubblelike structures with intriguing names like "coacervates" and "proteinoid microspheres"—entities which, said the texts, held great promise as possible protocells. For a while progress seemed encouraging, and one leading figure, biologist Dean Kenyon, voiced this widespread hope in his cleverly titled *Biochemical Predestination* (1969).

As new problems came to light year by year, however, the field stalled. Optimism about an early consensus on life's origin began to wane, and even Dean Kenyon began to have doubts about his previously published optimism.[31] The "prebiotic soup hypothesis," popularized by Miller's experiment, came under withering criticism from chemists for ignoring the role of competing and destructive cross-reactions with chemical ions that would be expected in any hypothetical ocean or pond. These reactions would have tied up or terminated any growing polymer-chain.[32] The more biologists learned about the new world of molecular biology, the more difficult it became to account for the delicate mutual dependence of the different parts of the system (the interactions between DNA, RNA, ribosomes, and proteins).

Chemists frequently cited other problems. For example, the Miller reactions, which required a strictly oxygen-free atmosphere, were called into question. Analysis of the deepest sediments on earth were showing the likely presence of oxygen during earth's earliest eras. In addition, scientists began noting that no progress was being made on how the correct kind of amino acids (exclusively left-handed) could be selected out of a random soup mixture that contained about 50 percent right-handed and 50 percent left-handed. Two other problems that persisted were how the right kind of linkages (called "peptide bonds") could be set up consistently and how the precise sequence of units (amino acids or nucleic acids) could be gained by a lotterylike process.[33]

By the late 1970s the discovery of a plausible chemical pathway to life seemed to some to be receding rather than coming closer. As early as 1973, just twenty years after his DNA milestone, Nobel laureate Sir Francis Crick had begun to toy with his own ideas of panspermia due to the difficulty of envisaging the origin of life on earth.[34] In his 1981 book, *Life Itself,* Crick reiterated his own idea of life being "seeded" here by aliens and explained why he entertained this scenario: "An honest man, armed with all the knowledge available to us now, could only state that in some sense, the origin of life appears at the moment to be almost a miracle, so many are the conditions which would have had to have been satisfied to get it going."[35]

This statement begs for rhetorical analysis with its multiple softeners and qualifiers, but I quote it only to demonstrate the drastic change in the rhetorical atmosphere since 1953. When a leading authority in biology deals with the festering problems of abiogenesis by turning to panspermia, it signals to skeptical observers that the conditions are right to organize a comprehensive attack. What emerged by the early 1980s was a revolutionary tension between the expectations the scientific world had projected through their public rhetoric and the reality of stalemate, revision, confession, pessimism, implausible scenarios, and rancor. An entire constellation of motivational symbols had developed and begun to interact, from the wrangling at Wistar, to the abruptly appearing fossils of Gould, to the shocking attacks of Grassé, to the weird ideas of Hoyle and Crick. Some of these were used immediately and kept on the rhetorical shelf as commonplaces. Francis Crick's quote above, for example, had such enduring value that it was repeated literally for decades. Denton used it in his chapter on the origin of life, and Michael Behe quoted it in his public lectures nearly twenty years later.

Rhetorical conditions were becoming right for a massive critique when Michael Denton began to write his book in 1980. At the outset of his dismissive review of Denton for *New Scientist*, philosopher of science Michael Ruse said, "In the past five years [1981–85], I have found myself reviewing an avalanche of books of this ilk, books whose authors think that something went badly wrong with science when, in his *On the Origin of Species*, Charles Darwin proposed his theory. . . ."[36] When Denton began to write his book, his problem was not where to find resources for rhetorical invention; rather, his challenge was to be wise in his selection and deployment of so much valuable new material alongside the older, classic lines of hostile evidence that he had accumulated. Like a wise military strategist, he knew that his book must avoid the appearance of a pathetic, quixotic lunge. It had to be constructed in such a way that its tone, theses, and varied lines of evidence acted in concert to overwhelm and intellectually defeat the reigning orthodoxy. When the startled defenders of Darwinism began to sense the severe intellectual danger in Denton's pummeling in 1985 and 1986, they unleashed a ferocious counterattack that is reminiscent of the great clashes of world history. We now turn to that momentous confrontation.

3

The Birth of Design

Denton Launches His Critique

To Darwin's skeptics, the coming of Michael Denton in 1985 produced the rhetorical analogue to the opening scenes of *Saving Private Ryan*. Darwinism had held the continent of scientific consensus for over a century. At last the time had come to wade ashore and establish the first beachheads of empirically based antievolutionism. The odds against the would-be "liberators" seemed so terrible as to border on the absurd. Everything hinged on the weapon held by the invaders—a 344-page book, *Evolution: A Theory in Crisis,* published in England (1985) and the United States (1986).[1] The author was relatively unknown—a British-educated biochemist and medical doctor laboring in the obscurity of the clinical chemistry department of Prince of Wales Hospital in Sydney, Australia.

As the invaders clambered up the cliffs towering over the beaches, they hurled Denton's explosive charges toward the pillboxes: *"Neither of the two fundamental axioms of Darwin's macroevolutionary theory—the concept of the continuity of nature . . . and the belief that all the adaptive design of life has resulted from a blind random process—have been validated by one single empirical discovery or scientific advance since 1859"* (italics mine).[2] MIT's Murray Eden with his Wistar colleague Schutzenberger joined the invading troops,[3] announcing that Denton "should be required reading for anyone who believes what he was taught in college about Darwinian evolution."[4] Paul MacLean, a former professor at the Yale Medical School and founder of the Brain Evolution Lab at the National Institute of Mental Health, exulted, *"Kant gave credit to Hume for arousing him from his 'dogmatic slumber.' This book promises to arouse an entire audience"* (italics mine).[5] Even the celebrated anthropologist Ashley Montagu praised Denton's breadth of knowledge, noted his "just and telling" criticisms, and welcomed his "valuable contribution."[6]

The Darwinian defenders were not asleep. Rhetorical bullets flew in savage counterattack:

> "Denton's book displays a vast ignorance about Darwin, evolution, and science in general."[7]
>
> "A specimen of creationism at its most subtle and up-to-date."[8]
>
> "No area escapes misrepresentation and distortion."[9]
>
> "a sham"[10]
>
> "fraught with distortions"[11]

A trio of prestigious evolutionists—Michael Ruse, Mark Ridley, and Niles Eldredge—wrote strongly negative reviews in prominent journals in an effort to crush Denton's credibility and repel his attack.[12] When the smoke cleared from the beach landings of 1985–86, the continent was still firmly in the hands of neo-Darwinism. Yet in the cliffs, here and there, certain pillboxes had fallen to the invaders, who were now spraying the countryside with sniper fire.

I trust the reader will bear with my Normandy metaphors, which of course are my own attempt to capture the wave of projection themes that powerfully chained out in the budding consciousness of early proponents of design when Denton's attack was published. The imagery fits well with Denton's comment in a letter in early 1987 that exemplified the Normandy fantasy scene: "I am totally committed to waging unceasing intellectual war on neo-Darwinian orthodoxy."[13] Denton was blunt, thorough, confident, and cognitively upsetting for anyone who began his reading with the assumption that Darwinian theory rested on ample confirming evidence. In a word (Denton's own word in the preface), the critique was *radical*—a notion I shall analyze later in the chapter.

It is impossible to trace the development of Design in the 1990s and beyond without reviewing the origin of that movement in the rhetorical invasion and bloody skirmishes that were triggered by Michael Denton in 1985. His *Evolution: A Theory in Crisis* (hereafter referred to as *Evolution*) is, to an astonishing degree, the initial impetus, inspiration, and even rationale of the movement. Together with *The Mystery of Life's Origin*,[14] a more technical critique focused on chemical evolution that preceded *Evolution* by a few months, Denton practically established the rhetorical matrix of values, communication styles, purposes, perspectives, assumptions, and beliefs that became the substance of the new Design genre.[15] These books also supplied a fund of rhetorical resources—lines of argument, phraseology, and especially patterns of evidence and anomalies gathered from many different biological fields. All of these outfitted the invading marines of the incipient Design Movement as they strategized their next sorties.

Denton's Development of the Core Thesis

Denton's core thesis in *Evolution* is a logical place to start an exploration of how Denton's criticism of Darwinism provided rhetorical resources for subsequent design proponents. His thesis is constructed in three steps. First, he establishes a *split between Darwin's two theories.* On the one hand is the "special theory" of speciation, called "microevolution," which is the generation of slightly different sister species. On the other hand is the "general theory" of the evolution of all life forms from common ancestors, called "macroevolution." After splitting "micro" from "macro," Denton shows how Darwin himself distinguished the two in the *Origin.*

His second step is a *happy concession,* acknowledging that the more modest theory of microevolution has good cause for full acceptance by biologists and the public at large. The rhetorical value of emphasizing this common ground with Darwin is obvious. In fact, Denton enthusiastically endorses microevolution in his fourth chapter, "A Partial Truth." He presents an array of evidence for the reality of microevolution but ends on a note of warning about the illegitimacy of extrapolating macroevolution from microevolution.

After Denton splits micro from macro and concedes micro, his third and final step is to *subject the macroevolutionary theory of Darwin to empirical investigation,* employing all relevant lines of evidence. He has eight separate evidence chapters covering an impressive range of topics—taxonomy, homology, fossils, hypothetical morphology-intermediates, and statistical analysis of random search processes (chapters 5 through 9 and 13). His own area of molecular biological evidence has two chapters (11 and 12) as well as one on the origin of life and one on the evidence of amino acid sequences in proteins.

All of these chapters call into question the two key Darwinian foundation stones of macroevolution—the mechanism (selection of random mutations) and the phenomenon of "biological continuity" (the interconnection of living things in unbroken linkages of descent). Denton's attack on Darwinian foundations boils down to the following question, which he uses repeatedly throughout his book: *Is there empirical evidence of transitions, or can we at least plausibly reconstruct a series of hypothetical intermediates?*

Denton shepherds all these lines of investigation toward a radical central thesis: *Macroevolution—continuous evolutionary development through the selection of random mutations—is not supported by findings in any area of biology. The theory is supported neither by empirical evidence nor by thought experiments, that is, by attempts at reconstructing plausible evolutionary pathways.* A rhetorical question arises naturally from this shocking thesis: If this is the true state of the evidence, then why does the scientific community tell the public that Darwin's theory is no longer a theory but a fact? One can scarcely imagine a greater discrepancy in overall assessment between Julian Huxley and Michael Denton!

To answer this, Denton presents a second major thesis—a Kuhnian corollary—in his final chapter: *It is the "priority of the paradigm" that renders these Darwinian problems and anomalies invisible.* Just as nature abhors a vacuum, said Denton in a published videotaped interview (1993), so science abhors a vacuum, and until a better naturalistic theory comes along, the Darwinian paradigm will have to do and will be treated as if it were factual. The rhetorical influence of Denton's arguments on Intelligent Design can hardly be overemphasized. His theses and the rhetorical logic used to establish them were absorbed directly into the intellectual bloodstream of Johnson, Behe, and other Design rhetors.

Anti-Narratives and the Taxonomy of Factual Narrative

In the persuasion of Design, narrative logic plays a central and powerful role, and this can be traced in Denton as well. In *Evolution,* he employed a narrative approach that became a model of argumentation for future Design critiques. Indeed, one can even say that Denton's core strategy was profoundly rooted in narrative but of a peculiar sort. *Evolution* pioneered a logic of anti-narrative. Let me describe his two distinct kinds of "anti-narrative."

Denton's first anti-narrative was the invention of a wholly new account of the Darwinian triumph, an account that would be aptly titled "A Skeptic's History of Darwinism." This new story line subverted and inverted the orthodox story of Darwinism's ascendancy, changing it from a triumph of truth to a descent into a new Dark Age, a mind-deadening tyranny of dogma.[16] Denton transformed virtually every dimension of the scientific script and featured his own favored system of biology—classical typology or the "theory of archetypes"—as the "explanatory key," the most cogent intellectual alternative to Darwinism. Typology thus dominated the antiplot, much more so than the usual opponent of Darwin, Genesis-based creationism. Given the antipathy of most biologists (and publishers) to creationism, the featuring of typology was one of the rhetorical keys by which Denton gained a hearing.

Into this revisionist narrative Denton plugged dozens of supporting scientific actors and micronarratives of their dissent, stating that their empirically based criticism had never been answered. Some of the dissenters, such as Cuvier, Owen, Agassiz, and Pictet, dated to the 1800s. Others, like Berkeley geneticist Goldschmidt, spoke out in the 1940s, and many were contemporary figures, such as Fred Hoyle and the doubters at Wistar and Alpbach.

This anti-narrative completely dominates the first three chapters. Chapter 1, "Genesis Rejected," is a charming retelling of the story of Darwin's voyage of discovery aboard the HMS *Beagle,* highlighting the actual process of internal "persuasion through observation" that Darwin experienced as he noted the variation of animals in the vicinity of the Galapagos Islands. As he pondered these patterns of apparent change, the old discontinuous picture of nature seemed less credible and at the same time the biblical framework seemed "increasingly obsolete in his mind."[17]

Chapter 2, "The Theory of Evolution," relates the gradual dawning of the idea of natural selection, Darwin's "elegant and beautifully simple" idea of how the variations between species that he had observed might have come about (42). Here Denton places Darwin in his historical context, as he traces the thought of materialistic origin of life back through Hume to Empedocles and the pre-Socratic philosophers and even to ancient Norse myths.

The heart of chapter 2 is Denton's review of both the positive evidentiary case that Darwin builds in his *Origin* and of Darwin's defense against anticipated objections and published criticisms. Here Denton thrusts before the reader Darwin's acute sensitivity to the lack of transitions or intermediates between different types. Denton stresses that Darwin's chief problem was that there was no empirical evidence for these transitional species nor could he "even . . . conjecture" the gradations by which they might arise (55–57). Altogether, Denton hammers home this point with nineteen quotes from the *Origin,* and the majority of these reveal Darwin's struggle to overcome these evidentiary difficulties. Darwin is portrayed sympathetically as an honest grappler with these issues: "Although convinced of the reality of evolution, nowhere, either in the *Origin* or in any of his other writings including his autobiography and letters, is he ever dogmatic or fanatical in his advocacy. Darwin always reveals himself to be a man of great common sense convinced, but still aware of the hypothetical nature, of his theory" (65). Denton's rhetorical strategy in chapter 2 is a clever twist. He pictures Darwin as a scientist who was highly sensitive to the gaps in evidence and who dealt with them in an earnest, forthright way. Denton has subtly recruited Darwin in a support role for his Normandy invasion.

Chapter 3, "From Darwin to Dogma," rounds out Denton's three-chapter opening "history-of-science" narrative. A tiny treatise (only nine pages), this chapter is one of the most forceful and effective chapters in the book. The goal of the chapter is to show how Darwin's theory became hardened over time into an unchallengeable axiom, when in fact none of its evidentiary difficulties were resolved. Besides the cultural trends in Victorian England that aided acceptance of Darwin's ideas, Denton pointed to the considerable success that science had achieved in explaining nature: "God came to be viewed increasingly as a distant and remote first cause, the architect of a clockwork universe which had continued from its creation to operate automatically without any need for further divine intervention." Denton added that it seemed "increasingly likely to most educated men . . . that all past phenomena would prove explicable in terms of presently operating processes and that the universe had gradually developed from a few elementary particles into its present state through operation of the basic laws of physics and chemistry. Thus, the universe came to be viewed in uniformitarian terms as a closed system and all the phenomena within it as essentially natural. Since its origin it had suffered no unnatural . . . interference" (71).

Although the formerly prevalent doctrine of typology fit the observed pattern of nature, this was no "scientific alternative" since in the typological system it was hard to see what "sort of natural process could have generated all the diversity of life on earth." To Darwin and many of his contemporaries, nature simply could not be *discontinuous* and yet at the same time reducible to scientific explanation. Denton concludes that it is "in retrospect perfectly easy to understand how Darwin's theory proved irresistible even though, as Darwin himself admitted, the actual empirical evidence was insufficient, and there was absolutely no evidence that any of the major divisions of nature had been crossed in a gradual manner. If nature was to be explained by natural processes, she had to be continuous" (73).

A bit later Denton employs narrative rationality again as he distinguishes facts from the various interpretations woven by scientists. Facts of nature had not changed with Darwin and the post-Darwin discoveries. Nature was just as discontinuous as ever. *What had changed, historically, was a massive shift in intellectual fashion*—that is, in the "intellectually respectable" interpretation given to nature. Denton creates a plausible inversion of the usual "progress narrative" of Darwinian triumph: "As the years passed after the Darwinian revolution, and as evolution became more and more consolidated into dogma, the gestalt of continuity imposed itself on every facet of biology. The discontinuities of nature could no longer be perceived. Consequently, debate slackened and there was less need to justify the idea of evolution by reference to the facts" (74).

On this view, it becomes clear why ferocious hostility was aimed at dissenters like paleontologist Schindewolf and geneticist Goldschmidt in the mid–twentieth century. Dissent becomes "by definition irrational and especially irritating if the dissenters claim to be presenting a rational critique." Denton adds, "It is ironic to reflect that while Darwin once considered it heretical to question the immutability of species, nowadays it is heretical to question the idea of evolution." In order to ratchet an already forceful chapter up to an appropriately smashing climax, Denton ends the chapter with a quote from Paul Feyerabend, the Berkeley philosopher-iconoclast whose anarchist vision of science is legendary. Called to Denton's witness stand, Feyerabend testifies to the power of metaphysical dogma to shape the image of truth, where "the stability achieved, the semblance of absolute truth, is nothing but the result of an absolute conformism." Such truth has come to function as myth, and the "myth is therefore of no objective relevance, it continues to exist solely as the result of the effort of the community of believers and of their leaders, be these now priests or Nobel Prize winners. Its 'success' is entirely man made." Thus, in Denton's retelling, Darwinism becomes a Feyerabendian case of mythology: Darwin's general theory "is still, as it was in Darwin's time, a highly speculative hypothesis *entirely without direct factual support* and very far from that self-evident axiom some of its more aggressive advocates would have us believe" (75–77, emphasis added).

The first three chapters function together as Denton's own new history of biology—a transformed metanarrative designed to overthrow and replace the one that is told by Darwinians.[18] This kind of narrative reappears also in several of his other chapters. For example, chapter 6 gives glimpses of the history of typology, starting with Linnaeus, founder of modern taxonomy, and continuing through such towering figures of the nineteenth century as French anatomist George Cuvier. The British anatomist Richard Owen, another giant, is brought in to speak against Darwin, and the paleontologist Francois Pictet testifies to his pause in accepting Darwin's theory. In chapter 7 on taxonomy Denton includes the contemporary story of Patterson's heresy and the dropping of evolutionary assumptions by contemporary scientists called "transformed cladists."

What is the rhetorical function of all this narrative reconstruction? Primarily, Denton is strengthening his case by embedding his own doubts in the seemingly forgotten tradition of empirically grounded dissent that stretches back to 1859. To use Kenneth Burke's term, Denton establishes himself as "consubstantial"[19] with these predecessors. His work becomes the latest chapter in the noble resistance saga of a long-lived, heroic but dogmatically suppressed, underground—a network of skeptics marked as truly scientific because of its respect for the overwhelming empirical evidence of the *discontinuity* of nature.

At this point, I feel compelled to point out something very important that we have just seen that will help us in the rest of our examination of Design. As I have described the complex system of stories and dramas that led up to Denton and appear in his book, unfolding before us has been a crisp three-part taxonomy of such stories. (Remember that all these stories inhabit the "factual level" that I introduced in my opening chapter, not the projection-theme level in which fact is mixed with faith or imagination.) Let me list and describe these three types of factual stories quickly.

The cosmology (or genesis) story. By "cosmology" I mean not just the study of the universe as a whole but origins in general, and "genesis" obviously does not refer to the biblical book but to a story of origins. Any such narrative that fills in the picture of the coherent historical development of any part of the universe has a goal—the fleshing out of the details of that narrative, whether they be steady processes or dramatic events, thereby satisfying our natural curiosity. Julian Huxley's oration at the Darwin Centennial contained a rare poetic sketch of such a genesis narrative. Many other examples could be cited, such as the story of the big bang, or of a prebiotic evolution scenario for the first cell, or of the origin, diversification, and ultimate extinction of the dinosaurs. We have begun to see already how Denton attacks the credibility of the Darwinian genesis story throughout his book.

The history-of-science story. Much of Denton's rhetorical work falls into this category. This is a clever strategy because all of our thinking about origins is rooted in a consciousness of *how* we arrived at our present enlightenment on genesis questions. In fact, in the minds of scientists and laypeople alike, there dwells an ever-changing, yet generally coherent, story line of the history of science. This is a narrative of how students of nature have progressed step-by-step in their understanding since the dawn of observation, reason, and research in the myriad scientific fields. In a typical Darwinian history-of-science story, the age of creationist belief (before 1859) is viewed somewhat as the Dark Age, punctuated at last by the dawn of biological evolution as it triumphed after Darwin's *Origin* swept the planet. This narrative includes several chapters, such as the transition to neo-Darwinism in the 1920s through the 1950s. Thus, there are also more specialized or narrow types of history-of-science narrative. One would be the story of a specific controversy, such as the triumph of plate tectonics theory in geology, while another would be the story of a scientific movement, such as Intelligent Design, which, of course, is the story of this book.

As we have seen, Denton spends much of his rhetorical energy dismantling the traditional Darwinian history-of-science story. He then erects in its place a new version—the skeptics' story. Yet we will see that anyone who would tell a history-of-science story must employ a third category of stories.

The personal or individual stories. Somewhat like limbs that shoot out from a tall, spindly pine tree, history-of-science narratives have many intersecting branches of individual scientists' narratives. By this I mean that any narrative of the progress of science will link many personal stories—stories of the lives, the struggles and epiphanies, the bold conjectures and discoveries, the triumphs and tragedies of individual scientists who contributed to the building of scientific knowledge. Such personal or individual narratives are embedded in the history-of-science narrative would of course include even those scientists who contributed indirectly by their failures. For example, Stephen Jay Gould's *Wonderful Life* tells of Charles Walcott's failure to realize the staggering importance of his discovery of the Burgess Shale fossils, shoehorning them instead into existing categories.[20] Fifty years later British researchers ascertained their significance. The Walcott story is thus embedded in and useful for the larger story of the Burgess Shale fossils.

One of the keys to success for writers who retell the history of science is to link together gripping, odd, or entertaining narratives of individual scientists who made significant contributions to a given field. Robert Jastrow does this in *God and the Astronomers,* as does Timothy Ferris in *The Whole Shebang.*[21] In many places, Ferris's history-of-modern-cosmology narrative takes the form of a string of individual scientists' stories, some touching, some amusing, some outrageous, but all utterly fascinating as they are told in vivid detail. Like Ferris, Denton adopted the wise strategy of embedding vivid individual stories in his inverted history-of-Darwinism narrative.

The Evidence and Discontinuity Motifs

Storytelling (or story demolition) is thus a master key to Dentonian rhetoric, and to the rhetoric of both Design and Darwinism. Yet stories and narrative rationality depend to a large extent on other artistic skills and rhetorical-symbolic strategies. The skill with which Denton decimates the credibility of the Darwinian genesis story depends significantly on his choice of key symbolic tools, such as his emphatic use of the word *evidence*. On seventeen occasions in the first three chapters alone, Denton employs the phrase "empirical evidence" or one of its cognates.[22] A tiny but noticeable rhetorical punch is delivered by the word *empirical,* and that is surely one reason Denton employs it so often. By means of his "empirical evidence" motif, which continues throughout the book, Denton projects a rhetorical ambience of scientific sensibility and competence. He creates an atmosphere of lively yet sober discussion, which hinges on an impartial review of the empirical evidence. He presents himself as radically committed to reporting the implications suggested by the evidence, no matter how unorthodox. This is one of the keys to the considerable success of his persuasion in the face of hopeless odds. By such symbolism, Denton seeks to align himself with the supreme value to which every scientist has pledged, namely, the priority of empirical investigation and the importance of going where the evidence leads.

It is the central theme of empirical evidence that is the bridge to the second kind of Dentonian narrative besides the retelling of the history of Darwinism. This second anti-narrative action, purely destructive in nature, is Denton's repeated degrading of the Darwinian genesis story. In nearly every chapter, Denton scrutinizes the Darwinian cosmological narrative of descent with modification, together with its vision of the continuity of all biological life forms, stretching backward across time in unbroken lineages to simpler animals and plants. Denton's strategy was to zoom in on this story line, focus on the gaps between different kinds of creatures, and submit them to a fine-grained inspection. As stated above, his rhetorical logic pivoted on two questions about the plausibility of these key transitions in evolutionary cosmology: (1) Is there good empirical evidence that these gaps were in fact crossed? Or (2) can we at least mentally conceive of plausible hypothetical intermediates to cross these gaps? This pair of questions is asked repeatedly throughout the extended criticism. The two questions eventually grow into their own respective chapters: "The Fossil Record," which is about empirical evidence, and "Bridging the Gaps," which is about hypothetical structural intermediates.

The stark contrast between nature's supposed continuity (claimed or theorized by evolutionists as historical reality) versus its discontinuity (demonstrated by empirical data, according to Denton) clearly dominates the book's argument. These two terms, "continuity" and "discontinuity," punctuate the text from the first chapter to the last. I was so struck with this pattern in one of my readings of Denton that I kept track of each usage of this pair of terms. In chapter 5 alone the "continuous/discontinuous" word cluster appears thirty-three times and the related "transitional/intermediate" word pair appears forty-three times. Denton's conclusion becomes overwhelmingly clear: Biology is now, and always has been, a *profoundly discontinuous phenomenon,* and this is made forceful by the sheer repetition of the terms in these two key word clusters. More importantly, these claims qualified Denton as a truly radical figure and made his book a rhetorical watershed in the history of Darwinism. In the years prior to Denton, a few secular critics—the Wistar dissenters and Pierre Grassé stand out—had published empirical attacks on the Darwinian *mechanism.* These questionings of the creative power of the mutation-selection mechanism were somewhat radical, but virtually no one outside of creationist circles had published a significant critique of the notion of macroevolutionary development or *continuity* itself.

Denton thus entered an already intense poker game, begun at the Wistar Symposium, and the onlookers gasped as he drastically raised the rhetorical ante. Employing erudition and an easy Socratic style, he questioned the existence of transitional branches in the evolutionary "tree of life" as well as Darwin's mechanism that was said to have driven and ramified the biological branches. Yet what was unprecedented, and extremely puzzling and frustrating to many reviewers, was that Denton "laid axe to tree" while at the same time dismissing the Book of Genesis as having any scientific relevance. As Mark Ridley wrote in the prestigious journal *Nature,* "Denton's purpose is purely destructive: he has no alternative to offer."[23] Actually, this "purely destructive" feature was foundational to his strategy. It was related to his ethos of radicalism.

Michael Denton's Ethos

Ethos—the communicator's credibility and reputation—is always central to rhetorical analysis in any context. Denton's case was unique, since he burst onto the scene with such a massive, detailed attack on Darwinism yet was a complete unknown in previous public discussions on evolution. Since ethos is always rooted in the details of a rhetor's biography and professional career, we must turn back to pick up the thread of Denton's own story where we left it earlier.

In the last chapter, I outlined the earliest dawning of doubt in Denton's mind as he completed his doctoral studies at King's College in London. Shortly after receiving his Ph.D. in biochemistry in 1974, he began to think about compiling a review of what others had already said about the empirical problems of Darwinism. As early as 1978, he had a synopsis of his assessment of the problems of neo-Darwinism and had begun to approach publishers. At that time, he wrote Arthur Koestler, organizer of the Alpbach Symposium, outlining his plans and expressing his frustration in his initial attempts at finding a publisher. Koestler sympathized, "I was much impressed by your synopsis and although I have nothing else to go by, I feel that your book may become a valuable contribution, throwing new light on an old controversy. I am not surprised by your difficulty in finding a publisher." Koestler said that his own publisher, Hutchinson's, was not a good prospect, since "your book, I am afraid, would be too technical"—and yet, ironically, Hutchinson's would actually become the eventual distributor in Great Britain of the Burnett Books edition of Denton's critique in 1985.[24]

To summarize his motivation and approach in this project, Denton included in the 1989 version of his professional resume a carefully worded paragraph about the writing of *Evolution* that bears repeating:

> For some years, I have been interested in the origin and evolution of life. Although accepting as all biologists must that the overall pattern of nature is best expressed by *an evolutionary tree of life,* I am sceptical that major evolutionary changes or macroevolution can be adequately accounted for in terms of the Darwinian model; that is by the gradual accumulation of small selectively advantageous mutations. I am unaware of any objective or quantitative evidence to support Darwinian claims. This scepticism was expressed in a book *Evolution: A Theory in Crisis . . .* which I wrote as an extra mural activity between 1978 and 1982, while a registrar in clinical chemistry in Hobart, Toronto and Sydney.

In effect, Denton had taken on a second registrar's job after leaving the hospital each day—rhetorical registrar of the scattered mutterings and murmurings of anti-Darwinian criticism spread over the previous 120 years.

Denton's scientific background was the foundation of his ethos as a scientific critic. His degrees and his toil as a biochemist established him as a true credentialed scientist. The back cover of the book (the 1986 Adler and Adler edition) lists Denton as an "Australian molecular biologist and medical doctor who has lived and worked in London, Toronto and Sydney, and who is best known for his biological research." It is not surprising that of the published reviews of *Evolution,* almost all of them (and all of the longer ones) mention Denton's credentials.[25] Of course, this is standard literary convention in reviews of works by relatively unknown authors. But I sense that most of these reviewers see Denton's scientific background as a highly relevant factor in the light of the radical thrust of his critique (whether they agree or not). This is especially true in the battleground of chapter 12, "A Biochemical Echo of Typology." In this chapter, the most controversial of his book, Denton argues that protein sequence analyses, far from supporting Darwinian claims, actually buttress the older notion of typology or archetypes. All of the reviewers who took positive note of this chapter mentioned Denton's credentials. Those who criticized the chapter did the opposite. They claimed the chapter cast a shadow of doubt upon his credentials and said or implied that Denton should know better.[26]

Denton's nonreligious orientation figured importantly in his ethos. Virtually every review noted that Denton was "not a creationist" and did not offer any biblical answers to the mystery of biology. This seemed to please many, although a few (Ruse, Eldredge, and Spieth) were puzzled or even irked about the probable confusion and damage that would be inflicted by Denton's book due to this odd quirk of his agnosticism.

In the case of an unknown author like Denton, ethos is largely established through the writing process itself. Denton's perceived flair for felicitous prose aided his ethos significantly. About half of the reviews praised the clarity of his writing. Typical were two British reviewers: Brian Inglis called the book "a remarkable (and agreeably readable) achievement," and Stuart Sutherland wrote, "Most of Dr. Denton's arguments are not new: what is new is the clarity with which he presents them." As one would expect, favorable reviewers generally praised his writing ability, while the negative ones often singled out stylistic or organizational flaws (e.g., Ruse, Shaffer, and Ridley). Perhaps the most revealing comment on style is in Walter Coombs's moderately critical review: "Denton pursues his avowed purpose, to critique the Darwinian model of evolution in a manner alternately *fascinating and tiresome*" (emphasis added). My suspicion is that Coombs's description of Denton as "tiresome" stems partly from his frequent recycling of certain word groups, such as the "continuity/discontinuity" or "transitional" word groups, and the "empirical evidence" cognate group discussed above.

Audience Adaptation and Goals

Denton had two main audiences in mind as he wrote *Evolution*. First, he wrote for a larger audience that included the educated lay public, mainly (but not exclusively) university-educated, having some interest in science generally or evolution in particular. This audience expectation was boosted by the fact that the book was published in Great Britain through a mainstream publisher (Burnett Books has also published Gould's books in Great Britain). A second anticipated audience—more specialized in science and thus more strategic in terms of influencing the future course of biology—could be pictured as a series of concentric circles. On the outside would be scientifically trained persons outside of biology; the next inner circle would be those working in fields relevant to the topic, such as biology and biochemistry; and at the center are the evolutionary specialists of all stripes in fields as diverse as zoology, paleontology, geology, and prebiotic chemistry.

A series of rhetorical goals is clearly implied by Denton's radical theses. In a general way he intended to influence all of his audiences' thought and belief toward fundamental doubt of the evolutionary paradigm. Such influence could be envisioned as ranging from tiny effects to huge, from as little as sowing seeds of incipient doubt about a few details all the way to a sudden, wholesale conversion like Behe's.

In the case of evolutionary specialists—the "inner circle"—Denton's goals would have to be limited. These are the scientists whose reputation and livelihood rested on the solidity of the paradigm and whose philosophical orientation was generally naturalistic or at least wary about any radical criticism of Darwinism. With this group, he sought to deflate dogmatism in the whole environment of Darwinian teaching and research. A minimal goal would be to coax Darwinists toward a greater willingness to admit evidentiary problems with the theory. More ambitious goals would be to create an openness to new and unorthodox ideas and to goad educators to begin revamping the way evolutionary biology is taught at all levels.

It appears from reviews that such goals were achieved very rarely or in rather slight measure among strongly convinced Darwinists. One of these rare occasions was the astonishingly positive review by Ashley Montagu mentioned at the outset of this chapter. Montagu's response was appreciative and conciliatory, and it invited Denton to the table to join the dialogue.

This is in sharp contrast with most Darwinists who sought to escort Denton abruptly off the stage. The most prominent examples of the latter approach are Niles Eldredge's brief review in the *Quarterly Review of Biology* and Michael Ruse's much longer one in *New Scientist*. In Ruse's rejection of Denton there are three major criticisms: First, many of his arguments are old. "He wheels out and dusts down all the standard arguments against evolutionism: supposed unbridgeable chasms between types or organisms; the impossibility of chance variations making the eye; the unlikelihood of spontaneous generation of new life forms." Second, Denton is confusing on just *what* he is attacking—the fact of evolution, the pathways, or the mechanism? If he is attacking the fact, then why "hardly a mention" of Darwin's "major source of support for the fact" in geographical distribution?

Third, in a fascinating misunderstanding, Ruse assumes that Denton is subtly arguing for the existence of God. This assumed conclusion of divine creation is taken (here Ruse quotes Denton) from the "sheer universality of perfection, the fact that we find an elegance and ingenuity of an absolutely transcending quality, which so mitigates against the idea of [chance]."[27] Ruse then asks, if God did this, how Denton "accounts for cancer. Either God was a bungler, or he deliberately allowed some cells to go astray, with horrific consequences. Neither option strikes me as very attractive." He then throws up his hands with transparent frustration and sighs, "Denton and I are in different paradigms, to use a trendy term of Thomas Kuhn."[28] Ruse's final comment is telling indeed in light of my hypothesis that the history of Design from Denton onward is thematically marked from the start by an implicit goal of a paradigm shift.

On the other hand, in terms of the broader audience, Denton did make some important inroads. Readers of his book in America were especially influenced by his arguments. It is an empirical fact that an extraordinary percentage of American professors who are now active in the Design Movement were either converted to skepticism or moved to activism by their reading of Denton.[29]

In other words, Behe and Johnson are just the tip of the iceberg of Dentonian conversions in academia. How did Denton manage in cases such as these to win a hearing and induce criticism on such a fundamental issue—considered "long settled" at the great universities? Are there any other keys besides the points discussed above—basic theses, anti-narrative strategy, and emphases on empirical evidence of discontinuity? To answer this, I must turn once again to Denton's rhetorical system and dig to a deeper level.

Rhetorical Strategy Once More: The Badge of Radicalism

I have already outlined key aspects of Denton's strategy, emphasizing the development of his theses and his narrative modes of criticism and advocacy. Now I will approach the question of strategy from a different direction, noting the recurring conceptual theme of radicalism that colors all of his rhetorical action. I need to illuminate the techniques by which he legitimized his radical call to revolt so that his book was polished into a highly persuasive document rather than stagnating as a curiosity of extremism.

A rhetorical analysis of Denton must reckon with his use of the term "radical." Denton did not conceal his radicalism. Rather, he transforms the term "radical" into a badge of bold intellectual independence. In fact, he states his theme of radical criticism explicitly in the two-page preface to his book. Keep in mind that by its very position in the text, this preface is crucial to his rhetorical strategy. After introducing the current controversy and tumult over evolution, he says that there are two philosophical approaches to this situation. One can "adopt the conservative position and view the difficulties as essentially trivial, merely puzzling anomalies, that will all be eventually reconciled somehow to the traditional framework. Alternatively, one can adopt a radical position and view the problems not as puzzles, but as counterinstances or paradoxes which will never be adequately explained within the orthodox framework, and indicative therefore of something fundamentally wrong with the currently accepted view of evolution." He then says that while most biologists who have written on evolution concede that the problems are serious, they nearly all take the conservative model that the ultimate solution will be found "by making only minor adjustments to the Darwinian framework." Denton, by contrast, takes a radical approach: "By presenting a systematic critique of the current Darwinian model, ranging from paleontology to molecular biology, I have tried to show why I believe that the problems are too severe and too intractable to offer any hope of resolution in terms of the orthodox Darwinian framework, and that consequently the conservative view is no longer tenable."

Denton's overture to his book strikes an adventurous, intellectually exciting note. He is careful not to surface immediately any quasi-creationist symbols (which appear later in the book), such as the lyrical descriptions of molecular machines as manifesting transcendent brilliance of design (see his chapter

"The Puzzle of Perfection"), or his hint about the rehabilitation of Paley's "argument for design." Instead, he begins with images of Darwinian tumult and emblazons the book's opening with a much more attractive intellectual symbol—*the courage of intellectual radicalism.* Indeed, Denton was not exaggerating; his position *is radical* in regard to prevailing orthodoxy. Of course, reviewers who suspected a covert creationist agenda characterized this claim to radicalism as a veneer to cover up a profoundly conservative move to sneak theology back into science.

Instances of radical rhetoric abound, especially in the closing chapter. One of these—the "explosive charge" quote hurled at Darwinists in the opening scene of that chapter—said that macroevolutionary claims have not been supported by one discovery since 1859. It is also not hard to hear a radical ring in his closing statement on mutations: "As the analogy deepens between organism and machine, as life at a molecular level takes on increasingly the appearance of a sophisticated technology and living organisms the appearance of advanced machines, *then the failure to simulate Darwinian evolution in artificial systems increasingly approaches a formal logical disproof of Darwinian claims*" (348, emphasis added). Here Wistar arises from the ashes with a rhetorical stinger attached. This is not the only place Denton invokes the "formal disproof" terminology. Earlier he had warned, "The fact that systems in every way analogous to living organisms cannot undergo evolution by pure trial and error and that their functional distribution invariably conforms to an improbable discontinuum comes, in my opinion, very close to a formal disproof of the whole Darwinian paradigm of nature. By what strange capacity do living organisms defy the laws of chance which are apparently obeyed by all analogous complex systems?" (315–16).

Several characteristics of Denton's writing render his radical position more plausible to the reader who is steeped in Darwinism. First, his radical barbs are mounted upon a fabric of *erudite richness* that stretches from cover to cover. My italicized phrase captures the sense of wide learning Denton projected in his book. I alluded to this quality earlier under "ethos," but here I refer not so much to writing style but more to content. He quotes frequently and effectively from an amazingly wide variety of sources. To read Denton is a highly educational experience. This struck many, including the prolific and well-read Montagu: "Imagine my surprise, then, when in reading Denton's critique of evolutionary theory, I found him to be a writer of the most astonishing range of knowledge in the natural sciences."[30]

Denton also fires a cluster of radical rhetorical rockets, each of them powered by the recalcitrance of the empirical data. Most of these are dazzling test cases, or Darwinian conundrums, which appear impossible to solve under the glare of "deep common sense." In his "Bridging the Gaps" chapter, Denton piles example upon example of complex organs, structures, or behaviors known to exist in nature that appear to defy any plausible step-by-step evolutionary scenario of development. One measure of the rhetorical strength of this chapter is the number of reviews (most of the published reviews, in fact) that cited this material as the most effective of Denton's book.

These reviews often summarized Denton's discussion of the difficulty of envisaging the pathway by which reptilian scales evolved into the complex engineering design found in the bird feather. Yet Denton's most dazzling test case is the bird lung mystery. Birds are said to have evolved from reptiles, but this poses an embarrassing problem due to differences in lung structure. Reptiles have a bellows-type lung, similar to that of humans and all mammals—air enters a branching, dead-end system; the airflow is reversed with each breath. All birds, however, have a fundamentally different kind of system. Air flows in, then breaks out into thousands of tiny parallel passageways (parabronchi) for the oxygen exchange, and then continues flowing in one direction through these parabronchi, and finally exits the lung. The bird lung is, thus, entirely unique in structure; it is a *circulatory* system (like the *cardiovascular* system of vertebrates).

After describing the two basic lung systems, bellows and flow-through, Denton asks what might have been the hypothetical intermediate forms by which a reptile's bellows or "dead-end" lung evolved into the completely different circulatory lung system of a bird. At the level of "deep common sense,"[31] the evolution of a bird's lung from a reptilian lung was portrayed as *incoherent*. It is virtually impossible to conceive a plausible step-by-step transitional scenario in thought experiments. What is rhetorically stunning about the entire "Bridging the Gaps" chapter, in which Denton discusses the bird lung problem, is that it amasses twenty-one such cases in which plausible intermediate structures cannot even be envisaged. Here again, Denton employs overwhelming "convergence of evidence," and his radicalism is rendered more plausible and intellectually inviting.

Dentonian Conversions: Implausibility and Vulnerability

In many ways, Denton's manifesto cleared a path for Intelligent Design. He galvanized and radicalized both of the principle rhetors of the coming movement, Phillip Johnson and Michael Behe. In a video interview, Behe was asked how he had become a skeptic of Darwinism. He replied that he had always assumed that Darwinism was true and never thought about the topic very much until he read Denton in 1987. At that point, said Behe, Denton revealed "very difficult problems for Darwinian evolution which I had never thought about and which no one in all my studies leading to my Ph.D. had bothered to mention. . . . I immediately recognized that they *were* difficult problems and I became angry that nobody had brought these up. I felt like I was being led down the garden path to a conclusion that didn't really have the evidential support that I thought it had." At that point, Behe took up Darwinism as his own research hobby, reviewing his own field of biochemistry through

Denton's critical gaze. Shortly thereafter, Behe designed and began teaching an introductory course on evolution at Lehigh, using *Evolution: A Theory in Crisis* as one of his texts.[32]

Behe and Johnson both report experiencing, through Denton, something akin to a "jolt of conversion," in the form of an intellectual awakening.[33] Which rhetorical factors were the most important to this sudden, forceful effect on their belief-structure? Denton's most important argument that won over academicians like Behe was that biology was now confronting a striking convergence of falsifying evidences in all its relevant branches of study.

Denton had no illusions about the difficulty of toppling Darwinism. One can hardly imagine a greater rhetorical challenge in science. Many lines of evidence had to be marshaled in for his ultimate radical act of paradigm demolition to become plausible. As a radical act *within science,* such criticism had to be argued with the firepower of overwhelming empirical evidence and careful articulation of the problem. This is just how Denton carried out his assault, and it explains his effectiveness as a genuinely radical rhetor. Yet his radical discourse was wisely grounded on a *conservative* value and mode of "doing good science." He gathered together under one cover all the major kinds of empirical anomalies and assessed them in one global vision. What scientifically minded person should complain about his appeal to the recalcitrant patterns in nature itself?

Denton's persuasion was all the more remarkable in view of the fact that his criticisms were devoted to destruction; he supplied no new cosmology. He did not lead readers to embrace any new belief about the *cause* of biological innovation—he himself was still looking for a replacement. In this sense, Denton defies Kuhn's historical model. He rejected the Darwinian paradigm himself, reverting his thinking back to a "pre-paradigm mode." He was inviting biologists, even evolutionists, to join him in this search for a new approach.

At the same time, in his chapter "The Priority of the Paradigm," he seemed somewhat resigned that professional biologists would cling to the current paradigm until a new causal mechanism of discontinuous biology was discovered. In this scenario, Denton's rhetoric functioned more as a spur to motivate scientists to discover the new paradigm. Thus, the radical nature of his critique is symbolically vindicated. In Kuhnian terms, Denton was *extremely radical;* he determined to be brutally honest in assessing the data, *even if it destroyed the current paradigm without furnishing an alternative.*

Of course, Behe and Johnson were reading Denton from a theistic perspective. They began any reassessment of evolutionary doctrine with a prior belief in God.[34] Their perspectives were similar; they viewed evolution as something that God *may have employed,* but they also possessed metaphysical space in their worldview that made it easier to entertain the radical skepticism of Denton. As Denton delineated the case for the biological realm as a "discontinuous reality," they considered the possible role of divine causation. For such newly converted intellectuals in America, the persuasion process did not stop there.

The Dentonian convergence of different lines of data convinced the converts that a rhetorical situation had emerged.

To those persuaded by Denton, this situation had two aspects—first, Darwinism's apparent evidentiary *implausibility,* and second, its *rhetorical vulnerability* in the hands of an effective critic. Once orthodox Darwinism was perceived as implausible and vulnerable, dissenting scholars were readily recruited, and there was a rhetorical multiplication effect. Furthermore, as individuals with prestigious academic appointments experienced Dentonian conversion, both intellect and heart were moved. Ideas flowed and emotions swirled—amazement, anger, disgust, or even contempt over a perceived intellectual betrayal—all of these and more were widely experienced. Michael Behe was not an isolated case in feeling (post-Denton) that he was grossly misled in previous teaching on evolution. These persons became much more of a threat than mere snipers in Normandy pillboxes. Those deployed at universities became creative activists who could and would exploit their academic niches and craft their own verbal attacks and invasions along the Normandy coastline.

I spoke earlier of Denton's rhetorical goals. One radical goal—inflicting widespread damage on Darwinian credibility—eventually took place, but in a much more indirect way he probably never envisioned: He persuaded a cluster of American intellectuals that his skepticism was justified. Denton ignited their imaginations and captivated their critical and rhetorical powers. Many would band together in the 1990s as a revolutionary movement whose main purpose was the toppling of Darwinian domination and the legitimization of intelligent design as a scientific hypothesis. To see how this unexpected turn of events came about, I must now tell the story of Phillip Johnson and his Berkeley-based campaign to put Darwin on trial.

4

The Virus Spreads

The Emergence of Phillip Johnson

The Aftermath of Denton

In American academia up through 1985, a broad rhetorical equilibrium prevailed in discourses on evolution. This sense of calm—a settled assurance about the reality of macroevolution—was beautifully expressed in 1976 by Cynthia Russett in her *Darwin in America,* a history of American intellectuals' response to Darwin. When she turned her attention in her final chapter to recent modifications in the Darwinian doctrine, she acknowledged that the violent phraseology Darwin used to describe natural selection had been revised: "The process need not involve any bloody competition at all." Then she added her assessment of the rhetorical situation at the time of her writing: "But in its essentials selection theory remains as it was one hundred years ago, and the essentials are beyond controversy. For many years it was possible to doubt the validity of Darwin's theory, but skepticism is not a tenable position today."[1]

The skepticism that appeared untenable to Russett in 1976 suddenly became highly tenable after the rise of Michael Denton nine years later. To adapt Stephen Jay Gould's famous terminology, the intellectual equilibrium was "punctuated and perturbed" by the abrupt appearance of a new hardy genus of dissent. While occasional traces of a secular doubt of Darwin can be seen in the evolutionary literature before 1985 (as shown in chapters 2 and 3), they had generated barely a ripple. There were a few other sophisticated attacks on evolution published at the same time as Denton's—notably *The Mystery of Life's Origin* by Charles Thaxton, Walter Bradley, and Roger Olsen; and Michael Pitman's *Adam and Evolution.* These books inflicted minor damage on the Darwinian edifice, yet together they produced a tiny fraction of the shake-up that Denton caused. It was primarily Denton's missile—with its broad target and comprehensive payload of secular arguments—which pierced the evolutionary calm.

Denton's attack was unprecedented in its rhetorical effectiveness. There are numerous reasons for this, many of which were discussed in the last chapter. Perhaps the greatest key to his success was simply that he managed to forge a rhetorical linkage of vast power that was previously unthinkable within mainstream biology. He claimed that *a consistent scientific empiricism*—quite apart from religious beliefs or motives—*will drive one toward doubt of (and opposition to) the Darwinian vein of evolutionary thought.* Portraying himself as anything but a creationist, he seized territory in the public discourse and began winning converts. Yet Denton stood out as a complex and enigmatic sort of revolutionary. He was palpably frustrated with his inability to furnish his followers with the ultimate weapon—a plausible mechanism of creation to replace natural selection. Nevertheless, he was convinced of the cogency of his critique and the timeliness of his call for scientific revolt.

Denton had dropped Kuhnian hints of a coming scientific revolution, and they were not at all subtle. The image of an incipient paradigm crisis was etched onto the closing chapter of *Evolution: A Theory in Crisis.* He described two earlier scientific crises—Copernicus's struggle against geocentrism and Lavoisier's against phlogiston—both of them classic cases used by Kuhn. He retold how these prerevolutionary paradigms had induced in scientists a mental blindness to glaring problems, and then he drew the parallel to evolutionary thought: "For the sceptic or indeed to anyone prepared to step out of the circle of Darwinian belief, it is not hard to find inversions of common sense in modern evolutionary thought which are strikingly reminiscent of the mental gymnastics of the phlogiston chemists or the medieval astronomers. . . . To the sceptic, the proposition that the genetic programmes of higher organisms . . . were composed by a purely random process is simply an affront to reason. But to the Darwinist the idea is accepted without a ripple of doubt—the paradigm takes precedence!"[2]

It is doubtful Denton thought that his book by itself would trigger the breakdown of biology's evolutionary paradigm.[3] However, his critique was clearly designed to help create a critical-rhetorical atmosphere for a revolution to gather steam. I have shown above two lines of anti-narrative he deployed as his weapons. One anti-narrative inverted the glorious story of the triumph of Darwin after 1859, and the other sought to shred the Darwinian "genesis tale" of origins—the step-by-tiny-step continuity of Darwinian development of new types of living things. His mission was to obliterate the credibility of the prevailing doctrine of *macroevolution driven by the engine of mutation selection.* How successful was he?

Denton's effectiveness can be evaluated historically at two levels. First, it can be pinpointed in retrospect as a historical watershed. Whether it stands as an *absolute turning point* in the rhetorical fortunes of Darwinism remains to be seen. This much is clear: Denton carved out the conceptual and suasory space[4] in which the forerunners of Intelligent Design would begin to toil and flourish in the coming decades. Through his massive rhetorical act, he stocked that suasory space with a panoply of resources—an unusual new perspective, along with passionately reformist motives, attitudes, and values that were absorbed and replicated in the Design Movement. Also, his narratives, models of style, and modes of argumentation and invention all shaped the rhetoric of future Design proponents. Above all, he thrust into the sphere of the general public a mass of data and several vivid test cases like the bird lung, which (to common sense at least) seemed quite difficult to reconcile with Darwinian gradualism. Many of his points became rhetorical commonplaces that were used over and over in later attacks on the Darwinian edifice.[5] Based on these lines of evidence, he also articulated a radical yet secular stance of skepticism toward macroevolution that was extremely appealing to many educated Americans who were already inclined to harbor some doubt about Darwinist scenarios but were not Genesis literalists.

Yet, on a different level, his book did not have any serious immediate impact on the course of biology. By all indications, the Darwinian paradigm weathered the shock quite well. Denton triggered no detectable decline of confidence among orthodox evolutionists. His criticism produced no retraction or softening of the rigid public stance of biology that *macroevolution by natural means is undeniably true.* Most importantly, it led to no conversions or defections among prominent evolutionary biologists. All published reviews by well-known scientists had panned *Evolution* as a strange, quixotic lunge and, above all, a "creationist tract."[6] The only prominent evolutionist with kind words for the book was Ashley Montagu, who nevertheless had chided Denton for going too far in his apparent questioning of common ancestry.[7]

Containing Denton: Spieth's Rhetorical Epidemiology

On the other hand, some biologists perceived Denton as a substantial threat. In the latter half of the 1980s, the book was selling widely enough and reviewed or quoted favorably in enough prominent newspapers and magazines to cause serious concern about its long-range potential to sow doubt about the scientific foundations of evolution. The few evolutionists who actually read the book had cause for worry. They had experienced firsthand Denton's effective marshaling of evidence from many different scientific fields and his confident, authoritative tone.[8] Geneticist Philip Spieth is typical of those who began to warn others about *Evolution's* siren call, noting the radical claim of Denton that "there is a crisis in evolutionary biology of fatal proportions." Spieth firmly rejected these arguments as "erroneous," "fallacious," and "spurious."[9]

Spieth and his fellow watchmen functioned as rhetorical epidemiologists, tracking Denton's flow of scientific persuasion as if it were an exotic intellectual infection—a mutant and savage virus to be reckoned with. This virus was tagged for destruction to keep it from spreading anti-Darwinism out of the quarantined huddles of religious creationists and into the unsuspecting university-educated public. Thus, Spieth ridiculed Denton's work as the sort of treatise that "could not pass the most sympathetic peer review"[10] and devoted much of his five-page review to a withering attack. Nevertheless, the geneticist fretted about the rhetorical power of *Evolution:* "The book . . . has the appearance of being strictly a book on biology. Intelligent laypersons reading Denton's book may think that they have encountered a scientific refutation of evolutionary biology." Sadly, a lay audience "cannot be expected to have the necessary expertise to avoid being deceived by the book's manifold abuses of evolutionary biology." It was critical, added Spieth, that Denton's secular antievolutionism never be permitted to lodge and fester in the universities. That could lead to intellectual conversions from which new outbreaks of the rhetorical infection could spring. The release of Japanese and French editions of *Evolution* in 1988 only served to underscore the danger of a global intellectual pandemic.[11]

Ironically, some of the most vigilant and hostile watchmen of Denton were self-identified religious believers. I turn once again to Spieth, the Berkeley geneticist quoted above, as a paragon of this odd *theistic indignation.* The lengthy review quoted above appeared in *Zygon,* a journal focused on the interaction between science and religion. Spieth, who is a Christian theist, first argued the science side of the issue, attacking Denton's notion of evolution as a result of mere "chance" or of "random search."[12] Then he moved into the religious dimension and said that the only crisis in evolution is a theological crisis for strict Genesis literalists. Yet as his tone of hostility subsided, he ended his review on a yearning, wistfully theological note. Contra Denton, he said, macroevolution had been convincingly established, but something very important still needed to be accomplished by scholars at the interface of biology and theology. Spieth suggested that scholars urgently need to integrate evolution with theological truths of God as ultimate Creator. Spieth gave several examples of earlier attempts, including that of the Jesuit priest and paleontologist Theilhard de Chardin with his Christian-evolutionary synthesis. Yet, he added, much more needed to be done to illuminate the relationship between Darwinian facts and theological truth: "We need good science and good theology. The two have operated too long in isolation. The time is ripe for a grand synthesis that will bring into register the complementary insights into human nature provided by modern biology and biblical theology."[13]

Little did Spieth suspect that just a few months after his review appeared in *Zygon,* his own friend and Berkeley colleague in the Law School, Phillip Johnson,[14] would become one of the most important American intellectuals captivated and energized by Denton's thesis. Furthermore, since Spieth was a staunch theistic evolutionist who accepted the basic accuracy of the story of macroevolution, he would have been horrified to learn just what kind of analysis Johnson would carry out on the relationship between Darwinism and religious belief. To properly introduce this new phase in the story of Intelligent Design, it may help to imaginatively recreate the historical situation in which Denton's disciple was recruited.

Phillip Johnson in England

In early October 1987, Berkeley law professor Phillip Johnson stepped down from the bus onto the London sidewalk and turned toward his visiting-faculty office, located just a few blocks away at the University College. As he strode through the chilly morning air, his thoughts turned once again to the lone frustration that marred the wonderful first weeks he and his wife, Kathie, had enjoyed as they settled into a year-long sabbatical in England. They loved England with its plethora of historical sites and museums; their hikes through the Welsh hills each weekend were glorious and exhilarating. Thankfully, he was free of all teaching duties—free to research and write on any topic. Yet, his first weeks of exploratory reading had led nowhere. He had read on insurance law and had toyed with educational issues, but nothing had seized his interest. He simply had not managed to settle on a professional research topic.

Following his usual route, Johnson passed by London's largest science book-store, and as he glanced at the storefront, he noticed a display of new books in the window. Pausing in front of the display, he noted two books on evolution—Richard Dawkins's *The Blind Watchmaker* and Denton's *Evolution: A Theory in Crisis.* His curiosity aroused, he entered the store, picked up copies of both books from a table near the door, and studied the dust jacket blurbs. The two biologists were apparently driving toward diametrically opposite conclusions. Sensing a delicious scientific dialectic, he bought both books and tucked them under his arm as he continued on to his office.

When he arrived, he settled down to a day of reading on a fresh topic, one on which he had read only sporadically—evolutionary biology. Although he was somewhat unfamiliar with the finer details of evolutionary theory, he was quite familiar as a legal scholar with the diverse modes and methods of persuasive argumentation. Within hours a fierce clash of argumentation over the plausibility of macroevolution had captivated his interest. A virtual debate over Darwinism unfolded at his desk between Denton and Dawkins. Denton the skeptic attacked macroevolution as empirically empty, a gossamer shell propped up by the sociological forces of a paradigm. Dawkins, the fervent believer and crusader, defended Darwinism as utterly compelling, supported by logical reasoning and even by his computer simulations called "biomorphs."

From Dawkins's vantage point, Darwinian evolution had vanquished all of its foes. It was logically impossible to deny.

Johnson, a former agnostic who had just ten years earlier embraced the Christian faith and joined a Presbyterian church in Berkeley, was fascinated with the strikingly different ways in which the two authors dealt with the whole notion of creation. Denton had mentioned the creationist explanation only briefly, in a relatively dismissive way, as if it were irrelevant to his search for a true explanation. In contrast, Dawkins singled out creation as the central thesis to be refuted. To do this, he focused on the famous "watchmaker" argument of William Paley in *Natural Theology* (1828, originally published in 1802).[15] Darwin himself as a young theology student at Cambridge University had read and enjoyed Paley's argument,[16] which rested on a famous central analogy: Just as a watch found on the ground would give evidence of having been constructed by a wise human creator, said Paley, so also (and even more) the consummate complexity of organic beings in nature pointed to a "design by intelligence"—that is, by God.

Dawkins said that Paley had at least one thing right. He had correctly singled out the key problem that evolution had to solve—the rise of sophisticated biological complexity. Dawkins even granted that Paley's argument was

> informed by the best biological scholarship of his day, but it is wrong, gloriously and utterly wrong. . . . All appearances to the contrary, the only watchmaker in nature is the blind forces of physics, albeit deployed in a very special way. A true watchmaker has foresight: he designs his cogs and springs, and plans their interconnections, with a future purpose in his mind's eye. Natural selection, the blind, unconscious, automatic process which Darwin discovered, and which we now know is the explanation for the existence and apparently purposeful form of all life, has no purpose in mind. It has no mind and no mind's eye. It does not plan for the future. It has no vision, no foresight, no sight at all. If it can be said to play the role of watchmaker in nature, it is the *blind* watchmaker.[17]

In fact, Dawkins saw Darwin's explanation as so crucial a blow to the plausibility of belief in God, he added that "although atheism might have been logically tenable before Darwin, Darwin made it possible to be an intellectually fulfilled atheist" (6). In other words, before Darwin, a potential atheist might be genuinely troubled by the stumbling block of biological design and its inference—a designer. The Darwinian revolution had cleared away that hurdle and had smoothed the path for the careful thinker to come to the inevitable conclusion of atheism or agnosticism.

Throughout the book, Dawkins explicated and illustrated again and again the "blind watchmaker" process—namely, random mutations, filtered by natural selection, "which is the very opposite of random" (41). He argued for this mechanism's ability to mimic any seemingly complicated work requiring a creator, even the production of the sonarlike navigational system of bats or the building of a human eye. By the mid-1980s, several rival theories in biology had appeared that somewhat downplayed the blind watchmaker mechanism. Among these were the neutralist genetic theory, transformed cladism, and Gould's theory of punctuated equilibrium. These seemed to pose a threat to classic Darwinism, calling for substantial remodeling of evolutionary theory. Dawkins took on each of these, arguing strenuously that each was a sham insofar as its being a credible challenge to Darwin. The blind watchmaker was fully vindicated in its confrontation with each rival.

Most importantly for Johnson, Dawkins wove into his arguments a variety of direct attacks on creation. In the chapter "Origins and Miracles," which dealt with the problem of the origin of life, he constructed what is probably his most crucial argument, which can be summed up in one sentence: "Creation explains nothing, because you have to explain where the creator came from." To give a sense of Dawkins's rhetorical approach and style in developing this point, I will quote two key paragraphs:

> So, cumulative selection can manufacture complexity while single-step selection cannot. But cumulative selection cannot work unless there is some minimal machinery of replication and replicator power, and the only machinery of replication that we know seems too complicated to have come into existence by means of anything less than many generations of cumulative selection! Some people see this as a fundamental flaw in the whole theory of the blind watchmaker. They see it as the ultimate proof that there must originally have been a designer, not a *blind* watchmaker but a far-sighted supernatural watchmaker. Maybe, it is argued, the Creator does not control the day-to-day succession of evolutionary events; maybe he did not frame the tiger and the lamb, maybe he did not make a tree, but he *did* set up the original machinery of replication and replicator power, the original machinery of DNA and protein that made cumulative selection, and hence all of evolution, possible.
>
> This is a transparently feeble argument, indeed it is obviously self-defeating. Organized complexity is the thing that we are having difficulty in explaining. Once we are allowed simply to *postulate* organized complexity, if only the organized complexity of the DNA/protein replicating engine, it is relatively easy to invoke it as a generator of yet more organized complexity. That indeed, is what most of this book is about. But of course any God capable of intelligently designing something as complex as the DNA/protein replicating machine must have been at least as complex and organized as that machine itself. Far more so if we suppose him *additionally* capable of such advanced functions as listening to prayers and forgiving sins. To explain the origin of the DNA/protein machine by invoking a supernatural Designer is to explain precisely nothing, for it leaves unexplained the origin of the Designer. You have to say something like 'God was always there', and if you allow yourself that kind of lazy way out, you might as well just say 'DNA was always there', or 'Life was always there', and be done with it. [141]

So for Dawkins, the proposal that an intelligent creator formed life in the beginning was vanquished on rather simple pragmatic grounds of informal reasoning. It would raise the unanswerable question "Where did God come from?" It is clear from the entire quote that Dawkins would declare the standard Judeo-Christian or monotheistic answer to his question "lazy," but it is unclear where the debate would proceed from there if the theist pointed out how biased Dawkins's informal logic was. The key observation that struck Johnson was the way Dawkins, throughout the book, deftly dodged such difficult areas as the fossil record and the Cambrian explosion and emphasized instead a vivid analogical-metaphorical understanding of natural selection. Here was a beckoning target for Johnson's own rhetorical analysis.

Johnson found that both books not only were superbly written prose but also constituted a pair of Herculean projects to convert the reader to opposite conclusions on a paramount question in science: *Just what is known with certainty about the origin and diversification of life?* Dawkins reassured the reader that humankind had come to secure and unshakable knowledge through Darwinian discoveries about how living things were developed by the powers pregnant in nature. Denton argued the opposite—that Darwinian theory had spectacularly flunked the empirical tests in regard to macroevolution and had become nothing more than a culturally and socially induced illusion. Both authors argued with vigor, creativity, and supreme confidence, conveying the impression that they viewed their own case as overwhelmingly persuasive. Johnson sniffed the rhetorical atmosphere of a courtroom trial and immediately began analyzing both sides in terms of the cogency of their arguments and the sufficiency of the evidence. He was amazed at the strength of Denton's case, and he recognized that *Evolution: A Theory in Crisis* was either "very, very wrong, or very, very important."[18]

At the end of the first day of reading, Johnson felt that here at last was a potential research topic for his sabbatical year. In addition to the dialectical clash, he was attracted to the subject because of its supreme sociocultural importance—its power as the supreme shaper of new ideas in the intellectual world. Virtually every trend in the university environment he inhabited was linked somehow to the reality of the Darwinian worldview. If this ultimate foundation proved false or even highly problematic, the universities and modern thought would undergo a vast intellectual upheaval.[19] Johnson glimpsed the dramatic supremacy and cultural centrality of the rising debate over macroevolution. (See chapter 1 for a discussion of these two qualities.)

In spite of his lack of formal training in science, Johnson felt confident that his own intellectual and professional skills equipped him more than adequately to evaluate this area. A graduate of Harvard and the University of Chicago Law School (placing first in his class), two of his earliest jobs out of law school were clerk to the chief justice of the California Supreme Court and then clerk to the chief justice of the United States Supreme Court, Earl Warren. After he started teaching at Berkeley in 1967, Johnson's primary field was criminal law and criminal procedure. Shortly after receiving tenure at Berkeley, he wrote a pair of law school texts on those two fields.

Johnson's law courses at Berkeley often revolved around the skilled analysis and construction of courtroom rhetoric. He taught criminal law students how to present their evidence and arguments in a convincing manner and how to spot flaws in opposing arguments. Professionally, he was a legal rhetorician—both as critic and strategist. As he assessed the arguments of Denton and Dawkins, he dissected the patterns of thinking they employed. He looked for hidden assumptions in their arguments; he analyzed their use of language—not just to clinch points but also to divert attention. He noted how assent was gained by redefining key terms. In short, Johnson was redeploying his old skills in a new area—he was becoming a rhetorical critic of evolutionary biology.

Within a week, Johnson had read both books through twice and had started to dig into the scientific literature on evolution at both the popular and technical levels. After the first few days of furious reading, he confided to his wife, "You know, I think I understand the problem with this whole field. But, fortunately, I'm too sensible to take it up professionally or to write about it. Nobody would believe me; I would be covered with ridicule and contempt. They would say, 'You're not a scientist, you're a law professor.' It would be absolutely foolhardy to take on this thing." Johnson adds, "Well, of course, that was irresistible. I started writing the next morning."[20]

That whimsical memory, which Johnson has repeated in countless published profiles, defines his chief rhetorical obstacle—to be taken seriously as a nonscientist, an outsider. One way to counter such criticism was through sheer thoroughness of preparation. From November 1987 through June 1988, Johnson carried out an intensive reading program. He began with the writings of George Gaylord Simpson and Ernst Mayer, both evolutionary biologists at Harvard and both esteemed as deans of that field during the twentieth century. He consumed all the published essays of Stephen Jay Gould and read the biology text of Futuyma and other widely used college-level texts on evolution. Then he turned to topical works on the history and philosophy of science and on key topics such as paleontology and chemical evolution. A significant portion of his time was spent perusing articles in the most prestigious science journals, especially the influential British journal *Nature* and its American counterpart *Science*. In his reading Johnson was absorbing necessary detail, but more importantly he was carrying out the same rhetorical analysis he had done on Denton and Dawkins.

Another important way Johnson dealt with his ethos problem—an assumed lack of scientific expertise and understanding—was through the cultivation of friendships with professionals in the various fields of evolutionary biology and then whenever possible having those experts check and evaluate his work. In the course of his reading in England, he stumbled upon the skeptical mutterings of Colin Patterson, one of the "transformed cladists" whom Dawkins had attacked. Patterson was an expert on fossil fish and served as a staff paleontologist and curator at the British Museum. One morning Johnson ventured out to Patterson's museum office and introduced himself, and that led to the two going to lunch and spending several hours together. Johnson explained his research project, probed Patterson's thinking, and asked for advice. By the end of their encounter, Patterson had agreed to help critique Johnson's work as he developed his research paper on evolution. During the spring of 1988 after receiving each draft from Johnson, Patterson would write back, encouraging him in his project and pointing out errors he had spotted. Over the coming years, this practice of submitting drafts to professional biologists would become one of the quality control methods of Johnson's writing on evolution. This not only helped build his own sense of self-confidence as a scientific critic, it also helped build a general respect for Johnson's critique as word spread of his practice of submitting to critical review.[21]

Johnson's New Berkeley Project: The Critique of Darwin

In August 1988 Johnson returned to Berkeley with a lengthy manuscript entitled "Science and Scientific Naturalism in the Evolution Controversy." His paper bore a certain resemblance to Denton's book, outlining multiple areas of evidentiary problems for macroevolution, but it featured several new controversies that had taken place during the 1980s, including court decisions in the United States on legislation dealing with evolution and creation in public schools. It also placed at center stage the galaxy of new evolutionary superstars, such as Gould and Eldredge in America and Richard Dawkins in England. The French zoologist Pierre Grassé, who was strangely absent in Denton's writing, was brought in by Johnson as a star witness—an acidic critic of Darwinism with superb credentials.

The main difference from Denton was visible in the paper's title. In the Johnsonian narrative of science, a crucial role was given to the *philosophy of naturalism*—the metaphysical conviction that nature is a closed system of causes and effects that cannot be affected by any outside factor like God. Johnson's placement of naturalism as part of the core of his thesis was an intellectual and rhetorical strategy somewhat different from that of Denton, who positioned naturalism in a much less prominent role. To Johnson, the presupposition of naturalism was central, not peripheral. He made it the central protagonist (or villain in the sense of a projection theme) on the Darwinist side in his explanatory metanarrative. Johnson argued that the ultimate foundation of the Darwinian worldview was not empirical evidence but metaphysical naturalism.

By the time he returned to Berkeley, Johnson had developed a basic strategy for advancing a credible critique of Darwinism. His chief difficulties would be to get the topic to be taken seriously and to avoid immediate dismissal as a religiously motivated proponent of "creation science."[22] To help deal with the latter obstacle, Johnson began to include in his earliest discourses a brief disclaimer identifying himself as a theist, thus admitting his worldview bias. To this admission he grafted two qualifiers: (1) He carefully excluded Genesis and biblical faith generally as relevant factors in testing the truth of Darwinism. Johnson's writings created a distance between his own position and that of Genesis literalists. (2) While admitting his theistic bias, Johnson pointed out that many in the field of evolutionary biology have their own equally strong religious bias in the other direction. In Johnson's rhetoric, the more dogmatic members of the latter group came to be described as the "Darwinian fundamentalists." The goal, proposed Johnson, is to see where the evidence pointed when the biases on all sides were acknowledged and efforts were made to minimize the distorting effect of these biases on scientific conclusions.

The metaphor Johnson used most frequently to sum up this first and foremost rhetorical goal was to "get the issue onto the table for discussion."[23] One may even identify this metaphor as a *key projection theme of design* in terms of the crucial initial goal to press toward. He often has stated that winning a given debate or argument on some issue was not nearly as important to his strategy as merely *getting the discussion started*—ideally in an intellectually unfettered setting, preferably on a university campus.

In August 1988 Johnson realized that one good way to get such a hearing for his paper on Darwinism was to arrange a faculty colloquium on his own campus. At Berkeley's colloquium office he filed a list of his invitees and sent them a copy of his paper. Included on this list were all of his fellow law faculty, several eminent philosophers, and professors in the departments of zoology, botany, genetics, and other related science fields. Cornell historian of biology William Provine, a staunch Darwinist, had just started to read Johnson's manuscript and suggested to him that Montgomery Slatkin, a friend of Provine's and a senior zoologist at Berkeley, would be an effective critic and should be invited.

Dozens of copies of Johnson's paper arrived in faculty boxes in early September. All members of the law faculty received a copy, and those who commented to Johnson were uniform in their general agreement with the validity of the thesis he was arguing.[24] Johnson enjoyed a sterling intellectual reputation in his department, and undoubtedly this helped make it possible for such a radical thesis to be considered in the first place. Another thread of Johnson's persona that made his paper seem in character to his long-time colleagues was his tendency ever since entering the faculty in the late 1960s to be "skeptical of authority, of conventional wisdom."[25]

The faculty seminar took place on 23 September 1988. Even though his response from law faculty was encouraging, only about four of the twenty individuals who converged on the faculty lounge at Boalt Hall to discuss Johnson's paper were from his own department. That did not bother Johnson; he was delighted that a number of prestigious Darwinists came, among them Montgomery Slatkin, botanist Thomas Duncan, visiting philosopher David Lyons from Cornell, philosopher Jeremy Waldron (with a joint appointment at Berkeley and Oxford), and renowned paleoanthropologist Sherwood Washburn. Fortunately, Johnson dictated notes of his recollections a few days later, which makes it possible to recreate a clear picture of his first oral performance as a rhetorical critic of Darwinism (see appendix 1).[26]

To start the meeting, Johnson said it was assumed that all of the attendees (as per his request) had read the complete text of his paper, and therefore he would only restate the paper's thesis briefly. He pointed his audience to the title, "which focuses attention on the difference between scientific evidence or scientific method on the one hand, and the philosophy of scientific naturalism on the other." Of this recap, Johnson recalls:

> My argument was that, although most people believe that an enormous amount of empirical evidence supports the general theory of evolution, *this is in fact an illusion.* Most people in the intellectual world are certain that evolution must be true because it is the only tenable naturalistic explanation for the development of complex life, or life in general, and it therefore *must be true* if non-naturalistic explanations such as creation are ruled ineligible for consideration. The evidence is then built up upon this pre-existing theoretical certainty based on philosophical presupposition. Non-evolutionary explanations of the evidence are not considered, and therefore the evidentiary support which seems to exist is the *product* of the cultural certainty rather than its *cause or support.* [See appendix 1, emphasis added.]

Johnson then suggested to the invitees that there were three ways in which his thesis could be attacked: (1) One could deny that the doctrine of evolution is in fact grounded upon "a conclusive presumption in favor of scientific naturalism." Johnson expressed his doubts "that anyone would want to deny so obvious a fact." (2) One could defend scientific naturalism on philosophical grounds, demonstrating that it is indeed a superior philosophy and we are therefore justified in taking it to be true. "The problem with taking this line," Johnson added, "is that it removes the question from the area of . . . scientific expertise, and puts it firmly in the camp of philosophy where scientists cannot claim to be experts." (3) Scientists could present compelling evidence that is "capable of proving the truth of the doctrine of evolution without support from the philosophical presupposition that only naturalistic explanations can be considered." Johnson said he doubted any such evidence exists, but he would be "eager to hear of any."[27]

Most of the criticism that followed took the second line of attack, supporting the inescapable validity of naturalism. Philosophers Lyon and Washburn aggressively argued that the presumption of naturalism "is philosophically valid and that any other approach was tantamount to irrationalism or mysticism." The philosophers, joined by the biologists, insisted on labeling the alternative to naturalism as "magic," which Johnson "considered a transparent effort to prejudice the inquiry from the start."

With the notion of creation labeled as "magic," the attackers said that proposing such "magic" as a legitimate possibility for study would make scientific inquiry impossible, since any such inquiry must presuppose uniform physical laws. Thus, Johnson's suggestion that science be open to nonnaturalistic explanations was an attack on scientific inquiry and "a form of irrationalism." Johnson has recalled his response: "Of course I denied this, claiming that my attachment to the scientific method was in fact greater than that of my critics, since I refused to rule certain hypotheses as out of bounds from the start. . . . I insisted that it may well be the case that science cannot answer all the questions that we would like to have it answer (such as the origin of life . . .). That is a real possibility, and every bit as plausible and eligible for serious intellectual consideration as the contrary assertion that there is a naturalistic explanation for everything" (see appendix 1).

The discussion turned for a while to the sections in Johnson's paper that argued that evidence for macroevolution was very weak if not nonexistent. Washburn tried to counter this by describing the willingness of scientists to modify their existing notions of evolutionary lineages and to redraw the human family tree where it was justified by the evidence. Johnson replied that this comment was irrelevant, since it showed only a readiness to revise the evolutionary narrative, not an openness to test the underlying assumptions of the paradigm. Slatkin explained that fruit fly experiments show that every characteristic of a fruit fly can be modified by inducing mutations. He conceded that "the resulting populations were infertile and composed of creatures which could not survive in the wild, but argued that all this was a problem only in the laboratory and would not impede evolution in the wild where populations are larger." Johnson summed up this part of the discussion with a note of private sarcasm, "In short, there was nothing new on the evidentiary line."

The third and final part of the discussion revolved around Johnson's agenda. Just what was he pushing for? Was he proposing that public schools teach creation science or religious doctrine of some kind? His law colleagues in particular, all of whom had told him before the colloquium that they were in substantial agreement with his paper,[28] pressed this question. Johnson, who considered this a very revealing moment, said in his memo of recollection, "I insisted that I would be content with a more modest science class of the kind outlined at the end of my paper, where there was candid teaching about the many problem areas that could lead to doubt about whether evolution is true. Some of my law colleagues then insisted that there could be no possible objection to reforming the curriculum along these lines, a point I labeled as naïve. . . . I noticed that the scientists present were staring stoney-faced at this suggestion from the lawyers that they would be willing to agree to a science class of the kind I proposed."[29] At this point the colloquium adjourned. All in attendance had been invited to Johnson's home for drinks and supper. Many participants went and extended the discussion into the night.

Was the colloquium a success in terms of Johnson's rhetorical strategy? He himself had no doubt that it was. He concludes his memo, "All told the session was remarkably calm, considering the controversial nature of my argument. I think that the scientists and Professor Washburn were baffled to see such a topic discussed in all seriousness in a law faculty, and I imagine they are passing the word now that some people over at our school have gone out of their minds. . . . *In all I consider this session a success from my point of view: A previously unthinkable claim—that the general theory of evolution is not true—was put out on the table for serious consideration and debated just as if it were any other academic topic*" (see appendix 1, emphasis added).

Johnson's own purpose for the seminar was, of course, not just to place the topic on the table but to test his paper's cogency and feel out the strengths and weaknesses of the attempts at refutation. That is why after restating his thesis, Johnson began by outlining the three possible avenues of attack. He wanted primarily to draw out the objections and only secondarily to answer them. At one point in his memo he even said, "I did not reply to these arguments particularly vigorously, however, being content to use this session for exploring the issues." The fact that most of the discussion revolved around a defense of *naturalistic thinking in science* was probably considered a practical and logical necessity by Johnson's critics in the meeting, but he viewed it crucial confirmation of his thesis.

It is rhetorically fascinating and instructive that a contention arose during the seminar when the philosophers charged Johnson with advocating "magic" in questioning scientific naturalism. These objectors felt Johnson was attempting to undercut the power and proven methodology that made science what it was—the *sine qua non* that made it uniquely powerful as an avenue to knowledge. Johnson, in reply, argued that the use of such a "stigma-word"[30] as *magic* was a rhetorical ploy and an unfair shifting of categories, and he sought to defend his view as more empirical than the naturalists' position. However, it appears that the use of such a stigma-word was a clever thrust and fairly effective as a rhetorical approach to counter Johnson's thesis. By the same token, Johnson himself had used a stigma-word that was equally forceful in his opening statement when he described belief in macroevolution as an *illusion*. We have seen other stigma-words above—*fundamentalist,* which was used by Johnson of Darwinian scientists who are dogmatic, and Denton's use of *cosmogenic myth* to describe Darwinian evolution.

The faculty session was Johnson's debut in a new rhetorical arena, but it was primarily a way of gathering critical response. In this regard, the unique venue of a faculty colloquium needs to be held in mind. It differs markedly from more mundane communication venues, such as political speeches, sermons, class lectures, or casual discussions. Johnson's faculty encounter was akin to a graduate seminar for which an author's works are read carefully during the week and then discussed around the table for an hour or more. Johnson's paper was fairly lengthy—about forty thousand words—so the vast majority of his suasory effort took place through his paper.[31] Nonetheless, Johnson's willingness to call for a seminar and to invite participants who would be strongly hostile toward his thesis became a rhetorical statement itself and thus a suasory means of multiplying the effect of the paper.

In comparison with Johnson's later writing and speaking, most of which began three years later, the 1988 paper and colloquium were like a core sample or microcosm of what would emerge as his chosen argumentative approach. It was this paper that grew, after extensive remodeling, into *Darwin on Trial*, and although many new scientific examples, stories, and subsidiary points were added over the next three years, his main thesis had been fully developed by this point. The Berkeley colloquium was to be reenacted dozens of times in the coming decade as Johnson cultivated diverse venues of engagement with American intellectuals and especially with Darwinian biologists and other scientists. Johnson found this part of his project sheer fun, yet it was a crucial part of his strategy as he created his own story of intellectual dissent.[32]

Johnson obviously was buoyed in his spirits not only by his sense of a successful seminar but also by a word of encouragement he received later: "The professor who presided over the colloquium, . . . who is very sophisticated in philosophical issues, told me privately afterwards that he was very impressed by the paper and even more impressed by my performance in the debate, particularly in view of the intellectual courage he thought it required to advance such an argument in front of hostile experts from the sciences" (see appendix 1). As a budding antievolutionist rhetor in a sophisticated university environment, Johnson's first clash was all he hoped for and more—a profitable and educational practice round with the bonus of substantial exhilaration. In the mind of any critic of a powerful orthodoxy in a field outside his own, such success breeds success. In retrospect, Johnson's Berkeley faculty seminar stands as a powerful historical symbol of things to come.

Gould vs. Johnson: The Campion Debate

Just fourteen months after the Berkeley colloquium, Phillip Johnson's plane began its descent into Logan Airport in Boston, carrying him on a collision course with Stephen Jay Gould. In a matter of hours Johnson would be meeting the prestigious Harvard evolutionist for the first time at a private gathering of experts called together to discuss the problem of "Science and Creationism in Public Schools." Johnson the Radical, anticipating some sort of exchange with Gould the Reformer, assumed that rhetorical sparks would fly. Yet to a law professor, this was the fun of the all-important work of engaging one of the world's renowned evolutionists in discussion. Such engagement played an important role in Johnson's short-range and long-range rhetorical strategies. What he did not quite expect was the ferocious attack and intense duel that would break out.

Gould had already established his reputation as one of the twentieth century's most prolific masters of scientific prose and was undoubtedly America's most popular and widely read spokesperson for evolution. His monthly columns in *Natural History* were periodically collected and published, producing a string of best-sellers, starting with *Ever Since Darwin, Hen's Teeth and Horse's Toes,* and *The Panda's Thumb.* By the late 1980s, his "punctuated equilibrium" proposal—an explanation of how macroevolutionary speciation could take place in a way that would leave exceedingly few transitional fossils—had been almost universally welcomed into the explanatory panoply of evolutionary biology. Because of his centrality in the debate over macroevolution (especially in connection with the problem of transitional fossils), Johnson had woven key strands of commentary and criticism of Gould into his own new book. Indeed, Gould was now the most prominent target of Johnson's analysis, dominating three of his book's twelve chapters and figuring centrally in several others.[33]

So Johnson was understandably excited when he heard in late November that Gould had decided to attend this informal, off-the-record meeting of a dozen scholars and professionals from various fields that had been convened to discuss the problem of creeping creationism. Johnson flew to Boston to attend the weekend meeting, and his shuttle sped him to the meeting place, the Campion Center, a Jesuit-run retreat on the west side of Boston. Glancing over the list of participants, Johnson relished the prospect of dialogue not only with Gould but with other creation/evolution luminaries as well, including the renowned paleontologist David Raup and theologian Langdon Gilkey, both from the University of Chicago.[34] Also attending were astronomer and historian Owen Gingerich, from the Harvard-Smithsonian Center, and Charles Haynes, an expert on the First Amendment who played a leading role in organizing the meeting.

Johnson had mailed to all attendees his Berkeley paper along with an eight-page point-by-point summary of his position (hereafter, the Campion Summary), which he had prepared specifically for the meeting. In this summary's title he suggested his own alternative agenda: "Science and Scientific Materialism in the Schools, the Universities, and the Public Broadcasting System."[35] This phraseology of course grew out of one of his most urgent rhetorical tactics—shifting from a "defense" of creation or creationism to the "offense" of placing a spotlight on the basis of our knowledge that Darwinism is essentially true.

For Johnson, the Campion meeting was a fitting way to cap off a very productive year since returning from England. To a large extent the past fourteen months had been an extension of the new biological research career he had launched in England. Continuing research was critical; he knew that his rhetorical project of fomenting radical doubt of Darwinism in the universities would be an intensely intellectual campaign. It would enjoy success only in proportion to the thoroughness and breadth of his own scholarship. Viewing the situation through the lens of Aristotle's classical triad of rhetorical resources—*ethos, pathos,* and *logos*—Johnson saw that his groundwork of impeccable and up-to-date research would be crucial in two ways. First, it was indispensable to effective *logos,* that is, rational argumentation, or "proof arising from rational linkages," of clear and established ideas.[36] Second, his research would greatly establish a positive *ethos*—*ethos* being the persuasive effect of the rhetor—in that it would enable people to view him as a trustworthy spokesperson even in an area outside his field of criminal law.[37]

In addition to research, the previous year was dominated by two labors of persuasion—dialoguing with scholars and rewriting his Berkeley paper into book form. The dialogue with scholars, which increased steadily during the year, began with a burst of correspondence with two colleagues who attended the Berkeley seminar—philosopher Jeremy Waldron and zoologist Montgomery Slatkin. Waldron wrote a short letter challenging Johnson to supply the "lawlike scientific principles" that would qualify Johnson's *own scientific research program* (based as it was on an openness to the possibility of some sort of creation) to function as a worthy and fruitful competitor to the Darwinian research program with *its* well-known principles. Johnson's reply to Waldron was a letter-essay that elaborated on the problematic nature of Darwinian assumptions.[38]

Because Waldron insinuated that Johnson was of one mind with the scientific creationists, Johnson inserted a disclaimer that stands as one of the earliest attempts to define the unique *ethos* of a Design critic:[39]

> I am not one of that party. . . . In common usage creationism has come to mean Biblical literalism, and in particular to imply the claim that the earth is no more than a few thousand years old. I have not personally studied the evidence regarding the age of the earth, or the metaphysical assumptions that underlie the methods of measurement, but I have no desire to quarrel with the generally accepted estimate of four or five billion years. More importantly, I am not concerned with whether the scientific evidence agrees with any dating derived from the Bible, nor do I assume that the literalists are reading the Bible correctly in its own terms, even without regard to any conflict with the scientific evidence. So I am no creationist, as that term is currently used, although I do believe that God can create. Let me call myself a "theist," in a perhaps futile attempt to avoid guilt by association.[40]

Montgomery Slatkin's letter to Johnson contained criticisms and questions similar to those of Waldron as well as a prophetic admonition. Slatkin warned Johnson that if his critique was thrust directly into the public's gaze, he might find himself swept up as the new leader of antievolutionism in the United States. This is the last thing Johnson would want, Slatkin pointed out, and he suggested that rather than having his paper published in book form, Johnson should turn it into a series of articles in academic journals.[41] Johnson was amazed at Slatkin's suggestion that he ought to quarantine himself from the rabble of popular creationism and immediately started rewriting his paper into book form and began looking for a publisher.

By late spring 1989 Johnson had completed his first book draft, heuristically titled "The Creation Question." He circulated the draft to biologists and other critical reviewers. As criticisms and suggestions poured in, Johnson kept tinkering with the text during the rest of 1989. Drafts were sent to several publishers who indicated interest. Just before the Campion meeting, a letter had arrived from a university publishing house informing Johnson that three critical reviewers, while impressed with his scholarship, could not come to a positive decision to publish. Earlier, to Johnson's astonishment, a major publisher turned his book down because his position was viewed as too much of a compromise; it was not controversial enough to generate interest.[42] In spite of the rejections (he expected it would be difficult to find a publisher), he felt fairly encouraged by his progress.

As the search for a publisher dragged on in late 1989, Johnson started to write a much shorter critique of Darwinism in hopes of eliciting written critical responses from prominent evolutionists and getting it into print. At about the same time as the Campion meeting, Johnson's shorter piece began to circulate, and it quickly elicited a mild response from Gareth Nelson at the American Museum of Natural History and feisty criticisms from William Provine at Cornell and Thomas Jukes at Berkeley. The resultant symposium-in-print, including Johnson's "Reply to My Critics," was published in *First Things* (October and November 1990) and then immediately reprinted by the Foundation for Thought and Ethics as *Evolution as Dogma: The Establishment of Naturalism.*[43]

In early December 1989 Johnson's shuttle pulled into the Campion Center and he was quickly settled into his guest room. Shortly afterward dinner was served, followed by the Friday evening session. All twelve participants except Gould were able to attend the opening session, which was designed for the participants to get acquainted and open up free-flowing communication. One by one, each person shared his background and worldview. It soon became obvious that a wide variety of beliefs, values, educational paths, career experiences, and philosophical foundations were represented around the table. By the end of the evening a mutual openness and cooperative spirit had been established.[44]

On Saturday morning as the participants were gathering for the second session, Johnson and Gould met briefly. Their chat was polite, but Gould signaled to Johnson that, having read the material shipped from Berkeley, his response to Johnson was going to be an urgent polemic. He told Johnson, "You're a creationist, and I've got to stop you."[45] As the morning session got underway, Johnson was first given an opportunity to summarize the gist of his Berkeley paper and the much shorter "Campion Summary." For over an hour Johnson reviewed point after point of his thesis. Near the end of his presentation, paleontologist David Raup briefly interjected his own evaluation of Johnson's work. He said that he had read the Berkeley paper and had even distributed it and discussed it with his students in one of his graduate seminars at the University of Chicago. Raup said he and his students agreed that Johnson's scholarship was fully accurate in its scientific detail and contained a clear understanding of macroevolution's anomalies and empirical gaps. In fact, said Raup, the various lines of evidence for Darwinian macroevolution were not nearly as strong as one would hope. The key point was clear—Raup had briefly but unmistakably certified the empirical quality of Johnson's critique.*

At this point, Gould immediately seized the floor and "donned the mantle of Darwin." Displaying agitation in voice and shaking bodily, he began to set the record straight.[46] In what one observer described as an "obliteration attack,"[47] Gould started pelting Johnson's thesis with vehement criticisms. Oddly, Gould argued that there is plenty of scientific evidence in the fossil record for Darwinian evolution and cited a number of fossil series that allegedly supported the validity of step-by-step Darwinian macroevolution. On this point, Gould was clearly backing away from the critical stance that had made him famous—that gradualistic neo-Darwinism was incapable of accounting for the rarity of transitional fossils.[48] On the contrary, Gould implied that the branches of evolutionary trees could be reasonably traced in the fossil record.

Very early in the attack, Johnson stepped in with strong rebuttals of a number of Gould's points, and immediately the two were engaged in a furiously paced seesaw debate that lasted for nearly an hour before a spellbound audience. The rhetorical purpose of Gould was clear—to so bury Johnson's criticism in a torrent of contrary evidence that the net effect would be to illegitimize both the logos and ethos of Johnson's critique while defending classic neo-Darwinism. However, in light of the *esprit de corps* that had been established among the Campion participants the previous evening, many felt the emotional intensity of Gould's all-out attack clashed with the spirit of the meeting and somewhat undermined his credibility.[49] Gould stayed on for the rest of the Saturday sessions and ate dinner with Johnson, after which he left to catch a plane for New York to appear on the *Today Show*. After he left, the subject of discussion among the remaining group was for some time the display of emotional intensity by Gould. Several said they had witnessed a side of Gould they had never seen before. The four participants who were sympathetic to Johnson's position said later they were impressed both by Gould's "technical finesse" in blitzing Johnson's thesis and also by Johnson's ability to contend all the key points Gould made. In the final analysis, many who attended described the private debate as a draw.[50]

As Johnson boarded his plane for California, he carried with him a psychological satisfaction—he had engaged Stephen Jay Gould and survived the baptismal blast. He also was gratified that he had become personally acquainted with key experts in a number of fields. Ongoing interaction with several of these would become pivotal to his project in coming years, especially with

*In my October 2000 interview with David Raup, he said, reflecting on the Campion meeting: "Phil Johnson's work is very good scholarship and, of course, this has been widely denied. He cannot be faulted; he did his homework and he understands 99 percent of evolutionary biology."

David Raup, whose quiet moral support and interactions became both an encouragement and a frustration (due to his public silence) in the years that lay ahead.[51]

Why and how is the Gould-Johnson debate significant? First, it gave both Johnson and Gould a more acute sense of the nature and scope of each other's rhetorical strategies with their inherent strengths and weaknesses. These were important insights in light of how central Gould had become in Johnson's critique. In this regard, the Campion debate was analogous to Johnson's Berkeley seminar in that it was useful for revealing the thought processes and rhetorical behavior of his opponents, clarifying the potential lines of attack against his thesis.

Second, it provided important background to Gould's lengthy attack on *Darwin on Trial* that appeared in *Scientific American* (July 1992). That review was mounted with virtually the same goal as the Campion debate—to overwhelmingly discredit Johnson as a knowledgeable and cogent critic of current evolutionary thought. Could Gould's perception of Johnson as a sufficient "menace"[52] to warrant a four-page bashing have been rooted in his meeting and debating Johnson personally at Campion?[53]

Third, the debate was valuable for Johnson because it revealed a unique species of rhetorical phenomenon in evolutionary response to radical empirical criticism. The Campion group had witnessed the spectacle of a reform-minded evolutionist (Gould) immediately distancing himself from previous criticism of neo-Darwinism and coming to the defense and vindication of the very gradualistic evolutionary scenarios he had previously disdained. It appears that conditions have to be just right for such an odd rhetorical act to occur, such as when academically respectable critics appear to be eroding confidence in the Darwinian paradigm or in its naturalistic base.

The fourth point, perhaps the most important, is that any revolutionary movement like Design thrives on *stories of engagement and success.* Such incidents serve as encouraging milestones along the path toward long-range goals—in this case, a scientific paradigm shift. These engagement stories function not only as evidence of progress but also as demonstrations of legitimacy. In the hands of leaders, they also serve as tools of motivation to spur on the rank and file of the movement.

Johnson was somewhat limited in his ability to use this incident in public communication. By common agreement the meeting in Boston was off the record. Yet he could (and did) mention it in private conversations and informal gatherings. It is in this context that many lines of projection themes began to chain out, and a mature rhetorical vision began to take shape. In retrospect, Johnson's Campion debate with Gould, coupled with Raup's quiet certification of Johnson's research, is the most important engagement story that circulated around the fledgling movement and energized its members in the period before *Darwin on Trial* was published.

Charles Thaxton and the Ad Hoc Origins Committee

Johnson's encounter with Gould became the occasion for a key meeting held two months later in February 1990 at an airport hotel in Portland, Oregon. About twenty-five individuals, many of whom comprised the earliest core of the Design Movement, gathered to hear Johnson report on the Boston meeting and discuss plans for the future.

This gathering was the latest in a series of irregularly scheduled meetings of like-minded skeptics of macroevolution that had taken place since 1987 under the name "Ad Hoc Origins Committee."[54] As early as 1987, when the Ad Hoc Origins Committee began to meet, the intellectual leadership of the committee was turned over to Charles Thaxton, the chemist and author of *The Mystery of Life's Origin*. Indeed, to understand the intellectual essence and rhetorical dynamics of the Ad Hoc Origins Committee, which was one of the forerunners to Intelligent Design, one must begin with Thaxton's "information-theory" critique of chemical evolution. This critique, first suggested in *The Mystery of Life's Origin,* was further developed in his later writings. Because of Thaxton's importance, I shall turn now to his role as one of the fathers of the Design Movement, which also will give important background to the February 1990 Ad Hoc Origins Committee meeting in Portland.

Charles Thaxton received his Ph.D. in chemistry from Iowa State University in 1968, and by the time he held a postdoctoral fellowship at Harvard in the history of science (1976), he had developed an avid interest in *chemical evolution,* also known as "abiogenesis"—the highly speculative field that sought to solve the enigma of how one-celled life got started on earth. In the late 1970s he met Walter Bradley and Roger Olsen, two other scientists who had similar research interests. After extensive discussion of the topic and their individual research interests, Thaxton decided to work with the other two to begin the writing of *The Mystery of Life's Origin* (hereafter referred to as *Mystery*) as a college-level comprehensive critique of abiogenesis research.

For a half century this field was dramatized and symbolized by the icon of Stanley Miller's glass-tube circulating apparatus and spark chamber, mentioned in chapter 2, which in December 1952 began to produce amino acids from sparking a mixture of hydrogen, ammonia, methane, and water. Because amino acids are joined together in living systems to form proteins, the apparatus was portrayed as having achieved a powerful simulation of mechanisms by which the building blocks of life could have been produced on the early earth. According to one "chemical soup" scenario, once concentrations of amino acids were sufficiently elevated in evaporating ponds, they combined to form chainlike macromolecules, which then eventually assembled themselves to form living organisms.[55]

Thaxton, Bradley, and Olsen viewed such scenarios as riddled with serious problems that were very rarely discussed (except among specialists in the field), and thus abiogenesis, as a scientific field, seemed overripe for a multifaceted critique.[56] The authors presented in *Mystery* detailed analyses of virtually all the relevant scientific topics of chemical evolution. Among their targets of analysis were the early earth atmosphere, geology and geochemistry, organic chemistry and Stanley Miller's "soup hypothesis," protocells (e.g., "proteinoid microspheres" and "coacervates"), information theory, entropy and the second law of thermodynamics, and even scientific questions about the protocol and relevance of recent experiments. Because *Mystery* was a moderately technical book, it was somewhat beyond the reach of the average reader.[57] For this reason, and because it was written with the assumption that the earth indeed had cooled about 4 billion years ago, *Mystery* did not fit well in the genre of scientific creationism. And yet its scientific thrust was one of cumulative, overwhelming skepticism about the plausibility of chemical evolution scenarios that graced the world's introductory biology texts.

This skeptical thrust is seen in a typical sentence from Thaxton's "Summary and Conclusion" chapter: "A major conclusion to be drawn from this work is that the undirected flow of energy through a primordial atmosphere and ocean is at present a woefully inadequate explanation for the incredible complexity associated with even simple living systems, and is probably wrong."[58] The very phrasing of this summary sentence signals departure from the polemic of a creationist text. Note the cautious yet confident tone contained in two key word clusters—"at present a woefully inadequate explanation" and "probably wrong." Such wording does justice to the severity of the problems and the lines of evidence already presented in the book, yet it leaves the door open slightly to the possibility of future discoveries resurrecting one's theoretical confidence. Due to the timing of *Mystery*—published in 1984, just a year before Denton's *Evolution: A Theory in Crisis*—these two books began working in concert as powerful midwives and sculptors of the emerging genre of Design.

How did the scientific establishment react to Thaxton's sophisticated skepticism? A few researchers in the field of abiogenesis hailed the book as a major contribution.[59] Through careful cultivation of contacts, Thaxton was able to secure several helpful endorsements. A foreword by San Francisco State University biologist Dean Kenyon added a modicum of rhetorical weight to the book. Kenyon had coauthored *Biochemical Predestination* (1969), an early, optimistic text on chemical evolution, but after reading scientific criticism of the prevailing theories, he changed his mind in the late 1970s and soon became an outspoken skeptic of his previous position.[60] The rhetorical legitimacy of *Mystery* also was reinforced by appreciative blurbs on the back cover from two celebrated scientists—Dartmouth space physicist Robert Jastrow and NYU chemist Robert Shapiro, who was writing his own critique of chemical evolution scenarios.[61]

Yet in its wake, *Mystery* triggered a wild cacophony of dissonant reverberations among scientists, ranging from high praise[62] to harsh criticism.[63] On the side of praise, Yale University epidemiologist James Jekel's comment is typical: "The volume as a whole is devastating to the relaxed acceptance of current theories of abiogenesis." On the critical side, the attack was not so much on the body of the book (which summarized the theory's scattered scientific problems) but rather on the epilogue. This controversial chapter discussed five alternatives to the current naturalistic chemical evolution scenarios: (1) *new natural laws*—science will yet discover the true naturalistic explanation, (2) *panspermia*—life drifted to earth from elsewhere, (3) *directed panspermia*—life was sent here by a civilization elsewhere in the cosmos, (4) *creation by a "creator within the cosmos"*—some other intelligence, perhaps non-carbon-based, engineered life, and (5) *creation by a "creator beyond the cosmos"*—life resulted from something like one of the classic theistic scenarios.[64]

Points four and five comprised *Mystery's* scientific heresy—its transgression of scientific decorum. Thaxton, Bradley, and Olsen had suggested two creation possibilities—either some creator within the universe (the idea of British scientist Fred Hoyle,[65] who was mentioned in chapter 2), or a creator who transcends or preexists the universe. Reviewers who lauded *Mystery* praised this section as helpful and provocative even though it was exploratory in nature. Those who panned the book usually were quite critical of this section.

In the period immediately following the publication of *Mystery* (1984–86), Thaxton began exploring the creation options in greater depth, focusing on the historical and philosophical reasons for which intelligent design had been banished from consideration within science during the previous two centuries. In 1986–88 he wrote a pair of papers to present his two emerging theses: (1) DNA and other information-bearing molecules were *prima facie evidence* of an intelligent designer because of their quantities of "specified complexity" embedded in the digital informational sequences, and (2) there is no good reason why science cannot consider the notion of "intelligent cause" or indeed any kind of cause for which humankind has "uniform sensory experience."[66]

In his rhetorical strategy to defend these points, Thaxton employed powerful weapons of analogy. For example, he argued that an observer beholding Mount Rushmore immediately perceives an "intelligent cause," even if he does not know the details about how or by whom the presidential faces were sculpted. So also the scientific observer may legitimately infer some sort of "intelligent cause" when looking at features of nature like DNA, which bear telling marks of design. Thaxton noted that one does not have to know the exact identity of the designer in order to come to a positive conclusion of design.

In another frequently presented analogy, Thaxton argued that just as SETI (Search for Extra-Terrestrial Intelligence) surveys radio emissions of stars looking for codelike sequences that would indicate intelligent origin, so scientists should have the academic freedom to be able to look at DNA and other information-rich systems in nature and consider the possibility of intelligent causation. The SETI analogy, which quickly came to function as a favorite commonplace in the rhetorical repertory of Design, proved especially helpful since SETI is seen as a valid scientific research program. For a period of time in the 1990s, NASA even funded some projects that scanned the night skies for radio messages that might be produced by distant civilizations. In other words, the "detection of intelligent design" is not a wild or illegitimate (or inherently religious) idea in science. Rather, it is currently being carried out by astronomers who are sifting inflowing galactic static by means of radio telescopes and code-recognizing computers.

In 1987 when Thaxton was developing these new ideas and starting to write his second paper, the Ad Hoc Origins Committee drew up plans for a scientific conference to discuss these ideas. Their chosen topic: *the mystery of the origin of genetic information.* Funding was obtained and a call for scientific papers went out in early 1988 to scholars from a wide variety of disciplines—mostly outside of the Ad Hoc Origins Committee. The three-day conference, entitled "Sources of Information Content of DNA," was directed by Thaxton, and it took place at a hotel in Tacoma, Washington, in June 1988 with about eighty persons present. One of the chief purposes was to explore and test Thaxton's theses—that the informational sequences of DNA yield a strong inference to intelligent causation and that scientists should have the freedom to consider the explanatory notion of *intelligent cause.*

Major impressions were made by two presenters who had not yet come to an intelligent cause conclusion but who were highly sympathetic with Thaxton's ideas. Hubert Yockey, an information theorist known for his analysis of the information content of DNA, presented a lengthy paper that supported a key contention of Thaxton's—that DNA was structurally or mathematically identical to human language.[67] An even more powerful impression was made by Michael Denton, who flew in from Australia to present a paper. His two presentations on the integrative complexity of proteins and the implausibility of their complex, hypersensitive structure arising via "chance mutational mechanisms" had probably the greatest impact of all on the conferees.[68]

Encouraged by the conference's results, Thaxton hurried to complete another book he had been laboring on for over ten years, *Of Pandas and People.* This book, which addresses not only chemical evolution but macroevolution as well, is highly significant in the history of Design. It took Thaxton's intelligent design notions and connected them with a wide range of evolutionary puzzles. More importantly, it was the first peer-reviewed book that advocated intelligent design as an alternative scientific model, legitimate for consideration alongside naturalistic evolutionary models. Of course, *Of Pandas and People* concluded in each chapter that the data were better explained by "design" than by Darwinian notions. Published in 1989 by Haughton Publishing in Dallas, it was substantially a product of the Foundation for Thought and Ethics, the same Dallas think tank that had produced *Mystery.*

With its semigloss pages and extensive color graphics, *Of Pandas and People* had the look and feel of an advanced high school text. Not surprisingly, it was created to serve in exactly that role—as a public high school supplement to biology texts, mainly for a teacher's use when dealing with the difficult topic of evolution. The text was written by a pair of teaching biologists—Dean Kenyon, the chemical evolution author whose change of mind was discussed earlier, and P. William Davis, who teaches biology at Hillsborough Community College in Tampa. The entire project was directed by Charles Thaxton, who served as editor and contributed a concluding "Word to the Teacher."

The scientific and rhetorical controversy surrounding *Of Pandas and People* has been considerable, including a negative jab in an article in *Newsweek* and mixed treatment in a front-page news article in the *Wall Street Journal*.[69] Despite the criticism, *Of Pandas and People* continued to draw increasing support as the Intelligent Design Movement developed. For example, Michael Behe assisted in the rewriting of a chapter on biochemistry in a revised edition of *Pandas*. The book stands as one of the milestones in the infancy of Design. Here at last emerged a published scientific proposal—a sketched-out explanatory system that could be viewed as pointing to a new heuristic paradigm.

Thaxton's work came to Johnson's attention (and vice versa) shortly after the Berkeley seminar of September 1988. The two met in person and quickly became close research colleagues, frequently exchanging information and ideas. When other members of the Ad Hoc Origins Committee read Johnson's Berkeley paper and his book draft, they were eager to hear Johnson present his views in person. Eventually a weekend gathering was scheduled for February 1990 in Portland. The main item on the agenda was to discuss the recent Campion meeting and the debate with Gould. Recollections of Johnson would be enriched by those of two other Campion participants who agreed to come to Portland.[70]

If the 1989 publication of Thaxton's *Of Pandas and People* symbolized the emergence of Design in print form, then the February 1990 Portland meeting symbolized the movement's social emergence. Here was a momentous convergence of like-minded persons, nearly all of whom were intent on cooperation and concerted action around a common set of ambitious goals in the arenas of research, discussion, and persuasion. With the coming together of diverse skeptics of Darwinism under the intellectual leadership of Charles Thaxton and Phillip Johnson, there appeared at last a realistic pathway toward the group's supreme goal—building a credible alternative to the Darwinian paradigm. During the weekend of meetings there was discussion of various projects that were just getting underway and others that the new movement would need to launch—networking, hosting of meetings and conferences, publications, communication with key outsiders, and continual clarification of the consensus on which the movement rested.

In Portland there was a visible deference to the newcomer Johnson. He was the center of interest as he reported at length on his journey to skepticism about macroevolution, his recollections of the Berkeley seminar and Campion meeting, and his progress in getting the book published. There was a quiet draft of Johnson to leadership of the group. Several factors led to Johnson assuming leadership of the group after Portland: his academic position at Berkeley, the broad and radical nature of his critique, his quick mind and knack at public speaking, his witty and jovial personality, his determination to carry the critique forward forcefully, and his network of interactions with scholars. This was a preview of Johnson's role during the 1990s. With attention focused on *Darwin on Trial* and its sequels, Johnson was universally recognized as the leader and spokesperson of Design. It was an informal role; there never was any formal hierarchical structure to the movement.

After Portland, Thaxton worked happily in Johnson's shadow, although there were times when the two would differ markedly over strategy. For example, in the two-year period before *Darwin on Trial* was published, Thaxton repeatedly urged Johnson to take a much more proactive role in his book on behalf of the intelligent cause inference. Instead of Johnson just criticizing Darwinism, why should he not also argue clearly on behalf of the legitimacy of science considering intelligent design? The time was right, in Thaxton's judgment, for Johnson to spread the very ideas that had been developed in Thaxton's recent papers—which also were debated at the 1988 Tacoma conference and woven into *Of Pandas and People.*

However, Johnson resisted any such additions to his own work. He felt that the time was not right or at least his book on Darwinism was not the right place for such a forthright advocacy of design, even though his book certainly contained abundant hints that creation (in the broadest sense) was well supported by the scientific evidence. In a letter to another attendee at the Portland meeting, he said,

> I am afraid of appearing to be trying to "put something over," by concealing my actual belief in a supernatural creator. It seems to me that I would get very quickly back to the supernatural in any event. If the DNA is the product of pre-existing intelligence because it employs specified complexity, then that intelligence is either itself the product of a naturalistic process like evolution, or it is a supernatural intelligence.
>
> The other main point in my thinking is that I am not as concerned as Charlie is to make my point within the category of "science." My consistent stance has been that science can consider the possibility of supernatural creation or ignore it as a matter of definition. If it takes the latter option, then it disqualifies itself from saying anything about creation and hence from making any absolute affirmations about naturalistic evolution. All science can say is that, if we rule out creation on an *a priori* basis, and similarly rule out any possibility of a non-evolved intelligence which guides the evolutionary process, then naturalistic evolution of the Darwinist kind is just about all we can plausibly conceive. In that case they have not met the challenge of creation at all, but have simply evaded it by the use of a philosophical definition.[71]

As a result of Johnson's reluctance to articulate the practical and philosophical bases upon which intelligent cause may be considered within science, the formal debut of an intelligent design proposal in mainstream U.S. media would have to wait another five years until the appearance of Behe's *Darwin's Black Box.*[72]

Finally, in late 1990 Johnson received an enthusiastic offer from Regnery Gateway to publish *Darwin on Trial* in 1991, and soon thereafter a joint publishing agreement was struck with InterVarsity Press. In early spring of 1991, galley proofs were heading out to potential reviewers and suppliers of endorsements. Anticipation was building steadily leading up to the announced release date of June 1991. The troops were marshaling for another assault on the Normandy beaches of Darwinism.

5

Putting Darwin on Trial

Johnson Transforms the Narrative

The Nature, Theses, and Strategy of *Darwin on Trial*

The importance of *Darwin on Trial* in the history of Intelligent Design can hardly be overstated. This book was not just Phillip Johnson's commentary on the flaws of contemporary biology nor merely a much-anticipated overhaul of his Berkeley paper. It was nothing less than an intellectually savage manifesto designed to overwhelm the opposition, to expose Darwinism as, in Johnson's stigma-word, a "pseudoscience."[1]

Another stigma-word, placed like a stinger at the end of his book's purpose statement, hints at the severity of his critique: "My purpose is to examine the scientific evidence on its own terms, being careful to distinguish the evidence itself from any religious or philosophical bias that might distort our interpretation of that evidence. I assume that the creation-scientists are biased by their precommitment to Biblical fundamentalism, and I will have very little to say about their position. The question I want to investigate is whether Darwinism is based upon a fair assessment of the scientific evidence, or *whether it is another kind of fundamentalism*" (emphasis added).[2]

By the end of the book, Johnson's answer is clear. He has indeed symbolically consigned Darwinism to the stigma-realm of fundamentalism. The new perspective—incongruous to say the least—was deliberately provocative, even shocking.[3] It pictures Johnson as standing between and above the two fundamentalist perspectives, equidistant and critically detached from both religious and Darwinian fundamentalists. This implicit positioning of Johnson in ideological space, in turn, was absorbed into Design's self-image and became a significant part of its developing rhetorical vision.[4]

Blurbs on a book cover often convey the vitality and energy of the book's message, and no reviewer better captured Johnson's aggressive persuasion than Michael Denton: "*Darwin on Trial* is unquestionably the best critique of Darwinism I have ever read. Professor Johnson combines a broad knowledge of biology with the incisive logic of a leading legal scholar to deliver a brilliant and devastating attack on the whole edifice of Darwinian belief. There is no doubt that this book will prove a severe embarrassment to the Darwinian establishment."[5]

Perhaps no biologist felt the book's engaging energy and sensed the imminent danger more than Stephen Jay Gould, who lashed back a year after its publication in a vitriolic and lengthy rebuttal. Gould's four-page review, appearing in *Scientific American,* charged that Johnson had produced a "very bad book" that "hardly deserves to be called a book at all." Clearly, Gould's purpose in this attack was not so much to review Johnson's book or even to defend macroevolution. His goal was primarily *personal*—to thoroughly demolish Johnson's credibility (ethos) as a competent commentator on Darwinism. In a verbal bludgeoning, which was rare for Gould, he strove to convey the exact opposite of David Raup's courageous endorsement of Johnson's scientific accuracy and quality, calling *Darwin on Trial* a "clumsy, repetitive abstract argument with no weighing of evidence, no careful reading of literature on all sides, no full citation of sources."

Gould's eighth paragraph began with an assault of *dismissive contempt*—a pithy rhetorical blow calculated to scorch and flatten Johnson: "The book, in short, is full of errors, badly argued, based on false criteria, and abysmally written." Gould then launched a meandering excursus of over two thousand words—a series of miniessays on Johnson's alleged errors and misunderstandings.[6] In a later chapter I will return to Gould's attack and describe Johnson's unsuccessful attempt to persuade *Scientific American* to let him publish a reply. At this point, I should simply note that Johnson saw the whole incident as a major step forward, and he quickly added an epilogue to *Darwin on Trial* (which first appeared in the 1993 revised edition) that described the clash and responded point-by-point to every one of Gould's criticisms.

Gould concluded his 1992 diatribe with rebukes reminiscent of a biblical prophet. Invoking powerful "Inherit the Wind" fantasies, he thundered, "Johnson's *grandiose claims,* backed by such poor support in fact and argument, recall a variety of phrases from a mutually favorite source: 'He that troubleth his own house shall inherit the wind' (Proverbs 11:29, and source for the famous play that dramatized the Scopes trial); 'They have sown the wind, and they shall reap the whirlwind' (Hosea 8:7)" (emphasis added).[7]

What were the "grandiose claims" Gould sought to annihilate? The delineation and rhetorical analysis of these claims and their persuasive presentation by the frames and tools of narrative is the main purpose of this chapter and the one that follows. Even though Johnson's book was half the length of Denton's, his themes, critique, and rhetorical strategy were similar. In fact, Denton and Johnson had had over two years of friendly correspondence by the time *Darwin on Trial* was published. At one point, Denton visited Johnson in Berkeley for several days, and as they hiked along Johnson's favorite nature trails near San Francisco, Denton gave him advice on confronting the bitter rhetorical onslaught that awaited him.[8]

Like Denton, Johnson affirmed Darwinian *microevolution* as respectable science, but he relentlessly attacked *macroevolution* as a thoroughly counterfactual enterprise. Here Johnson and Denton share a common negative thesis—non-Darwinian (or anti-Darwinian) inferences spring naturally from the relevant scientific data. Where Johnson differs from Denton is in his incorporation and extensive development of *three additional theses*. Let me list all four:

1. Johnson's negative thesis (shared with Denton) can be restated: *Biological and paleontological evidences and other scientific data, with very few exceptions, tend to falsify the Darwinian story of macroevolution and its chemical origin-of-life prelude.* I call this thesis, focused on scientific evidence, *T–1.*

2. An additional thesis (not shared with Denton) asserts: *Darwinian macroevolution, as a comprehensive truth claim, is ultimately grounded on the philosophical assumption of naturalism.* Johnson describes naturalism (or materialism) as a philosophy that "assumes the entire realm of nature to be a closed system of material causes and effects, which cannot be influenced by anything from 'outside.'"[9] This thesis, focused on the philosophical base of Darwinism, I call *T–2.*

3. A second additional thesis, more in the vein of rhetorical analysis, is a charge of "shoddy rhetoric": *When Darwinism is brought into question, it is routinely protected by empty labels, semantic manipulations, and faulty logic.* This is an underappreciated thrust of Johnson's persuasion, yet it is quite prominent throughout *Darwin on Trial.* This thesis will be labeled *T–3.*

4. Johnson's final thesis is stunningly radical and also a major conclusion drawn from the other three: *Therefore, Darwinism functions as the central cosmological myth of modern culture—as the centerpiece of a quasi-religious system that is known to be true* a priori, *rather than as a scientific hypothesis that must submit to rigorous testing.* The religious-mythological functions of Darwinism and its corollary of shying away from testing are major recurring themes of *Darwin on Trial.* I will label this thesis *T–4.*

I will return to these four theses periodically when I flesh out the strategy of the book. But first I need to probe Johnson's overarching purpose, and to do that it will help to connect *Darwin on Trial* to its historical context. I have

described the quietly deteriorating rhetorical situation within evolutionary
biology in the latter half of the twentieth century. There arose a simmering tension between the evolutionary consensus and the annoying stream of
questions, puzzles, anomalies, and murmurs of dissent against the reigning
orthodoxy. I then showed how Denton (1985) symbolically captured much
of this stream of dissent and portrayed its flow as a whitewater narrative of
an incipient paradigm crisis. Denton's massive tributary was joined in the
late 1980s by other rhetorical rivulets, especially Thaxton, Bradley, and Olsen
(Mystery of Life's Origin) and Johnson himself *(Evolution as Dogma)*, although
by this time many others had contributed (Norman Macbeth, Pierre Grassé,
Hubert Yockey, Michael Pitman, Robert Augos and George Stanciu, Thomas
Bethell, and Robert Shapiro).[10] By 1991 this river of rhetoric had carved out
a proactive nucleus of several dozen American dissenters, all of whom were
skeptical of Darwinism (agreeing with T–1) but also were dissatisfied with
the approach of scientific creationism. Under the leadership of Thaxton and
Johnson, this group of skeptics wielded *Darwin on Trial* as their new weapon
of choice as they enlisted in a new campaign to undermine the most powerful
scientific paradigm of all time.[11]

In chapter 1, I described Design's rhetorical vision and its projection themes
of an ultimate paradigm conquest of Design over Darwinism. I suggested that
this dramatic vision which fueled the skeptics' efforts was driven by the two
key qualities of *cultural centrality* and *dramatic supremacy.* The idea that "a reliable creation narrative is *culturally central*" is not controversial. Most observers recognize the function of naturalistic macroevolution as a keystone of the
Western worldview and its myriad secularized subcultures and socioeconomic
practices. Implicit recognition of this centrality is revealed in Steven Jay Gould's
vehement and trembling reaction to Johnson's thesis at Campion.

If this kingpin of truth were to erode or be dislodged, the impact on the
world's intellectual and social history would be extensive and even unprecedented, compared with earlier paradigm shifts. This expectation is the essence
of *dramatic supremacy.*[12] Part of the supreme drama is that such a change would
almost certainly mark a major reversal in the intellectual (and hence institutional) marginalization of theism that has proceeded unchecked in the last two
centuries of "scientific enlightenment." Recall Gould's pointed summation of
his own motives when he first met Johnson in December 1989. He said to
Johnson, "You are a creationist, and I've got to stop you." Clearly a great deal
more is at stake than the objective truth about creation, evolution, or design,
and all of this comes to be implied or embodied in the notion of a paradigm
shift and revolution.[13]

Related to dramatic supremacy and cultural centrality is Design's driving
motivation—*reformative urgency.* Johnson wrote in 1990 to Notre Dame
philosopher Alvin Plantinga, "My goal . . . is to legitimate the critique of
Darwinism and its philosophical background so that others, particularly in
the scientific world, can step forward and finish the job."[14] This is an early

glimmer of what later became articulated and symbolized as Johnson's *wedge strategy*, in which he sees himself functioning as the "thin edge of the wedge" with other Design scholars following in his wake, extending and deepening the critique.[15] Yet Johnson has frequently said that his even bigger long-range goal is to help open up higher education *in all spheres* to the possibility that scientific materialism may not be true, and thereby to "legitimate a theistic perspective in the universities."[16] This global dimension of the critique is a key to understanding the expansion of the rhetorical vision of Intelligent Design during the 1990s.

Rhetorical gusts and gales whip briskly at exactly this point. Defenders of the academic status quo who read these words are surely tempted to protest what they perceive as the dark political motives of Johnson and his theistic cohorts to legitimate a theistic perspective. They may suspect an effort to take over university education, which they believe would stifle and even devastate the sciences.[17] Theistic advocates of Design would reply that the issue is not the religious party *taking control* but rather being given a chair at the intellectual table after being shut out in a long exile.

Perhaps the common ground that fair-minded persons on both sides can press toward at this point is the establishing of *pluralistic forums for unfettered rhetoric*—occasions for reasoned discussion between opposing points of view with all worldview assumptions placed on the table for inspection. In fact, such rhetoric of metaphysical engagement and negotiation began in the wake of *Darwin on Trial*, as seen in several interdisciplinary conferences held at American universities to discuss Johnson's shocking theses. The stories of these conferences (one of which I will tell in chapter 7) have enhanced Design's self-image and its positive image in the public and on university campuses. They have also shaped and matured its rhetorical vision.[18]

Because most change in America's science-education system (in terms of content—what shall be taught) trickles down from the universities rather than going the other direction, Johnson primarily targeted his manifesto at what he calls the "high university world."[19] In this specialized world, a theistic perspective is generally seen as having very little, if anything, to contribute to intellectual discourse. Theism is viewed as a personal quirk or preference—tolerable as long as it stays a private matter.[20] In Johnson's view, most intellectuals in universities like Harvard and Yale (both of which have divinity schools) actually possess very little respect for theology as a legitimate academic field.

Furthermore, if a Christian professor working in a scientific field wishes to integrate his research with his Christian perspective, says Johnson, he typically tracks the secular thinking in his area and adds some "excusable god-talk"[21] but rarely challenges naturalistic assumptions. Of course, university professors in the arts and literature, as well as in social sciences and hard sciences, are free in the West to hold theistic beliefs and function as members of a religious community both on and off campus. However, according to Design theorists,

*these same theists often do not have the freedom to connect their academic work
with their deepest convictions about the existence of a deity who transcends and
rules the universe.*[22]

In my judgment, this last statement not only is an accurate depiction of
Johnson's view of academia; it also corresponds generally to the *social reality*
of western universities at the start of the twenty-first century. Theology, as a
system of belief or of rational investigation, holds only the most tenuous and
marginal place in Western universities. Primarily, its visibility in that arena
is only as an object of study—a persistent sociocultural phenomenon to be
analyzed. For example, I was amazed once to hear a brilliant and renowned
rhetorician whom I respect very highly describe the issue of God's existence
as a nonrhetorical issue, implying that it is a purely subjective (that is, non-
rational) issue, one that cannot really be argued at all.[23] Furthermore, if
this situation implies certain unwritten rules about epistemic constraints in
academic discourse, then *Darwin on Trial* constituted Johnson's massive act
of dissent against those rules and their underlying intellectual rationale. As a
manifesto, it went well beyond the point of "comprehensive critique" of the
dominant paradigm. It constituted a brief for the rejection of the status quo,
namely, *submitting to naturalism as an inviolable starting point of all research
and teaching in academia.*

Johnson's new path for biology was not as clear. *Darwin on Trial* itself did
not present a well-developed case for the consideration of intelligent causes.
As I said earlier, this second stage—the explicit case for design—began in
1996 with the proliferation of the arguments of Behe, Dembski, and many
others.[24] *Darwin on Trial* prepared the way for those later discourses by
plowing the ground and sowing doubt about the claims of Darwinism. It
heightened the existing tension and encouraged readers to initiate their
own reassessment.[25] With these strategic purposes in view, Johnson aimed
at a target audience that was similar to Denton's—a system of concentric
circles with scientists at the center, ranging out to university graduates and
the general lay public with reading skills to handle Johnson's moderately
advanced vocabulary.[26]

Johnson's overall rhetorical strategy was not dependent solely on his
writing. Rather, he placed a high priority on his public speaking as a key to
legitimizing his critique. Immediately after the publication of *Darwin on
Trial* and in the twelve years since, he took well over one hundred oppor-
tunities to speak in college or university venues.[27] One of the goals of his
public speaking was causing students and professors to think, *There is more to
this issue than I realized,*[28] and thereby generate sufficient curiosity for them
to read his critique or at least engage the issues he was raising. As Johnson
crafted *Darwin on Trial* to "arouse from dogmatic slumber," he realized that
two formidable obstacles stood in his way, and he had to devise a plan to
overcome them.

Johnson's Obstacles: Creationism and Competence

Phillip Johnson's critique did not sail into a rhetorical vacuum; rather, it had to lurch over smoke-filled battlefields pockmarked with craters left by attacks and counterattacks between creation scientists and the defenders of evolution. As a result, the suasory atmosphere around *Darwin on Trial* was thick with negative emotions—many of them deep and bitter.[29] Johnson himself acknowledged to a colleague, "The great problem is in the sciences. The evolutionary biologists are bitterly hostile, as might be expected, and other scientists are reluctant to go out on a limb on something outside their field."[30]

By 1991 there had built up an acrid and enduring cluster of feelings, attitudes, and emotions that surged within evolutionary biologists as they encountered the Genesis-based creationism that dominated the earlier debate. Evolutionists' hostility ranged on the mild end from amusement, derision, disapproval, frustration, and nervousness to the more severe reactions on the other end such as disgust, anger, deep resentment, and loathing.[31]

Undoubtedly this emotional sensitivity to creationism produced a major rhetorical obstacle, especially among readers with strong evolutionary convictions. Johnson had to forestall an early rejection or even an automatic dismissal before he could gain a hearing. Scientists who harbored such negative attitudes could perceive in Johnson's sharp attack on Darwinism that his critique was cut from the same cloth as "creation science." This would quickly shut down a hearing.

The situation was complicated by the fact that Johnson never tried to hide his own metaphysical assumptions as a Christian, including his belief in a basic (but nonfundamentalist) idea of creation—that the biosphere owes its existence ultimately to the will and action of the Creator. Because of this acknowledged bias, Johnson had to brace himself for the charge that he was religiously driven, thus incapable of dealing objectively with the evidence.[32] Johnson uses a systematic plan to surmount this "creationist obstacle." In chapter seven, I will discuss this strategy in the narrative setting of hostile book reviews. I simply note here that Johnson's book contains many direct statements as well as verbal clues that distance him from Genesis literalism and picture the speed and timing of creation as less important issues that distract from the main questions.

In addition to the possible perception of creationism, a second obstacle loomed—questions about the competence of a legal scholar in criticizing an area of science outside his own field of criminal law. A common question in media interviews with Johnson was, "What is a law professor doing, writing about problems in evolutionary biology?"[33]

Johnson confronted this "competence obstacle" both with explicit remarks addressing this concern and with scholarly qualities within the book that project competence. (Again, I will return to this obstacle in the next chapter.) The most persuasive strategy for establishing competence of the author lay within the substance of the book itself. Thus, perceptions of Johnson's excellence as a critic are projected primarily by his apparent breadth and depth of research and his accuracy of understanding, especially when dealing with the scientific evidence. In addition to conveying the sense of "having done one's homework,"[34] the author who is criticizing a field other than his own is helped enormously if he displays the qualities of a nimble, robust intellect and a communicative vibrancy. By nearly unanimous consent, one of Johnson's strongest qualities as a rhetorician is his blend of logic, insight, use of metaphors and analogies, and wit, all of which project the wisdom, intellectual excitement, and verve in the author's mind and personality. Many reviewers, including several who were generally negative, commented positively on his vigorous writing style and clever turns of phrasing.[35]

Surmounting the two rhetorical obstacles is part of the strategy of *Darwin on Trial,* yet it is only preparatory for the meat of Johnson's argument. His critique was crafted not only as a case of persuasion by evidence but, more astonishingly, as a matrix of compelling stories.

Johnson's System of Narrative Arguments

Johnson's most compelling dimension of rhetoric is his intricate system of narratives, which entails the destruction of old stories and construction of new ones. Each narrative Johnson tells has a purpose that supports one or more of the theses outlined above (T–1 through T–4).[36] It is helpful here to recall my observation (in chapter 3) that there are three basic types of factual narrative: (1) genesis or cosmological stories, (2) history-of-science stories, and (3) personal or individual stories. Working with these three basic types as a "story toolkit," I see within *Darwin on Trial* three veins of narrative argument.

One vein of Johnson's narrative argument works with the first type of story—the evolutionary cosmology. Johnson seeks to question, degrade, or destroy that story. In other words, his argument is a shredding of the Darwinian genesis narrative. Let's call this the "anti-genesis" genre of argument. Such a line of argument involves many coordinated acts, such as storytelling, the exposure of bad reasoning, or the setting forth of evidence that plagues the orthodox story line of evolutionary descent. These credibility problems gradually overwhelm as they accumulate.[37]

Amazingly, in the place of the dissolved genesis story, Johnson puts almost *nothing*—that is, no detailed or specified cosmology at all. The reader is given only a vague or minimal replacement story line—a generalized act (creation)[38] and an agent (the creator). In chapter 1, when he identifies his own bias that a creating agent—God—exists, he states that in his investigation he has not prejudged whether that agent has created using evolutionary mechanisms or more direct, interventional methods. "I do not exempt myself from the general rule that bias must be acknowledged and examined. I am a philosophical theist and a Christian. I believe that a God exists who could create out of nothing if He wanted to do so, but who might have chosen to work through a natural evolutionary process instead."[39]

Eight chapters later he refers to his theistic cosmology in the most general of terms:

> Why not consider the possibility that life is what it so evidently seems to be, the product of *creative intelligence?* Science would not come to an end, because the task would remain of deciphering the languages in which genetic information is communicated, and in general finding out how the whole system *works.* What scientists would lose is not an inspiring research program, but the illusion of total mastery of nature. They would have to face the possibility that beyond the natural world there is further reality which transcends science.[40] [emphasis added]

Johnson implies a tentative agnosticism over how creation happened. This view is portrayed as a proper, honorable, and intellectually honest position given the present fragmentary state of knowledge. In Johnson's prototype of Design strategy, the timing of creation is not so much unknown as it is deliberately not discussed.[41]

A second vein of narrative argument is destruction of the triumphant history-of-science story in the particular arena he is targeting—evolutionary biology. I shall refer to these narrative arguments, which invert and retell the history of evolutionary biology (HEB), as the *anti-HEB* type. (The prefix *anti* seems doubly appropriate since it can imply both "opposition" and "replacement.") To build these arguments, Johnson writes into his script two kinds of guest appearances: first, Darwinists whose statements and acts reveal the philosophical or religious foundations of their thought or who exhibit shoddy arguments (supporting theses T–2, T–3, and T–4), and second, respected scientists, such as Darwin himself or even Gould, who expressed doubts about some aspect of the orthodox theory or who alluded to serious problems in the scientific evidence (supporting T–1).

As some narrative arguments undermine the old HEB narrative, others build in its place a new history-of-science (HOS) narrative, which carves out a place for design theory. I will call this third type of argument *new-HOS*. In this revision of history, normal roles are reversed. Darwinism becomes a dogma-driven, deteriorating paradigm. Within biology, the heroes of the new story are the scientists, like Michael Denton, Pierre Grassé, or Murray Eden, and the mathematicians at Wistar, all of whom had the intellectual courage to raise radical questions and propose heretical ideas. The emergence and growth of Design's proposed paradigm thus function as the goal and projected vision that are implied in most of the new-HOS narrative arguments. As a result, this sort of argument has a very intimate and dynamic connection with the more imaginative type of story seen earlier—the projection themes that picture ultimate paradigm triumph.[42]

We should note in passing that all three of Johnson's narrative arguments incorporate and are powered by numerous personal narratives—the third key story type in the toolkit of stories.[43] There are two kinds of personal stories. One revolves around a single individual whose actions or words (and reactions to them) either undermine old master-narratives[44] or contribute to new ones. Another kind is an "incident narrative" which revolves not around a person but a single incident—a dramatic scene that has some effect on a master-narrative but involves several persons whose interaction constructs the meaning. A good example of an incident narrative in *Darwin on Trial* is the 1981 British Museum brouhaha over unsettling captions placed in a new display about evolution. This one story with commentary takes up sixty percent of the text of chapter 11![45]

Earlier I pictured the relationship between specific stories and a larger narrative that links them together as a tree with limbs jutting out. Another use of the tree metaphor is to see the tree as the scheme that organizes the whole "macronarrative" of Johnson's presentation of his case in *Darwin on Trial*. I am suggesting that the spectacle of a leading intellectual figure on the Berkeley campus, clothed and in his right mind, presenting publicly his case against the factual status of Darwinism, was widely perceived as a significant story in and of itself.[46] It is the rare spectacle of a scientific manifesto with the potential to do serious damage to the reigning orthodoxy. Johnson's inclusion of an epilogue in 1993 summarizing the response of leading Darwinists underscores the book as an "event."

Let me clarify what I mean about the book as a "story" itself. In *Darwin on Trial,* Johnson does not tell in any detail the story of how he was led to doubt Darwin. Here and there one can pick up hints of that story, of course, but the actual biographical story of Johnson in 1987–1991 (which I told in the previous chapter) was one that lived in the rhetorical consciousness of the Ad Hoc Origins Committee, not in the pages of *Darwin on Trial*. Rather, Johnson's manifesto is a "trial story"—somewhat like the summarization of evidence and analysis coming from the lips of a prosecuting attorney before a vast grand jury. Johnson's book—with eight chapters summarizing the state of the scientific evidence and four chapters on the philosophical, religious, and educational dimensions of the debate—is a running commentary on *why* and *how* he has come to doubt virtually everything the Darwinists are telling America about macroevolution. As the product of a nonfundamentalist intellectual, his book's body of argumentation counts as a provocative narrative event in the context of the long-running global debate over creation and evolution. *Darwin on Trial,* then, is an exotic "tree" among the forest of other commentaries in the creation-evolution debate. In my analysis of *Darwin on Trial,* I will occasionally apply the tree metaphor to the book itself as a massive *communication event.* Johnson's barrage of narratives, quotes, and analyses constitute the dozens of tree limbs.

Science writers do not hesitate to tell stories; they are among the most important tools of their trade.[47] So it is not unusual for Johnson to build his case upon narrative bedrock. In fact, in all his books he uses storytelling frequently with persuasive effect. His five other books after *Darwin on Trial* are heavily populated with fascinating and instructive stories from scientific, legal, and educational controversies.[48] The point, of course, is not the quantity of narrative discourse used by Johnson but *how his narrative arguments work*—how they advance his theses. One of his most probing narrative questions in *Darwin on Trial* is, "How, historically, did we arrive at this 'confident conclusion' of Darwinian evolution?" Now we are ready to turn to Johnson's opening chapters to see how he answers that question.

6

The Matrix of Stories in *Darwin on Trial*

Johnson Argues His Case

We have seen how Phillip Johnson's book functioned as a lean manifesto, designed to demolish the "Bible versus science" stereotype that had dominated the debate over evolution. By the time the reader arrives at chapter 12, Johnson is concluding that Darwinism, in final analysis, fits Popper's criteria as a "pseudoscience." For readers who opened the book with a relaxed acceptance of Darwinian claims, Johnson's thesis was shocking, and for some it seemed unthinkable. He argued that naturalistic evolution of all life from simple precursors, while posing as robust science, is actually a species of fundamentalism with its arguments and proofs running upon the rails of dubious metaphysical assumptions. The central question that hovers over the book's network of argumentation is, *What is the basis for many scientists' supreme confidence in the doctrine that scientific laws and chance were sufficient to account for the rise of all the complexity and diversity of life?* The contention of *Darwin on Trial* is that as one scrapes away the layer of topsoil, one discovers metaphysical naturalism, not empirical evidence, laid bare as the bedrock of this confidence.

In the previous chapter, I sketched in broad strokes how Johnson built his case for such a shocking thesis with its attendant radical reordering of scientific consciousness. Now I turn to the pattern of muscle and sinew in *Darwin on Trial*—the actual matrix of stories that constitutes Johnson's flow of argument. For brevity's sake, I will restrict myself primarily to the four opening chapters, in which much of the evidentiary force of his case is developed.

The Legal Stories

It was logical and prudent for Johnson, a legal scholar, to begin his critique of Darwinism on his own turf, reviewing the major twentieth-century court cases that grappled with the evolution controversy. In chapter 1, the Scopes trial of 1925 is briefly retold so as to puncture the "Inherit the Wind" legend.[1] Then Johnson quickly moves to his second and much more important story, the 1987 Supreme Court case, *Edwards v. Aguillard.* In this case, the high court ruled unconstitutional, by a seven to two vote, a 1981 Louisiana statute that required creation science to be taught in public schools alongside evolution science. This case is the opening fulcrum of Johnson's critique. It introduces all four of his theses and establishes his own voice as a legal-intellectual critic. In the process, he points out the problems with one of the best-known legal chapters in the modern story of Darwinism.

As background to this legal story, Johnson starts with a description of the changing situation leading up to the 1987 decision. He says that in the modern era, creationists'

> goal was no longer to suppress the teaching of evolution, but to get a fair hearing for their own viewpoint. If there is a case to be made for both sides of a scientific controversy, why should public school students, for example, hear only one side? Creation-scientists emphasized that they wanted to present only the scientific arguments in the schools; the Bible itself was not to be taught.

To balance things, Johnson points out the vehement denial of the legitimacy of creationism. "Mainstream science does not agree that there are two sides to the controversy, *and regards creation science as a fraud. Equal time for creation science in biology classes, the Darwinists like to say, is like equal time for the theory that it is the stork that brings babies*" (emphasis added).[2]

A federal judge blocked the Louisiana statute, holding it to be an "establishment of religion." When the Supreme Court upheld this decision, the majority opinion written by Justice William Brennan stressed the purpose of the Louisiana law "was clearly to advance the religious viewpoint that a supernatural being created humankind." In the dissenting opinion, Scalia said the people "are entitled, as a secular matter, to have whatever scientific evidence there may be against evolution presented in their schools, just as Mr. Scopes was entitled to present whatever scientific evidence there was for it."

Johnson then tries to establish his own detached and judiciously balanced perspective, saying, "Both Justice Brennan [writing for the majority] and Justice Scalia [writing for the minority] were in a sense right." The drama that begins here pictures Johnson entering a complex conversation, listening to the experts, and injecting new insights into a deadlocked discourse. This is a microcosm of the dramatic action inherent in the entire book. Johnson's comments throughout the chapters are designed to portray himself as not only brandishing a manifesto but also doing so from a well-informed perspective. Here is Johnson's critical stance in relation to both Scalia and Brennan:

> The Constitution excludes religious advocacy from public school classrooms, and to say that a supernatural being created mankind is certainly to advocate a religious position. On the other hand, the Louisiana legislature had acted on the premise that legitimate scientific objections to "evolution" were being suppressed. Some might doubt that such objections exist, but the Supreme Court could not overrule the legislature's judgment on a disputed scientific question, especially considering that the state had been given no opportunity to show what balanced treatment would mean in practice. In addition, the creation-scientists were arguing that the teaching of evolution itself had a religious objective, namely to discredit the idea that a supernatural being created mankind. Taking all this into account, Justice Scalia thought that the Constitution permitted the legislature to give people offended by the allegedly dogmatic teaching of evolution a fair opportunity to reply. [7]

Johnson acts here in the role of a wise counselor who wants to bring sanity to a complex situation. He deals with questions of fact, but he also illuminates issues of procedure, worldview entanglements, and the relationship between metaphysics and the claims of science.

In the course of discussing the Supreme Court case and pointing out the complexity of the issues, Johnson has raised central questions he poses for the current controversy: *Are there true scientific problems for Darwinism? Does the current teaching of Darwinism have a religious edge to it, seeking to discredit the idea of a creator?*

Instead of going directly into the evidence here, Johnson first speaks as a legal-rhetorical scholar and observes the persuasive advantage obtained by the way key terms like "science" and "religion" are used:

> If we say that naturalistic evolution is *science,* and supernatural creation is *religion,* the effect is not very different from saying that the former is true and the latter is fantasy. When the doctrines of science are taught as fact, then whatever those doctrines exclude cannot be true. By use of labels, objections to naturalistic evolution can be dismissed without a fair hearing. [7]

Such arguments over labeling are common in *Darwin on Trial* and feed into Johnson's T–3 thesis, which focuses on manipulation of terminology to protect Darwinism from questioning.

Johnson adds a coda to the Supreme Court story—the "friend of the court" argument submitted in that case by the National Academy of Sciences (hereafter referred to as "the Academy"). He quotes from the brief to show how the Academy rejects the creationists' tactic of "negative argumentation," saying that it is a "dilution" of science that is "antithetical to the scientific method."

The argument in the Academy brief supplies an inviting target for Johnson, and he moves in to expose its illogic. His reply is worth quoting to get the feel of Johnson's rhetorical style in hunting down and seizing upon Darwinian misstatement:

> The Academy thus defined "science" in such a way that advocates of supernatural creation may neither argue for their own position nor dispute the claims of the scientific establishment. That may be one way to win an argument but it is not satisfying to anyone who thinks it possible that God really did have something to do with creating mankind, or that some of the claims that scientists make under the heading of "evolution" may be false.
>
> I approach the creation-evolution dispute not as a scientist but as a professor of law, which means among other things that I know something about the ways that words are used in arguments. *What first drew my attention to the question was the way the rules of argument seemed to be structured to make it impossible to question whether what we are being told about evolution is really true.* For example, the Academy's rule against negative argument automatically eliminates the possibility that science has not discovered how complex organisms could have developed. However wrong the current answer may be, it stands until a better answer arrives. It is as if a criminal defendant were not allowed to present an alibi unless he could also show who did commit the crime.[3] [emphasis added]

Johnson observes here a "protection by illogic," which is at the heart of T–3. He also has sketched his role as an outsider who is trained in arguments and who spots an oddity about the way rules of argument have been structured to make it impossible to ask key questions concerning the factual truth of Darwinism. His role as critic of Darwinism is plainly set forth. As a law professor he is trained in the ways that words are used in arguments; his training as a rhetorician has equipped him for critique. This statement begins to erect the "tree" of *Darwin on Trial* itself—a massive and unique communication-story.[4]

The Opening Swirl of Darwinian Controversies

Johnson quickly moves from legal stories to other controversies that cast doubt on evolutionary theory. Johnson first sums up the religious tendencies of modern Darwinists:

> The literature of Darwinism is full of anti-theistic conclusions, such as that the universe was not designed and has no purpose, and that we humans are the product of blind natural processes that care nothing about us. What is more, these statements are not presented as personal opinions but as the logical implications of evolutionary science. [8–9]

This is a strategic statement. The explicit religious function of Darwinism—*to wean educated persons from belief in the supernatural*—is not just a key thesis (T–4) of *Darwin on Trial;* it is a predominant theme of all Johnson's writing and public speaking since 1991.[5] This religious function is exemplified by George Gaylord Simpson, "Man is the result of a purposeless and natural process that did not have him in mind."[6]

Johnson immediately places another witness on his stand—Oxford biologist Richard Dawkins, who claimed Darwin made it possible to be an "intellectually fulfilled atheist." Then Johnson links this exultation with the disgust with which Dawkins viewed unbelievers in Darwinism: "It is absolutely safe to say that if you meet somebody who claims not to believe in evolution, that person is ignorant, stupid, or insane (or wicked, but I'd rather not consider that)" (9). Johnson, quick to squeeze comic effect from juicy quotes, adds,

> We must therefore believe in evolution or go to the madhouse, but what precisely is it that we are required to believe? "Evolution" can mean anything from the uncontroversial statement that bacteria "evolve" resistance to antibiotics to the grand metaphysical claim that the universe and mankind "evolved" entirely by purposeless mechanical forces. A word that elastic is likely to mislead, by implying that we know as much about the grand claim as we do about the small one. [9]

At this point Johnson has advanced another one of his most pointed criticisms of Darwinism (part of T–3)—the shifting back and forth in definitions of evolution from weak (e.g., microevolution or change in gene frequencies) to strong (strictly naturalistic common ancestry of all life forms).[7] Johnson returns to this point many times, but the most important summary of this theme actually comes in chapter 12. There, in three key paragraphs, he portrays this switching of meaning as a species of semantic sleight of hand—a verbal shell game. Johnson pictures this process as a situation that is repeated constantly in the world of Darwinian science and is thus typical.[8] I will jump into the middle of this summary:

> If critics are sophisticated enough to see that population variations have nothing to do with major transformations, Darwinists can disavow the argument from microevolution and point to *relationship* as the "fact of evolution." Or they can turn to biogeography, and point out that species on offshore islands closely resemble those on the nearby mainland. Because "evolution" means so many different things, almost any example will do. The trick is always to prove one of the modest meanings of the term, and treat it as proof of the complete metaphysical system.
>
> Manipulation of terminology also allows natural selection to appear and disappear on command. . . . The fact of evolution therefore remains unquestioned, even if there is a certain amount of healthy debate about the theory. Once the critics have been distracted, the Blind Watchmaker can reenter by the

back door. Darwinists will explain that no biologist doubts the importance of selection, because nothing else was available to shape the adaptive features of the phenotypes.[9]

What is the persuasive effect of revealing such maneuverings? Johnson exposes here a deceptive habit, creating a new sensitivity to a pattern of communicative manipulation. It also casts the current theory in a highly vulnerable light. Its defenders are like clever defense lawyers, using word elasticity to deflect and divert hard questions.

A Controversy Story: Colin Patterson

Johnson inflicts his most important early damage through two controversy stories.[10] The first of these is about Colin Patterson, the British scientist who helped critique Johnson's early draft in London. I briefly discussed this account in chapter 2. It possesses shock value and is probably the most powerful narrative in chapter 1 of Johnson's book. Outside of creationist circles, few people were aware that a renowned British paleontologist had visited gatherings of U.S. evolutionists in 1981 and was greeted by silence wherever he asked his famous question, "Can you tell me anything you know about evolution, any one thing . . . that is true?"[11] Johnson wisely qualifies the story of Patterson's speech at the American Museum by telling that he "came under heavy fire from Darwinists after somebody circulated a bootleg transcript of the lecture, and he eventually disavowed the whole business." Yet, this softener only heightens the drama of the story, and Johnson's use of Patterson (in spite of the Britisher's disavowal) is vindicated in the "research notes" in which he adds that during his meeting with Patterson in London, none of the critical points were retracted.

Crucial to the story's power (besides the legendary question) are two of Patterson's provocative comments from the museum speech. First, Patterson said in his speech that evolutionists increasingly talk like creationists in that "they point to a fact but cannot provide an explanation of the means." Second, he said that both evolution and creation are forms of "anti-knowledge."[12] That is, they "are concepts that seem to imply true information, but do not."[13] Patterson's second point paves the way for Johnson to restate his "label critique": "We can point to a mystery and call it 'evolution,' but this is only a label. The important question is not whether scientists have agreed on a label, but how much they know about how complex living beings like ourselves came into existence." This charge of employing labels in place of real knowledge is a common theme in Johnson.[14] In fact, both Patterson and Johnson seem to agree that evolution functions as a totalizing verbal symbol (what Burke called a "god term"[15]), and yet ironically it is empty—it has no real empirical substance or clarity. This emphasis on word content and labeling is a new rhetorical strategy that Darwinists had not previously encountered—*critique by dogged rhetorical criticism.*

The Patterson story has a potential awakening effect on the reader—awakening believers in Darwinism to a heretical parallelism between evolution and creation. Patterson considers both to be explanatory systems that are now equally agnostic on the exact mechanism of creation. This point is clearly part of anti-HEB; it casts a shadow on the current HEB narrative. The outlines of a new story appear in which Darwinists are working away, blind to their growing similarity to creationism until they are rudely jarred awake to this point by their restless, disgruntled colleague from Great Britain.

As a noncreationist, Patterson plays a role that is similar (though on a much smaller scale) to Michael Denton's role and to the quietly supportive role of David Raup. In common rhetorical action all three fulfill what Kenneth Burke described as the "prophet" role in science. In Burke's metaphor, with the decline of the church, the clergy now play a relatively minor cultural role. Science and the universities have taken over the bulk of the "priesthood function" of maintaining the proper worldview. So in science the "priests devote their efforts to maintaining the vestigial structure; the prophets seek new perspectives whereby this vestigial structure may be criticized and a new one established in its place."[16] In light of this perspective, we might describe "controversy stories" like Patterson's as "prophet narratives."

The Kristol-Gould Controversy Story

The other significant controversy story is the Kristol-Gould exchange. Irving Kristol, a social theorist, had proposed a mild corrective in a *New York Times* essay: "[I]f evolution were taught more cautiously, as a conglomerate idea consisting of conflicting hypotheses rather than as an unchallengeable certainty, it would be far less controversial." He then said religious fundamentalists were not "far off the mark when they assert that evolution . . . has an unwarranted anti-religious edge."[17]

Stephen Jay Gould, far from welcoming this corrective, criticized Kristol. He denied that evolutionary science is anti-religious. Most important of all, said Gould, Kristol ignores the all-important fact/theory distinction. There are conflicting hypotheses about the exact mechanism of evolution, "but evolution is also a *fact* of nature, as well established as the fact that the earth revolves around the sun."[18] This reply is vital for Johnson's strategy. Later Johnson devotes two entire chapters to an analysis of this very fact/theory distinction and attacks Gould's proofs that evolution is a fact.

When Gould compares evolution's factualness with the fact that the earth revolves around the sun, Johnson responds mildly, "As an outside observer who enjoys following the literature of evolution and its conflicts, I have become accustomed to seeing this sort of evasive response to criticism" (11). However,

later in the book when Gould compares the "fact of evolution" with the fact
that apples fall down, not up, Johnson attempts a verbal puncture and argumentative blowout: "The analogy is spurious. We observe directly that apples
fall when dropped, but we do not observe a common ancestor for modern apes
and humans. What we do observe is that apes and humans are physically and
biochemically more like each other than they are like rabbits, snakes, or trees.
The ape-like common ancestor is a hypothesis in a *theory,* which purports to
explain how these greater and lesser similarities came about. The theory is
plausible, especially to a philosophical materialist, but it may nonetheless be
false. The true explanation for natural relationships may be something more
mysterious" (66–67).

Johnson's introduction of Gould in chapter 1 is very strategic, since the late
Harvard professor practically dominates Johnson's courtroom thereafter.[19] In
order to characterize Gould's disillusionment with the standard gradualistic cosmological story, Johnson quotes a key statement in a 1980 paper. In this article
Gould focused not on the common ancestry thesis (held by all evolutionists)
but on the Darwinian picture of how the branching took place—by gradual
accumulation of adaptive changes via mutation and selection. Gould concluded
that the neo-Darwinian synthesis, "as a general proposition, is effectively dead,
despite its persistence as textbook orthodoxy" (11). (Johnson says that Gould,
having colorfully dismissed the Darwinian mechanism, might be expected to
warm up to Kristol's idea to teach evolution more cautiously.)

Gould's "effectively dead" quote is a massive gift to Johnson (and to Design
generally). It serves simultaneously to advance all three of his narrative arguments—anti-genesis, anti-HEB, and new-HOS. As with Patterson's speech,
this shock-quote tends to undermine the reader's confidence in the textbook
story. It records Gould admitting at least one damaging point: The Darwinian
scenario of the rise of life's diversity is now defunct as a general theory. Johnson
finds this quote so rhetorically useful, he recycles it twice in another chapter.[20]
Of course, Gould saw his own "punctuated equilibrium" concept as playing
a significant transitional role in the forthcoming new and general theory of
evolution that he said would soon emerge. However, as Johnson explains in
later chapters, the hope languished. No new theory arose.

Natural Selection

Near the end of chapter 1, a key skeptical refrain begins that is repeated in
many variations throughout the book, usually in the form of two questions:
*"How much do evolutionists really know about the process whereby all living things
evolved from microbial ancestors? Specifically, do they really know what they have
been claiming to know—that it was a mindless process?"*[21] If *Darwin on Trial*
is a communication event that is pictured as a tree, then this pair of probing
questions helps to form the trunk of that tree.

This emphasis on the alleged ignorance of the "how" of evolution makes it logical for Johnson first to tackle the two sides of neo-Darwinism's creation mechanism—*natural selection,* which sifts out and adds up beneficial *mutations.* He critiques this mechanism with a pair of chapters, one on selection, one on mutations.

As the lead-off chapter in Johnson's seven-chapter overview of scientific evidence (chapters 2 through 8), chapter 2 on natural selection is supremely important.[22] Some universal common ground exists on natural selection. Even creationists agree that selection is a real process in nature (16, 68–69). All recognize it as a historical (narrative) process at the level of local populations. The question at issue is whether this process can be extrapolated beyond the tinkering and tweaking within a type (microevolution). Is selection the creative agent that produced the entire biosphere? Creationists and Design theorists view natural selection only as a conservative process—weeding out the genetically unfit, maintaining the integrity of the genetic information of a given kind. At most, it may also create new varieties within a species, or it may lead to the production of sister species (again, microevolution) within a genus.[23]

However, most evolutionists since Darwin, especially after the neo-Darwinian synthesis emerged, viewed natural selection in Darwin's own dramatic role—as a fantastically powerful natural creator. Besides weeding out the genetically damaged and producing diversity, it also creates new organs and organisms and drives animals and plants to higher and higher levels of complexity. In a classic Darwinian nutshell, *natural selection drives macroevolution.*

In opposition to the old creationist metanarrative, evolutionists have vigorously argued that selection mimics the work of an intelligent designer—it acts as the creator. As stated in chapter 4, Dawkins called selection the "blind watchmaker."[24] Darwin himself wrote of selection in anthropomorphic terms, describing the mechanism as a vigilant breeder who was daily and hourly scrutinizing all the variations, adding up the positive variants, and weeding out the less advantageous ones, steadily and insensibly creating new and improved species out of the older species that were less fit.[25]

For believers in natural selection's creative power, Johnson's second chapter is jarringly iconoclastic. I have sometimes experienced firsthand how this section of *Darwin on Trial* could upset convinced Darwinists. For example, after sending a copy to one of the most renowned Darwinian theorists in the world, I received the following in reply:

> Thanks very much for the copy of *Darwin on Trial.* . . . I'm sorry to say I don't share your enthusiasm for the book. I only read the first thirty or so pages, by which point it had become clear that Mr. Johnson and I see things so differently that further reading would succeed only in raising my blood pressure. My own view has long been that any modern religious philosophy must somehow reconcile itself with the theory of evolution by natural selection, and I seriously

doubt that anything Mr. Johnson could say would change that view. In any event, his first twelve thousand words didn't have that effect.[26]

Because of biologists' sensitivity on this topic, it was especially important that Johnson manifest his scientific competence in this chapter. He sought to do that by sophisticated discussions and the projection of a thorough and subtle knowledge of the topic. An example of this projection of competence comes at the end of the chapter, where he moves into moderately technical terrain and discusses three subsidiary concepts connected with selection—pleiotropy (important in the defense of selection), kinship selection, and sexual selection.

At the outset of the chapter, Johnson lists five ways that the creative power of selection is said to be defended. Selection is (1) demonstrated by the analogy of animal breeders (Darwin's argument), or (2) shown true by definition—a tautology, or (3) demonstrated as a deductive argument, or (4) proven by clear-cut evidence, or (5) demonstrated true—as philosophical necessity.

Prosecution by Embarrassment

Of the five ways of defending selection, only one has pure nobility—the fourth way, shown by clear evidence. The other four ways of confirming the creative power of selection are much less impressive or even embarrassing. Johnson's basic chapter strategy is a special kind of anti-genesis argument—shredding natural selection by embarrassment. For example, Darwin's "breeder analogy" is disposed of rather easily, as we shall see later. The other dubious defenses of selection (two, three, and five) are virtually stigmatized by being listed as philosophical in nature rather than empirical. Natural selection "as deduction" is somewhat more respectable, but the other two—"as tautology" and "as philosophical necessity"—seem deeply suspect in their very description. Johnson symbolically lays bare the bedrock issues of philosophy, exposing the "tautology" concept as a label of unique embarrassment, since it is a statement that restates in the predicate what is already in the subject. Hence, "survival of the fittest," if the fittest be judged simply by the quality of survival, becomes—to coin a word—the survival of the "surviviest." A narratively rich nuance of shame and irony is made even richer in Johnson's listing and quoting Darwinists who acknowledged that selection *is tautological,* and yet who showed little or no worry or embarrassment on this point. Johnson repeats this narrative point later in chapter 9: Darwinists are undisturbed by selection as tautology.[27]

Johnson injects another "controversy narrative"—curiously similar to the Colin Patterson story—into the tautology discussion. Sir Karl Popper, the late celebrated philosopher who is so central to chapter 12, is described here as writing "that Darwinism is not really a scientific theory because natural selection is an all-purpose explanation which can account for anything, and which therefore explains nothing." Then the plot thickens—Popper recanted. He "backed away from this position after he was besieged by indignant Darwinist protests, but he had plenty of justification for taking it." The parallel with the Patterson story is obvious and striking. This is another case of Johnson building the tree of a new-HOS narrative. Limb by limb he attaches factual and fascinating personal micronarratives to his new story of how the scientific world began to wake up to Darwinism's fatal flaws. To add insult to narrative injury, Johnson quotes Popper's own 1983 discussion of his previous (but by then recanted) criticism, citing in self-defense several leading biologists who "formulate the theory in such a way *that it amounts to the tautology that those organisms that leave most offspring leave most offspring*" (emphasis added).[28] It almost seems that Popper recanted in letter but not in spirit.

As I turn from "selection as tautology" to Johnson's opening refutation, "selection as analogy," it is wise to retrace the connection with my narrative theory. Natural selection, in the Darwinian vision, is an easily pictured theory and story, which (if true) implies a sweeping, inexorable drama of steady transformation of simple to complex, filling out the myriad branches of the tree of life. It offers a universal and potentially plausible fine-grained cosmological story. I say "potentially plausible" in the sense of Johnson's argument. To establish his own objectivity and fairness, he first describes quickly this exact Darwinian cosmological narrative. Then he adds, "That is, all this can happen if the theory is true. Darwin could not point to impressive examples of natural selection in action and so he had to rely heavily on an argument by analogy."[29]

Johnson then explains what he sees as fatal differences between intelligent breeders and natural selection. Breeders, pursuing conscious goals, *intelligently* guide their process, and even then, the potential for variability runs out and the process reaches its natural limit.[30] This argument (undermining Darwin's analogy) has its own persuasive value as an anti-genesis argument, but there lurks here an equally important anti-HEB argument: Johnson has just said that Darwin could not provide direct evidence for natural selection, and his breeder analogy was his main persuasive tool with which he convinced his own generation. The embarrassing implication is clear: The scientific world of the late 1800s was convinced (as history-of-science fact) not by direct evidence of selection's power but by an argument that is now shown to be problematic.[31]

Scientific Evidence for Selection

Natural selection's most important pillar of support is evidence showing selection in action. Here Johnson has two rhetorical tactics. First, he invokes French zoologist Pierre Grassé three times in the chapter as a star witness against the creative role of selection.[32] Then he lists and critiques Douglas Futuyma's six evidences for selection in action. The most famous of these, the peppered moth observations in England, may be the most frequently cited evidence for evolution, but Johnson subjects it to a severe anti-genesis critique and portrays it as no evidence at all for macroevolution.[33] In fact, referring to the whole list of six observations, Johnson says, "None of the 'proofs' provides any persuasive reason for believing that natural selection can produce new species, new organs, or other major changes, or even minor changes that are permanent. . . . That larger birds have an advantage over smaller birds in high winds or droughts has no tendency whatever to prove that similar factors caused birds to come into existence in the first place."

To support his point, he inserts a powerful quote from Grassé: "The 'evolution in action' of J. Huxley and other biologists is simply the observation of demographic facts, local fluctuations of genotypes, geographic distributions. . . . Fluctuation as a result of circumstances, with prior modification of the genome, does not imply evolution, and we have tangible proof of this in many panchronic species [i.e., living fossils that remain unchanged for millions of years]."[34]

Of course, Johnson agrees, and he asks a series of questions that projects a pathos of head-shaking amazement at Darwinists' gullible cognitive processes. His questions have the ring of a prosecutor speaking common sense to the grand jury, and they imply that something more than scientific theorizing is going on:

> Why do other people, including experts whose intelligence and intellectual integrity I respect, think that evidence of local population fluctuations confirms the hypothesis that natural selection has the capacity to work engineering marvels, to construct wonders like the eye and the wing? . . . Everyone who studies the [peppered moth] experiment also knows that it has nothing to do with the origin of any species, or even any variety, because dark and white moths were present throughout the experiment. . . . How could intelligent people have been so gullible as to imagine that the Kettlewell experiment in any way supported the ambitious claims of Darwinism?[35]

This remarkable paragraph is Johnson's transition to the chapter's finale, in which he concludes that natural selection *is* a "philosophical necessity." This section is his answer to his question about Darwinists' gullibility. He is not just casting doubt on Darwinian science (T–1); he is establishing naturalism as its ultimate foundation (T–2): For those who believe Darwinism must be true, "there is no need to test the theory itself, for there is no respectable alternative to test it against. Any persons who say the theory itself is inadequately supported can be vanquished by the question 'Darwin's Bulldog' T. H. Huxley used to ask the doubters in Darwin's time: What is your alternative?" (28). This is the topic he raised earlier in connection with the NAS brief to the Supreme Court, the "prohibition of negative argumentation." He closes the chapter by surveying several technical concepts that are used to deal with empirical situations that seem difficult to explain. Ultimately, he points out, natural selection must work with what mutation serves up.

Mutations Great and Small

The drama of Johnson's third chapter boils down to a three-way scientific battle between (1) proponents of change driven by "small mutations," versus (2) those who proposed the need of "great mutations" (known as "saltations"), versus (3) those like Gould who "split the difference."[36] Saltations are described in the opening paragraph as "sudden leaps by which a new type of organism appears in a single generation." They were spurned by Darwin as equivalent to a miracle.[37] The debate over saltations—an HEB story line that stretches back to Darwin's time and beyond—serves as Johnson's revelatory tool, focusing attention on the perennial difficulty of envisaging how a string of tiny mutations could build up complex organs such as eyes and wings. Recall that the problem of "hypothetical intermediates" was so rhetorically potent that Denton spent his "Bridging the Gaps" chapter surveying organs that defy scenarios of step-by-step development through intermediate stages. Johnson devotes his chapter on mutation to the same problem, but he uses a different approach from Denton's—the historical debate over saltations, which includes many touching personal stories.

Scene 1: Darwin and His Contemporaries

In scene 1, Darwin and his contemporaries are the actors on Johnson's stage. Even though Darwin knew nothing about mutation as such, his theory of natural selection depended upon something similar—*variation*. In Darwin's own words, which Johnson quotes early in the chapter: "Natural selection can act only by the preservation and accumulation of infinitesimally small inherited modifications, each profitable to the preserved being." Johnson says repeatedly in this section that the key to Darwin's theory is his "uncompromising philosophical materialism, which made it truly scientific in the sense that it did not invoke any mystical or supernatural forces that are inaccessible to scientific investigation. To achieve a fully materialistic theory Darwin had to explain every complex characteristic or major transformation as the cumulative product of a great many tiny steps."[38]

Dramatic tension arises when T. H. Huxley suddenly appears on the stage, warning Darwin not to rigidly dismiss saltations. In a letter to Darwin he said, "You have loaded yourself with an unnecessary difficulty in adopting *natura non facit saltum* [nature does not make a jump] so unreservedly."[39] The tension between Darwin and Huxley over jumps (saltations) is rooted in the gaps in the fossil record and in the intricate complexity of organs themselves. Johnson, the scene's narrator, then explains the problem with the step-by-tiny-step production of complex organs:

> The eye and the wing are the most common illustrations, but it would be misleading to give the impression that either is a special case; human and animal bodies are literally packed with similar marvels. How can such things be built up by "infinitesimally small inherited variations, *each* profitable to the preserved being?" The first step towards a new function—such as vision or ability to fly—would not necessarily provide any advantage unless the other parts required for the function appeared at the same time. As an analogy, imagine a medieval alchemist producing by chance a silicon microchip; in the absence of a supporting computer technology the prodigious invention would be useless and he would throw it away. [34]

Then Johnson digresses from scene 1 to discuss a few organs (the eye, wing, feather, and bird lung) that have triggered controversy. Here Johnson stumbles rhetorically, blunting what should have been a piercing argument. On the one hand, he effectively employs scientists' comments (Dawkins, Mayr, Gould, and even Denton) about the puzzles and problems entailed in the evolution of key organs. But when Johnson turns to the crucial bird-lung problem, he spends no time explaining the odd morphology that is the heart of the puzzle. He never describes the flow-through structure of the bird lung (in comparison to the familiar bellows lung of all other land vertebrates). Instead, he vaguely refers to a "distinctive avian lung" and then quotes Denton to illuminate the problem. However, that quote uses a technical term, *parabronchi,* which is only meaningful if one already understands the bird-lung structure. Thus, the problem of the hypothetical structural change from one lung form to the other is never clearly sketched. If one had not read Denton on this mystery (or had not been instructed elsewhere), Johnson's discussion, far from being a clear and dramatic example, winds up as technical noise about an obscure topic.

Immediately following this stumble, however, Johnson makes one of the most effective statements in his entire book, which connects with my narrative perspective on rhetoric:

> Whether one finds the gradualist scenarios for the development of complex systems plausible involves an element of subjective judgment. It is a matter of objective fact, however, that these scenarios are speculation. Bird and bat wings appear in the fossil record already developed, and no one has ever confirmed by experiment that the gradual evolution of wings and eyes is possible. This absence of historical or experimental confirmation is presumably what Gould had in mind when he wrote that "These tales, in the 'just-so' tradition of evolutionary natural history, do not prove anything." Are we dealing here with science or with rationalist versions of Kipling's fables? [36]

Johnson resumes shredding the evolutionary cosmology as a scenario that is constructed out of speculations and just-so stories that are even mocked by Gould.

What follows next is the chapter's pivotal text. Having framed the general problem, Johnson again calls Darwin to the witness stand to supply the effective test of falsification. This is the quote that became famous later in the strategy of Behe: *"If it could be demonstrated that any complex organ existed which could not possibly have been formed by numerous, successive, slight modifications, my theory would absolutely break down"* (emphasis added).[40]

Scene 2: Goldschmidt and the Wistar Institute

Unlike Behe, who follows this quote with examples of molecular machines that could not be formed by numerous slight modifications, Johnson follows the quote with personal narratives of scientists who felt the theory had broken down on this basis. The chapter's drama lifts the curtain on scene 2, dominated by a new star, Richard Goldschmidt, a mid–twentieth-century geneticist at Berkeley who "issued a famous challenge to the neo-Darwinists, listing a series of complex structures from mammalian hair to hemoglobin that he thought could not have been produced by the accumulation and selection of small mutations." Johnson enriches this controversy story with tender pathos as he relates the "savage ridicule" of Goldschmidt for his heretical notions—something even Gould compared with the "Two Minute Hate" that is directed at Emmanuel Goldstein in Orwell's *1984*.

Johnson concedes the validity of evolutionists' criticism of Goldschmidt's "hopeful monster" ideas,[41] but he points out that disproving the creative role of macromutations *does not prove the opposite,* that evolution happened by the accumulation of tiny mutations. This is one of the best examples of Johnson's practical reasoning and argumentative logic—the muscle of his logos. In fact, after a very effective discussion of the problems with advantageous micromutations (calling Dawkins to the stand as a key witness),[42] he then casts doubt on the plausibility of the Darwinian genesis narrative in all of its implicit requirements:

> The probability of Darwinist evolution depends upon the quantity of favorable micromutations required to create complex organs and organisms, *the frequency with which such favorable micromutations occur just where and when they are needed,* the efficacy of natural selection in preserving the slight improvements with sufficient consistency to permit the benefits to accumulate, and the time allowed by the fossil record for all this to have happened. *Unless we can make calculations taking all these factors into account, we have no way of knowing whether evolution by micromutation is more or less improbable than evolution by macromutation.* [38, emphasis added]

After this classic anti-genesis narrative argument, Johnson picks up on his own keyword, "calculations," to introduce his next controversy story—the conference at the Wistar Institute in Philadelphia.[43] This story, which was told in chapter 2, is as important as the Goldschmidt narrative. Johnson says the Wistar "mathematicians did try to make the calculations, and the result was a rather acrimonious confrontation." The action of this story is revealing and humorous—to the rhetorical disadvantage of Darwinism. For example, when mathematician D. S. Ulam argued the mathematical improbability of the evolution of the eye by an accumulation of mutations, two Darwinists—Sir Peter Medawar and C. H. Waddington—said that "Ulam was doing his science backwards; the fact was that the eye *had* evolved and therefore the mathematical difficulties must be only apparent" (38).

The embarrassment for Darwinian rationality reached a climax when Ernst Mayr concluded, "Somehow or other by adjusting these figures we will come out all right. We are comforted by the fact that evolution has occurred" (39). (Johnson considers it unnecessary to point out to the reader the question-begging logic of Mayr.) Later in the Wistar confrontation, the discussion became heated when the French mathematician Schutzenberger said there is "a considerable gap in the neo-Darwinian theory of evolution, and we believe this gap to be of such a nature that it cannot be bridged within the current conception of biology" (39). Immediately, the Darwinists accused him of pointing toward creationism, but he vehemently denied it.

Of course, the mathematicians function as "prophets" in Burke's sense, and the defenders of Darwinian orthodoxy are the "priests." The rhetorical effect is both destructive and constructive: First, the Wistar incident hovers as a nimbus cloud of trouble, casting early shadows of deterioration on the HEB narrative. Second, it also serves as a limb being attached to the tree of the new-HOS being constructed. Later in the chapter Johnson casts the Wistar exchange as a typical situation that symbolizes what he calls a "principle at work"—the logical incoherence and circularity of Darwinist thinking:

> The prevailing assumption in evolutionary science seems to be that speculative possibilities, without experimental confirmation, are all that is really necessary. The *principle at work* is the same one that Waddington, Medawar, and Mayr invoked when challenged by the mathematicians. Nature must have provided whatever evolution had to have, because otherwise evolution wouldn't have happened. It follows that if evolution required macromutations then macromutations must be possible, or if macromutations are impossible, then evolution must not have required them. The theory itself provides whatever supporting evidence is essential.[44] [emphasis added]

Scene 3: Stephen Jay Gould

Scene 3 features another major point of view ("split the difference") and revolves around its chief advocate—Gould himself. His solution, a middle position, tries to introduce a modest level of macromutation into Darwinism. Johnson spends several pages on Gould's writings, making points that invite rhetorical analysis. Overall, the discussion serves to highlight Gould's own doubts about the orthodox theory. For example, Gould is quoted as saying it is too hard to "invent a reasonable sequence of intermediate forms—that is, viable, functioning organisms—between ancestors and descendants in major structural transitions." In fact, we will have to accept "many cases of discontinuous transition in macroevolution." Gould's comment reinforces the fact that serious problems exist with "mutations great and small."[45]

The Fossil Problem

Of the thirteen chapters in the 1993 edition of *Darwin on Trial*, the one on fossils is the most rich in narrative, as well as the most laden with narrative-type arguments. It also is probably the most rhetorically powerful chapter in the entire science section. I make that evaluation based on my sense that the fossils are the most easily grasped and interesting part of the debate.

Darwin's Fossil Struggle

Johnson begins his labyrinth of narratives by whisking us back to Victorian England and letting us listen in on Darwin's fossil musings and those of his contemporaries. Johnson's opening sketch on the shifts and trends in Darwin's day weaves together four historical points. After noting that (1) "Darwin's most formidable opponents were not clergymen, but fossil experts," he sketches (2) the geological theory of "multiple catastrophes" developed in Darwin's youth by French scientist Cuvier. He tells of (3) the revolution in geology, whereby Cuvier's catastrophism was pushed out by Lyell's uniformitarian ideas—that natural features were the result of "the slow working over immense time of everyday forces." He notes (4) the puzzle that Lyell "had great difficulty accepting biological evolution," even though some evolutionary theory seemed an inevitable extension of Lyell's geology.[46]

Lyell and others were troubled by the fact that the biological world had basic divisions, each with a unique structural plan, and there was scarcely a hint in nature of intermediate types. Johnson repeats Darwin's question, "Why, if species have descended from other species by insensibly fine gradations, do we not everywhere see innumerable transitional forms? Why is not all nature in confusion instead of the species being, as we see them, well defined?" (46). Darwin's own answer was his theory of extinction, based on survival of the fittest. In this view, said Darwin, "if we look at each species as descended from some other unknown form, both the parent and all the transitional varieties will generally have been exterminated by the very process of formation and perfection of the new form" (46). This theoretical process, which David Raup calls "competitive replacement,"[47] would imply that appearances in the living world—an array of stable species—would not reveal the evolutionary history behind them. As Johnson puts it, "The links between the discontinuous groups that once existed have vanished due to maladaptation." Yet those maladapted intermediate forms would be expected in one place—the fossil record.

The opening discussion raises expectation and leads the reader to anticipate that in the fossil record, scientists should find the crucial evidence of this "world of the past" where nature was continually transforming her species. Having turned the spotlight onto the fossil data, Johnson places Darwin on the witness stand again, where the British naturalist acknowledges that his theory implied that "the number of intermediate and transitional links, between all living and extinct species, must have been inconceivably great."[48] Johnson then summarizes the situation in Darwin's day, saying that scientists, rather than "continually uncovering fossil evidence of transitional forms," actually discovered species "which appeared suddenly rather than at the end of a chain of evolutionary links."

Even more damaging is Darwin's confession that such evidence is "the most obvious and gravest objection which can be urged against my theory," adding that this is why "all the most eminent paleontologists . . . and all our greatest geologists . . . have unanimously, often vehemently, maintained the immutability of species."[49]

Darwin's explanation of the problem and defense is summarized by Johnson: "Darwin argued eloquently that the fossil problem, although concededly serious, was not fatal to his theory. His main point was that the fossil record is extremely imperfect. Fossils are preserved only in special circumstances, and thus the various fossil beds of the world probably reflect not a continuous record but rather pictures of relatively brief periods separated from each other by wide intervals of time." Johnson adds more of Darwin's defensive points and quotes twice more from the *Origin* to illustrate the pressure Darwin felt from the lack of fossil transitions. Johnson reflects on Darwin's fossil confessions with interwoven logos and pathos. He pictures Darwin as deeply puzzled to the point of desperation over nature's opposition to the discovery of transitional forms, yet keeping faith in his theory in the teeth of misleading fossil evidence:

> Darwin did as well with the fossil problem as the discouraging facts allowed, but to some questions he had to respond frankly that "I can give no satisfactory answer," and there is a hint of desperation in his writing at times, as in the following sentence: "Nature may almost be said to have guarded against the frequent discovery of her transitional or linking forms." But Darwin never lost faith in his theory; the only puzzle was how to account for the plainly misleading aspects of the fossil record. [47]

The purpose of the review of Darwin's interaction with the fossil problem is to lead the reader into a jarring revision. Fossil evidence, in Darwin's experience, was a major problem to be dealt with, not a mother lode of data that led him toward his conclusions. Johnson points out, "Opposition to Darwin's theory could hardly be attributed to religious prejudice when the skeptics included the leading paleontologists and geologists of the day. Darwin's defense of the theory against the fossil evidence was not unreasonable, but the point is, it was a *defense.* Very possibly the fossil beds are mere snapshots of moments in geological time, with sufficient time and space between them for a lot of evolution to be going on in the gaps" (47–48).

Then comes the conclusion: "Darwin's arguments could establish at most that the fossil problem was not fatal; they could not turn the absence of confirming evidence into an asset" (48). Johnson's tour of Darwin's fossil struggle works to toss away any misconceptions about the role of fossils at the dawn of the Darwinian era; it is a narrative argument that is both anti-HEB and new-HOS. Employing my tree metaphor, Johnson uses Darwin's discussion of the fossil problem to cut down one tree—the textbook narrative of paleontology as a plus for Darwin—and he erects a new narrative tree in its place, with Darwin himself now as a reluctant limb.

What Really Happened

At this point Johnson begins a highly unusual discussion, employing complex layers of narrative imagination. First, he takes the approach of musing with the reader about "what might have happened" in regard to the evaluation of Darwinism. In what might be called an "O. Henry twist," it turns out that each of the musings in this discussion turns out to be *what really happened!*[50]

After pages of "what might have happened" scenarios (meaning, in his twist, *what really happened*), Johnson shifts gears and prepares for the deepest rhetorical thrust of the entire book. He returns to the conventional narrative of the history of Darwinian success, and for several paragraphs he summarizes "what supposedly happened." He describes the textbook story of one fossil triumph after another, culminating in a bold description excerpted from a leading textbook: "The fossil record suggests that macroevolution is indeed gradual, paced at a rate that leads to the conclusion that it is based on hundreds or thousands of gene substitutions no different in kind from the ones examined in our case histories" (49–50).

True to the spirit of a manifesto, Johnson abruptly crushes this fossil summary as utterly fraudulent: *"But that last sentence is false, and has long been known to paleontologists to be false"* (50, emphasis added). By boldly and authoritatively exposing a deceitful fossil summary found in a textbook, he implicitly tears the covers away from Darwinian propaganda that is cruelly misleading the public. This is one of the strongest anti-genesis and anti-HEB bombshells in the book, and it is an extremely important moment in the development of Johnson's T–1 thesis. It is the chapter's turning point, thrusting the reader forcefully into a rhetorical barrage that follows, designed to discredit the textbook fossil narratives of both the genesis and HEB types.[51]

A Fossil Primer

Johnson's fossil argumentation is peppered with bird's-eye surveys of current scientific controversies. These accounts are extremely cogent; they not only powerfully summarize the state of the evidence, they also paint the human dimension in science. Johnson manages to integrate the logical-evidential factors with those that are ideological and sociological—such as biases, professional boundary squabbles, quirks of personality, interpersonal rivalries, pride of ownership, and more. This is Johnson's goal. He is telling not just a story about fossils but rather a complex social drama about contemporary scientists facing an embarrassing and frustrating situation and groping toward a clever solution.[52]

The chapter's fossil primer (with Gould's words in italics) crackles with rhetorical energy:

> Gould and Eldredge proposed a new theory . . . to deal with an embarrassing fact: the fossil record today on the whole looks very much as it did in 1859, despite the fact that an enormous amount of fossil hunting has gone on in the intervening years. In the words of Gould:
>
> *The history of most fossil species includes two features particularly inconsistent with gradualism:*
>
> *1. Stasis. Most species exhibit no directional change during their tenure on earth. They appear in the fossil record looking pretty much the same as when they disappear; morphological change is usually limited and directionless.*
>
> *2. Sudden appearance. In any local area, a species does not arise gradually by the steady transformation of its ancestors; it appears all at once and "fully formed."*
>
> In short, if evolution means the gradual change of one kind of organism into another kind, the outstanding characteristic of the fossil record is the absence of evidence for evolution.[53]

This famous quote is extraordinarily important both in modern paleontology and in Johnson's strategy. The fossil facts of sudden appearance and stasis become the rails upon which the train of Johnson's critique runs. Johnson says there are ways by which "sudden appearance" can be explained, but "stasis—the consistent absence of fundamental directional change—is positively documented." To show that stasis is the norm, not the exception, he reports what Stephen Stanley found in the Bighorn Basin in Wyoming (Stanley was a codeveloper of punctuated equilibrium). Paleontologists expected to find fossil transitions because of the continuous record of deposits covering millions of years. Yet expectations were not met; "the fossil record does not convincingly document a single transition from one species to another" (51).

Johnson explains that in order to accept punctuationalism (punctuated equilibrium), one would have to accept the following scenario as plausible: A few members of a species migrate to a peripheral, isolated area where mutations produce rapid structural change.[54] Because the isolated breeding group is so tiny and "selection pressures" are so stringent at the edge of the range, the inherited changes spread rapidly. When the evolutionary change is complete, the tiny new isolated species eventually spreads out and multiplies in number, increasing its chance of fossilization after this spurt of evolution has been completed. This explains why so few transitional forms are found in the fossil record. The key changes occurred rapidly, in tiny populations. How possible is this scenario?

To answer this question, Johnson assesses the mixed reaction to this theory from evolutionary biologists. Many feel they have much better conventional ways to explain sudden appearance and stasis. Johnson then applies a dialectic tinged with satire: "If Darwinism enjoys the status of an *a priori* truth, then the problem presented by the fossil record is how Darwinist evolution always happened in such a manner as to escape detection. If, on the other hand, Darwinism is a scientific hypothesis which can be confirmed or falsified by fossil evidence, then the really important thing about the punctuationalism controversy is not the solution Gould, Eldredge, and Stanley proposed but the problem to which they drew attention" (54). Johnson acknowledges that punctuated equilibrium may have played some role in development of new species, but he asks how it could account for "discontinuities between the major groups—phyla, classes, orders—[that] are not only pervasive, but in many cases immense. Was there never anything but invisible peripheral isolates in between?" (54). Next he illustrates this point with two major fossil anomalies, which constitute the next section of the chapter.

The Cambrian Explosion and the Great Extinctions

The fact that Johnson spends nearly a third of the chapter on these two controversial anomalies shows their vital place in his anti-genesis narrative strategy. The Cambrian explosion is his first illustration of the difficulty in resorting to punctuated equilibrium to explain the gaps between phyla. There is no need to summarize here in detail Johnson's description of this most dramatic event in all the fossil record, when virtually all the multicellular phyla appeared at once—in a geological instant—in strata dated about 540 million years ago. In the course of Johnson's description, he scatters several key quotes. One of the best is Dawkins's amazement at the Cambrian event: "It is as though they were just planted there, without any evolutionary history." Even Darwin's own puzzlement is employed in this anti-narrative:

"The case at present must remain inexplicable, and may be truly urged as a valid argument against the views here entertained" (54). The central puzzle here is that in Cambrian strata, virtually all of the major categories of living things—the phyla—appear abruptly. Below, in the pre-Cambrian rock, one never finds transitional ancestors to the Cambrian phyla. Except for the unrelated Ediacaran species and sponges, this was a world of bacteria and algae.[55] Another mystery Johnson summarizes is what happens after the Cambrian: "*No new phyla evolved thereafter.* Many species exist today which are absent from the rocks of the remote past, but these all fit within general taxonomic categories present at the outset" (55, emphasis added).

As usual, Gould is the linchpin to Johnson's narrative strategy. Gould's celebrated book *Wonderful Life,* which tells the story of the Cambrian fossils pulled from the Burgess Shale in Western Canada, is used as a springboard. In his book, Gould inverts the traditional "cone of increasing complexity." Instead of the gradual evolution of greater and greater disparity in complex body plans over the various epochs of earth history, the picture is one of a sudden appearance of all major kinds, followed by their gradual disappearance through extinction. That is, instead of an evolutionary cone that fills out as one ascends the time axis (a cone with the point at the bottom), the picture is inverted: the cone's wide base—representing the sudden appearance of disparate categories of animals in the Cambrian—is in the distant past and gets narrower and narrower as one approaches the present.

Thus, the fossils of the Burgess Shale in Canada, now housed primarily in the Smithsonian Musuem, are enlisted as star witnesses in a controversy story. According to Gould, the Canadian fossils destroyed the old explanation and left one good one—the "fast-transition" theory. He says the transitional fossils "really didn't exist, at least as complex invertebrates easily linked to their descendants," and that such animals can evolve much, much faster than we ever imagined before now (55).

Johnson says (in support of T–3) that Gould's words "fast-transition" are a label attached to a mystery. Siding with the orthodox Darwinists in their skepticism of Gould, Johnson muses, "An orthodox Darwinist would answer that a direct leap from unicellular organisms to 25 to 50 complex animal phyla without a long succession of transitional intermediates is not the sort of thing for which a plausible genetic mechanism exists, to put it mildly" (56).

Johnson includes a discussion of the problem posed by the "great dyings"— massive global extinctions by which most species have gone extinct (and not by Darwin's theory of competitive replacement). He then caps the section with a situation report that summarizes the ongoing erosion of the Darwinian narrative: "*The point to remember, however, is that the fossil problem for Darwinism is getting worse all the time*" (emphasis added).[56]

The Real Reasons for Punctuated Equilibrium

"We can tell tales of improvement for some groups, but in honest moments we must admit that the history of complex life is more a story of multifarious variations about a set of basic designs than a saga of accumulating excellence" (57–58, emphasis added). This now legendary quote by Gould, which helps to open the last section, is touching in its gentle honesty and eloquent irony. In fact, the closing section is so filled with such rich situational narrative and is so steeped in irony and understanding pathos that it tends to engender a feeling of sympathy for the deep frustration experienced by paleontologists in the decades before the rise of punctuated equilibrium.

One of Johnson's main goals here is to drag into the spotlight the true situational or professional reasons for punctuated equilibrium. Gould and Eldredge, again, are the main witnesses. They candidly tell about constructing their theory in part to allow young paleontologists to publish their findings—to report what they were seeing in the sedimentary rock. Otherwise, if a graduate student pursuing a Ph.D. went out looking for gradual evolution in a certain lineage and instead found that the species' structure persisted unchanged for millions of years, then the findings were useless—they were considered unpublishable failures.

To illustrate the situation, Johnson tells of a German paleontologist, Otto Schindewolf, who went to the saltationist extreme of suggesting that the first bird hatched from a reptilian egg. George Gaylord Simpson, while panning Schindewolf's book that contained this suggestion, admitted the bizarre idea arose from the German scientist's "thorough knowledge of the fossil record." Eldredge makes a similar point: "Either you stick to conventional theory despite the rather poor fit of the fossils, or you focus on the empirics and say that saltation looks like a reasonable model of the evolutionary process—in which case you must embrace a set of rather dubious biological propositions." Johnson quips, "Paleontology, it seems, is a discipline in which it is sometimes unseemly to 'focus on the empirics'" (60).

Punctuated equilibrium met a practical need—something had to be done to free up the researchers to tabulate what they were seeing. In view of all the normal challenges and difficulties in doing graduate work in this field, including the pressure for results in scientific research, a theory was needed that would allow the researchers to report what they saw, which was *sudden appearance* and *stasis*. What was needed, says Johnson, was "a theory that was saltationist enough to allow the paleontologists to publish, but gradualistic enough to appease the Darwinists." Out comes Johnson's knife: "Punctuated equilibrium accomplishes this feat of statesmanship by making the process of change inherently invisible. You can imagine those peripheral isolates changing as much and as fast as you like, because no one will ever see them" (61).

Johnson cites other embarrassing quotes, such as when Gould described the "extreme rarity" of transitional fossils as the "trade secret" of paleontology.

The chapter ends with Darwinian gradualism in rhetorical tatters and with punctuated equilibrium characterized as a desperation measure. The setting is ripe for Johnson's transitional question, "If there are so many problems with Darwinism, and no satisfactory alternative within the framework of evolution, why not reevaluate the framework? What makes our scientists so absolutely certain that everything really did evolve from simple beginnings?" (62). This propels the reader into Johnson's critical review of Gould's "fact/theory" essays in chapters 5 and 6.

Deconstructing Stephen Jay Gould

In the empirical evidence chapters we have reviewed (chapters 1 through 4), Johnson has laid out the substance of his critique. In later chapters other key topics are covered, such as the "molecular clock" (chapter 7) and the "origin of life" (chapter 8). Yet in terms of fundamental points or lines of argument, relatively little new appears after chapter 4. In the remainder of the science section, for example, he continues to shred the Darwinian genesis-narrative and further plagues the HEB narrative with problems, embarrassments, unexpected surprises, controversy stories, and the like. He also builds, limb by limb, a Design version of the modern history of science.

My analysis of the first four chapters made clear the preeminence of Stephen Jay Gould in Johnson's critique. Gould functions as an implicit prophet, and yet his role is a tragic one, creating new kinds of problems for Darwinian credibility in the course of his bold attempt to rescue the ship. In fact, Gould's career-long running critique of Darwinian gradualism could be joined together as one long controversy story. Space does not permit a close analysis of the next two chapters (5 and 6), in which Johnson critiques Gould's key distinction—evolution as fact/theory. These chapters make Johnson's critique just as radical as anything we saw previously in Denton. Johnson strides into the "holy of holies" in evolutionary biology—*common ancestry as fact*—and commits intellectual desecration. He views this notion as overwhelmingly lacking in empirical support, and he casts Gould's defense of evolution as fact as something of a desperation act, a last stand in the defense of a crumbling paradigm. If this picture is so, a rhetorical bargain looms: Defeat Gould's three proofs of the fact of evolution, and the notion of common ancestry is brought into question.

I shall select for brief comment just two of Gould's three arguments for "the fact of evolution." (These two are so highly vulnerable, one wonders why he ventured to use them.) The first of these—"microevolution points to macroevolution"—is punctured with pithy rejoinders. Johnson says it was Gould himself who wrote "that even the first step toward macroevolution (speciation) requires more than the accumulation of micromutations." Johnson accuses other Darwinists of not so much ignoring this problem but rather resorting "to bad philosophy to evade it" (part of his T–3 thesis).

For example, Mark Ridley says the proof of macroevolution is just "observed microevolution added to the philosophical doctrine of uniformitarianism which . . . underlies all science." Johnson replies: "But what sort of proof is this? If our philosophy demands that small changes add up to big ones, then the scientific evidence is irrelevant. Scientists . . . do not assume that the rules which govern activity at one level of magnitude necessarily apply at all other levels. . . . What the Darwinists need to supply is not an arbitrary philosophical principle, but a scientific theory of how macroevolution can occur" (69).

Gould's second proof of "the fact of evolution" is to show that structures exist that are poorly designed or clumsily thrown together from preexisting structures. This "argument from imperfection" concludes that a wise creator would not have done such inferior designing. Johnson characterizes this argument as logically flawed, since it is fundamentally *theological* in nature—speculating, for example, that God would not have created the appendages of all vertebrate animals with similar bone structures. Johnson criticizes such theological argumentation—"God would not have done it this way"—as a woefully inadequate means to establish the sufficiency of a materialistic scientific theory. Note that the entry of Darwinian scientists into theological speculation connects with my T–4 thesis—Darwinism is said to function as much as a religious system as it does as a scientific one.[57] Chapter 6, which is a review of the putative evidence for the rising evolution of different classes of vertebrates (fish, to amphibians, to reptiles, to mammals, to humankind), employs an anti-genesis strategy that is filled with current discussions of the evidence along with vivid controversy stories. In this section Johnson appears to have scored points with many readers by making a few concessions to Darwinists.[58]

The Summation Completed

The most important chapters outside the science section are 9 and 12, both on the philosophy of science, and the epilogue (13), which surveys critical response to his first edition. Chapter 9 is especially important since it is the chapter in which he firmly argues his T–2 thesis that naturalism is the bedrock upon which Darwinian evolution rests. Johnson's discussion of Thomas Kuhn's vision of the history of science as a series of paradigms is a significant part of this section. Given Johnson's narrative approach, Kuhn's ideas are important, and Johnson, while urging caution in applying theories like Kuhn's, finds the Kuhnian analysis "an illuminating picture of the methodology of Darwinism." Johnson summarizes Kuhn's steps of a paradigm shift, leading from normal science, to struggle with anomalies, and finally into crisis. Kuhn's doctrine implies that a prevailing paradigm is deeper and broader than a hypothesis, which can be discarded if it fails a test. Paradigms are much more fundamental; they are ways of looking at the world, or some part of it. Because *ad hoc* hypotheses can be employed to explain anomalies and protect the paradigm, it is much more difficult to do damage to a paradigm without a new paradigm.

Has Johnson rhetorically fallen into "Denton's limbo" at this point? Did Johnson hope that his book would produce a rare (the first?) exception to the Kuhnian pattern he described? Without a new paradigm, what is the hope of success? Where is Johnson headed?

Johnson was laboring to provoke a *falsification crisis*. He was training a vast and diverse audience in his own school of the rhetoric of science so as to create a common consciousness in which Darwinian evolution was seen as badly in need of outside audit. This is the theme of Johnson's twelfth chapter, and the other chapters (7 through 11) also are sprinkled with analyses that call for a start in this testing of the bedrock notions of the theory.

Falsification, however, is not the same as construction of an alternative research system. Johnson believed in 1991 that an embryonic paradigm was in the wings. Thaxton's intelligent design notions were fairly well developed at this point and had been in print for two years,[59] and several young scholars allied to this new framework were finishing their doctorates. Some of their dissertations were already addressing the questions that would be foundational to the construction of a design paradigm.[60] The conceptual base of the "inference to design" was emerging; arguments were being honed and supporting data was being analyzed. Johnson was the forerunner, and he paved the way by heaping up anomalies and dramatizing the simmering crisis. Yet it would be the task of his new cohorts to carry the argument to the next stage.

7

The Roaring Nineties

David Takes On Goliath

An Overview of the 1990s

As *Darwin on Trial* and its author began to circulate on college campuses in the fall of 1991, two important rhetorical processes began to accelerate. One was Johnson's energetic engagement with his critics. He immediately took on a heavy schedule of speeches, conferences, and debates, and his public oratory quickly became one of his most effective ways to influence university audiences.[1] A vivid characterization comes from Lynn Vincent, writing for *World*:

> Once evolutionists read his book, they were eager to sink their teeth into Mr. Johnson, whom they saw as a middle-aged, Harvard-educated dilettante sticking his unscientific nose where it didn't belong. Critics lined up to debate him. But once engaged, his adversaries found him to be both ruthlessly intelligent and maddeningly congenial. With his agreeable, favorite-uncle face, wire-rimmed specs, and a perpetual smile in his voice, it was hard not to like Mr. Johnson as he shredded their arguments. And, of all things, he even wanted to be friends when the debates were through.
>
> "I've been overplayed as a controversialist," said Mr. Johnson, who sees such bridge-building as his greatest strength. . . ."I see myself as a person who tries to build alliances and friendships. To win the debate, you have to carry both the moral high ground and the intellectual high ground rather than try to win by any sort of power tactics. That's really what we're trying to teach people."[2]

The second key rhetorical process was his recruitment and grooming of new, bright revolutionaries. He especially sought colleagues with academic qualifications who could collaborate in research, criticism, conceptualization, and persuasion. As the decade wore on, the ranks of Design began to swell with newly recruited researchers and public advocates who began to publish their own books and articles, which in turn generated fresh cycles of engagement with other critics on an ever-widening set of issues.

Working together, these two recycling processes (engagement—recruitment—new publication—further engagement) transformed Intelligent Design from a committee of rebel outsiders into a tight-knit, aggressive network of several hundred activists, many of whom began carrying out sorties of persuasion at their own universities. By the end of these "Roaring Nineties," Design texts and speeches were exuding confidence about the future. To the younger members, a paradigm shift seemed not just possible within their lifetime but inevitable. David felt certain that he could defeat the aging Goliath through a relentless critique of the vulnerable points of the Darwinian edifice and through a flow of books and articles enriched with new arguments and concepts that putatively established Design's practical explanatory power. This chapter and the two that follow will trace these vital interwoven processes during the 1990s as they propelled Intelligent Design to the brink of a serious challenge to Darwinian domination within U.S. science and education.

After surveying first the historical highlights of the first part of this period, 1991–96, I will place my central focus on Michael Behe, telling of his entry into the Design Movement (before it was even called by that name). The next chapter will trace the story of his book, *Darwin's Black Box,* which exercised its own unprecedented persuasion, triggered the creation and sharing of projection themes, and fed into the movement story. These cascading processes, accelerated by Behe's book and lectures, worked in concert with similar rippling that spread out in smaller measure from the half-dozen other critiques emanating from the Design community during the 1990s.[3] All of these narrative processes are important to my analysis because they shaped the rhetorical vision of the young movement.

The first of these two processes, Johnson's forceful engagement with scholars in evolutionary biology and others in academia, was sparked primarily through *Darwin on Trial* and to a lesser extent through his five later books.[4] As shown in chapter 4, Johnson already had interacted with many scholars before *Darwin on Trial* was published, but the pace accelerated greatly in the latter half of 1991. By early 1992 such interaction occupied much of Johnson's spare time when he was not teaching law at Berkeley's Boalt Hall. In the last chapter I stressed that the epilogue of the 1993 edition of *Darwin on Trial* summarized the most important early reactions. Here Johnson reported the remarkable concession of Michael Ruse, described later in this chapter. He also responded in detail to Stephen Jay Gould's 1992 review in *Scientific American.* In fact, in his research notes to the epilogue he pled guilty to one scientific error in a footnote; beyond that, he defended—item by item—the accuracy of every statement that had been attacked by Gould.

Most reviews of *Darwin on Trial* from evolutionists were similar to Gould's, seeking more to discredit Johnson as a competent critic than to engage with his book's main criticism of macroevolution and the creative power of natural selection. Johnson had expected this, and he relished the opportunity to craft rebuttals. Some of Johnson's most powerful encounters with critics grew out of his use of other channels for persuasion—a plethora of articles, videotapes, conferences, symposia, and campus lectures. All either contained or fostered a wide variety of reactions and interactions with academicians. Some interactions were face-to-face; others were developed through e-mail correspondence. Johnson used many of these encounters later as illustrative material in his writing. The stream of reaction played a major role in Johnson's rhetoric; his writings after 1993 are generously sprinkled with interaction vignettes.[5]

Some of the most powerful interaction stories that enhanced Design's credibility are found in published videotapes.[6] Undoubtedly the most popular and rhetorically useful videotape was Johnson's 1994 debate at Stanford University with Cornell's historian of biology, William Provine. This debate advanced Design two ways. First, it was one more vehicle for laying out Johnson's evidentiary case against macroevolution. Second, Provine's dismissal of the mirage of free will and his repeated sneering at Johnson's belief in God served to illustrate the thesis that Darwinism functions just as much as an antitheistic philosophical frame of belief as it does as a scientific frame of research.[7]

Johnson's colleagues began to speak on college campuses frequently in the latter half of the 1990s, especially in the wake of Behe's *Darwin's Black Box* (1996). Some lectures lived on as videotapes, such as Behe's "Irreducible Complexity," a lecture at Princeton University with a lively question-answer period involving members of the biology department. Such lectures and videos reinforced the persuasive effect of the Design books, enabling Johnson's individual rhetoric and his interaction story to expand into a much broader "Design Movement Story." This composite story, with its tales of successful or revelatory interactions with critics, proved to be one of the most powerful weapons in the maturing rhetoric of Design.

Design had to cope with one major interaction story that was more negative than positive. That was the two-hour PBS "Firing Line Debate," which was watched by hundreds of thousands of Americans during the month of December 1997. In this debate, filmed in a packed auditorium on the campus of Seton Hall University in New Jersey, Michael Ruse, Eugenie Scott, and two other Darwinists managed to rhetorically outscore the Intelligent Design team of Johnson, Behe, David Berlinski, and William Buckley.[8] While Johnson and especially Behe proved very capable in the debate, two factors hindered the persuasiveness of the Design performance. First, biologist Kenneth Miller of Brown University, a skilled Darwinian orator, spoke with crisp and impervious confidence, and he repeatedly and enthusiastically brandished posters filled with fossil lineages that he claimed powerfully illustrated macroevolutionary descent.

Second, the Darwinists pictured macroevolution as *something God may well have established.* Thus, the hard, atheistic knife-edge of Darwinism that was so blatant and damaging in William Provine's presentation at Stanford was transformed here into a softened, theologically sensitive velvet pillow. This difference was symbolized by Barry Lynn, the lawyer-clergyman on the Darwinist side, when he quoted John 1:1—"In the beginning was the Word"—and then said that God's "word" was "Evolve!"[9] Within Intelligent Design's assessment and private chaining of projection themes, this debate was an occasion for introspection and licking of wounds. Those in Johnson's e-mail discussion group saw it as a frustrating and embarrassing (but rare) setback—something to learn from and to keep in mind in preparing for future debates.[10]

So far I have attempted an overview of the 1990s with my focus on the rhetorically pivotal process of engagement with critics. But the second rhetorical process—recruitment of new revolutionaries—was every bit as important as the interaction with outsiders. Obviously, a robust numerical expansion of Design and an augmenting of its conceptual richness will greatly increase its potential for long-range rhetorical success. Michael Behe was the supreme example of a powerful new ally who made huge contributions, yet he is no isolated case. Another powerful recruit, philosopher and logician David Berlinski (who participated in the PBS debate), first burst onto the scene in June 1996 when his article "The Deniable Darwin" appeared in *Commentary* magazine. This article argued the shocking thesis that Darwinism had not yet risen to the level of a true scientific theory. It triggered an extraordinary outpouring of vehement response from both sides of the evolution debate, and I will weave some of his story into the section on the emergence of Behe in the next chapter.

Behe and Berlinski's addition coincided with a steady influx of other academicians into Design. Many were attracted through Johnson's conferences and college speaking tours, organized by Christian Leadership Ministries and the Veritas Forum.[11] Others such as philosopher Alvin Plantinga, the renowned chemist Henry Schaeffer, and several professors at Yale and Princeton had become skeptical of both Darwinism and chemical evolution through reading the works of Michael Denton and Charles Thaxton. They became early allies of Johnson.

Several dozen scholars decided to support Johnson publicly after a crucial turn of events in August 1992. *Scientific American* had just published Gould's July 1992 review, "Impeaching a Self-Appointed Judge," which I discussed in chapter 5. Johnson immediately asked *Scientific American* for space to reply to Gould's attack, but his request was denied. Ironically, Gould's bashing of Johnson coupled with the magazine's denial of a chance to reply proved to be a major step forward in building Johnson's stature as a critic. It handed Johnson his own rhetorically useful story of unfair treatment at the hands of the "biological thought police."

Because of this turn of events, the Ad Hoc Origins Committee obtained a grant to send out (in 1993) a copy of Johnson's reply to Gould—a four-page essay called "The Religion of the Blind Watchmaker"—directly to mailing lists of five thousand university science professors.[12] The Ad Hoc Origins Committee cover letter for that mailing (see appendix 3), signed by forty-five professors, along with Johnson's essay reply, serve as paragons of the emerging communication genre I have called the "Intelligent Design subgenre."[13]

Besides the adherents drawn from professorial ranks at colleges and universities,[14] four younger scholars were finishing their doctorates in the early 1990s at prestigious institutions: Steven Meyer at Cambridge University, William Dembski[15] and Paul Nelson at the University of Chicago, and Jonathan Wells at the University of California at Berkeley. Dubbed "the four horsemen,"[16] this group worked in close partnership, honing each other's ideas and collaborating on research and writing. In fact, after 1995 many of the Design ideas and texts went through an internal scrutiny from Phillip Johnson's entire Internet discussion group. This Internet village, which grew from 75 members in 1995 to over 200 in 2003, was the developmental womb of much of the rhetoric of design in this period.

The importance of this sort of community in the development of revolutionary ideas and texts is captured in Hull's parallel observation of Darwin's own community in *Darwin and His Critics:* "But I believe that there is also a kernel of an original truth in Kuhn's work—the role of scientific communities in the reception of scientific theories. I do not know whether this truth belongs in the philosophy of science or in the sociology of science and do not much care. It is an interesting feature of science which needs investigating. New ideas, like mutations, stand a better chance of surviving and being promulgated if they develop for a time in an isolated community. Then, after they have been established firmly in it, they are injected more massively into the larger community as a whole."[17]

Just as in Darwin's day, the close collaborations between the four horsemen and the interaction within Johnson's Internet community played a major role in evaluating progress, discussing strategy, and developing the new ideas before they were injected into the larger public. The actual design theory texts of the four horsemen (and several other new rhetors) made their publishing debuts in three key books: *Darwinism: Science or Philosophy?* (ed. Buell and Hearn, 1993), *The Creation Hypothesis* (ed. Moreland, 1994), and *Mere Creation* (ed. Dembski, 1998). In chapter 9 I will profile the conceptual-rhetorical contributions of the most important horseman, William Dembski, whose "explanatory filter" became the central heuristic framework for the detection of intelligent design in any phenomenon or object.

In my survey of the 1990s, I have discussed the engagement-recruitment processes that spiraled outward and also profiled key rhetorical texts and battles. Some were critically important for chaining out projection themes of paradigm conquest in the twenty-first century. These visions, coupled with a steadily increasing flow of fresh rhetoric that embodied new evidence and concepts and generated unprecedented positive interaction, augmented the confident "David versus Goliath" drama inherent in the rhetorical vision of Design. Now I turn to the first of the major flashpoints in this Davidic drama—the published responses to *Darwin on Trial.*

Jukes and Hull on Phillip E. Johnson

Robert Pennock (1999) and Kenneth Miller (1999) critique Johnson and Behe at length in their books on Design. Adding many shorter published discussions and reviews, Johnson's critics' total published interaction amounts to several hundred pages.[18] Many of the later Pennock-Miller criticisms of Johnson are found in seed form in the earliest published reviews. Some, such as Thomas Jukes's "Random Walking" column in *The Journal of Molecular Evolution,* were hostile in tone, attacking Johnson's character. Jukes described Johnson as sounding "truculent and bombastic"[19] in newspaper interviews. In general Jukes vilified his opponent as a buffoon, an uncomprehending incompetent: "Johnson does not have enough comprehension of evolution to discuss intelligently any article in this or any other issue of *The Journal of Molecular Evolution.*" Even law schools are subjected to Jukes's ridicule. He quotes a friend's wisecrack: "Indeed, if necessary to win the case, the truth must be hidden. A successful lawyer like Johnson probably has strong natural instincts in this direction, reinforced by training at the best law schools." About half of the review takes aim at Johnson's chapter 7 on molecular evidence and alleges several errors of understanding.[20]

Not long afterward, in the same "Random Walking" column, Johnson replied: "This Journal has published a polemic by Thomas Jukes which denounced me, the entire legal profession and anyone else who dares to question the Darwinian account of evolution. Readers should be aware that this overwrought attack does not controvert any of the important points in my book, *Darwin on Trial.*" A detailed review of Johnson's reply is unnecessary, but it is relevant to my narrative rhetorical questions to note how Johnson used the space allotted him to respond to Jukes's vicious attack. He first recaps his book's overall argument step-by-step and then clarifies the purpose of chapter 7, which Jukes had so heavily criticized. Johnson defends his own understanding of the molecular evidence, saying the pattern of hierarchy, or "groups within groups," is not confirmation of the Darwinian explanation. "The logical and imaginative appeal of Darwinism justifies treating the theory respectfully as a hypothesis, but not shielding it as a sacred dogma from possible disconfirmation." He adds, "I predict that, if readers will examine the book for themselves, they will find it hard to identify any important scientific errors. That Jukes had to rely upon newspaper quotes and irrelevancies speaks for itself. On the other hand, I welcome any critical comments that readers can supply. The book will be used in many college courses and other educational programs, in this country and around the world. I want it to be as accurate as possible."[21]

Johnson's strategy in this reply is fourfold. First, he took the opportunity to present a pithy abstract of his entire book, beginning with an offense rather than defense. Second, he ably defended his chapter on molecular evidence, which Jukes attacked. Third, Johnson modeled in his style and tone a *patient, calm, scholarly attitude*—the opposite of the truculence and bombast Jukes claimed were typical of his ethos. Fourth, Johnson also strove to prompt readers to give his book a hearing—encouraging this response by inviting readers to identify any errors and by saying that the book would be used in many college courses around the world.

The most important book reviews of *Darwin on Trial* were Gould's review and the review by the Darwinian philosopher of science at Northwestern University, David Hull. Hull's review appeared in the British journal *Nature*, which is widely regarded as the most prestigious science journal in the world. It is noteworthy and relatively rare when any book is singled out in the pages of *Nature* for attack and wholesale rejection; Hull's review of *Darwin on Trial* is one of those rare cases.[22] Hull's tone was much calmer and more reflective than Jukes's, but his final evaluation was just as dismissive: "If Johnson had written a *religiously motivated* criticism of thermodynamics, quantum theory or plate tectonics, it might have been worth reading, but *I cannot imagine why anyone would want to read yet another rehash of creationist objections to evolutionary theory*" (emphasis added).[23]

It was predictable that evolutionists would target Johnson's religious frame of reference. In an attempt to overcome the "creationist obstacle," he often wrote of his own nonliteralist view that *somehow* God was behind the creation of living things, yet he professed openness to what science may show about *how* the creation took place. Hull turns Johnson's theistic admission into a rhetorical weapon. He opens with a lengthy (305-word) introduction on scientific "fallibilism," in which he reviews how scientists from time to time "have to backtrack to re-evaluate some tenet that they had all taken for granted for years." There are genuine disagreements and questions, and this applies even to "proper scientific method." He adds, "Because scientific knowledge is so interdependent, rarely is evidence totally independent of the hypotheses that it is used to test. As dogmatic as scientists can be at times, they are committed to fallibilism. Not only may they be mistaken, but quite frequently they are." He goes on to say that Johnson "exploits the complex fallibilist character of science in general and evolutionary biology in particular *to urge equal time for his religious convictions*. He is not a creationist in the sense that he thinks God miraculously created all species in six days some ten thousand years ago, but he does believe that any purely naturalistic explanation of the creation of life on Earth and the emergence of human beings is inadequate. Instead, as a 'philosophical theist and a Christian' he believes that God brought all living things into being to further His own purpose, possibly by creation from nothing, pos-

sibly in the way that contemporary evolutionists claim."[24] Of course, by saying
that Johnson is urging "equal time for his religious convictions," the reader of
the review would likely infer (erroneously) that Johnson's religious convictions
are a substantial part of the content of his discussion of Darwinism.

A similar misleading and far more sarcastic jab is found near the end of
the review. Hull claims that "Johnson's problem is that he does not like the
answers that he hears. He wants evolutionary biologists to include reference
to God in their professional writings in the way that he, I presume, does in
his."[25] Ironically, this comment can be viewed two ways. It can be seen equally
as (1) a vicious rhetorical cheap shot or (2) a weighty symbol of the substantial
quandary Johnson's rhetoric created for evolutionists.

The "cheap shot" side is obvious. Johnson clearly would be satisfied by
the confession of ignorance of truth about origins; he voiced no demand that
biologists figure God into their theories. He might conceivably reply to Hull
that different branches of academia analyze very different areas of reality. Since
laws are human constructions (leaving aside the question whether natural law
exists, stemming from God), it is absurd to suggest that legal scholars address
the relevance of God in their study of the rise of these human constructions.
On the other hand, biologists face a different reality, which came into existence
outside of human observation but necessarily arose either with or without the
substantial participation of a mind—a designing agent.[26] The knowledge-claim
of Darwinian biology, says Johnson, is that we have compelling evidence of
purely naturalistic mechanisms that can accomplish the task of building highly
complex and information-rich systems. This claim of science is precisely the
locus where the notion of a *designer* of biological complexity is branded as "decisively refuted" or "outmoded." It is exactly this claim that Johnson argues is
completely empty of empirical support. Thus, the potential *God question,* by
the nature of subject matter and prevailing theory, is thrust squarely into the
center of the debate by evolutionary biologists themselves. Therefore, to suggest
that Johnson must be referring to God in his research or that he wants biologists to do the same is to ignore the huge differences in the nature of subject
matter and in the inherent foundational issues.

Yet my explanation of Hull's cheap shot also has revealed how very weighty
is this rhetorical issue. What Johnson proposes, and indeed what Intelligent
Design insists on, is a peculiar kind of revolution of thought in regard to
macroevolution—openness to the *possibility* of a creator of some kind. The
weightiness, of course, is partly due to the fact that this issue is exceedingly
central to human cultures. At the same time, it is the biologists who must
grapple with this issue (along with physicists and cosmologists who now face
the implications of a "fine-tuned universe"). This much is clear: If enough
influential biologists were persuaded that something were fundamentally amiss
in this set of naturalistic claims, it would require a top-to-bottom reevaluation
and restructuring of their entire paradigm. Whereas physicists would only
have to face the possible "design" of boundary conditions of the universe, the

biologists would also have to reckon with something closer to home—the notion of possible "intelligent intervention" along the course of the universe's development. This latter concept (intelligent design through intervention) would veer even closer to the ideas of classic theism than the former possibility of "boundary-conditions design" in physics (which may merely imply a deistic notion of God). So Hull's comment is an implicit balk at the mere thought of biology undertaking such a revamping of its conceptual foundations and horizons. In Hull's case (as with all others who have no theistic belief), this paradigm remodeling is not just unpalatable; it is (at least pre-Behe and pre-Dembski) quite literally unthinkable.

Between his two comments about theism, Hull hurls three feisty rebuttals at Johnson's theses. The first two, scientific in nature, are brief; the third, highly theological, dominates the review. Hull's first reply lists the main scientific objections of Johnson and implies that they are old, weak arguments and that some are mere caricatures (such as the case of natural selection being a tautology).[27]

In his second rebuttal, Hull discusses four scientists whom Johnson quoted as star witnesses against Darwinism's credibility. Grassé and Goldschmidt are just mentioned; the other two, Popper and Patterson, are discussed briefly. Hull's argument here takes on an informal logic that is utterly intriguing because it is so profoundly personal or sociological or even narrative type. In a nutshell, he says that these four, despite their vehement criticism of Darwinism, did not express any doubts that evolution has occurred. Their doubt went only to the *how* of evolution, not the *fact* of evolution.[28] This narrative argument leads to Hull's rhetorical question: "Why are even the scientists who are most sceptical of darwinian [sic] versions of evolutionary theory nevertheless forced to accept evolution?" He then gives Johnson's answer: It is due to their "commitment to naturalism. Evolution is the only viable naturalistic explanation for the structure of the living world, and scientists are committed to naturalistic explanations."

This is Hull's transition to his lengthy third argument, which is unabashedly theological. One might even call it an "antitheological argument" because its abstract could be phrased, "God's agency is impractical in science and totally unthinkable in biology." Hull's first point—naturalism is unavoidable—is based on a practical claim that "scientists have no choice." He uses *reductio ad absurdum* to show why scientists cannot entertain the idea of divine causation: "Once they allow references to God or miraculous forces to explain the first origin of life or the evolution of the human species, they have no way of limiting this sort of explanation. Why does the Earth have a magnetic field, why do organisms use only *laevo* amino acids, why is the savings and loan industry in such trouble? It is easy enough to answer that these phenomena are all part of God's great plan, but in the absence of some partially independent knowledge of God and His intentions, such explanations are no less vacuous than the usual parodies of the principle of survival of the fittest."[29]

For those in sympathy with Hull, this hypothetical nightmare in science (in which every phenomenon can be claimed to be a manifestation of God's will) is not just an argument; it is also the tender shoot of a projection theme. The plot line of this implied nightmare can be replayed and readapted in any academic department, in any area of research, and in each fantasy of this type. The villain (the Design proponent) effectively shuts down science by his inane "God willed it!" explanation. It is a dreadful vision of absurdity and irrationality, of the death of science.

Johnson replies to Hull's point in his research notes of *Darwin on Trial:*

> This is a caricature of theistic rationality, of course. Theists do not throw up their hands and refer everything to God's great plan, but they do recognize that attempts to explain all of reality in totally naturalistic terms may leave out something of importance. Thus they reject the routine non sequiturs of scientism which pervade the Darwinist literature: because science cannot study a cosmic purpose, the cosmos must have no purpose; because science cannot make value judgments, values must be purely subjective; because science cannot study God, only purposeless material forces can have been involved in biological creation; and so on.[30]

Johnson has effectively deflated Hull's *reductio* here, but a question may linger for his reader: "Can the theory of design explain precisely when and where it will consider intelligent design as a possible cause? Specifically, what kinds of phenomena may trigger or invite this explanatory option?" It fell to Behe and Dembski to begin to answer these questions later in the 1990s.

The second part of Hull's theological argument uses a different line of antitheological reasoning: God's agency is *morally unthinkable in the realm of biology* because such a deity cannot be efficient, or kind, or good at all. Rather, he is "almost diabolical." I shall quote this section of Hull to convey the full rhetorical effect as he piled example upon example:

> What kind of God can one infer from the sort of phenomena epitomized by the species on Darwin's Galapagos Islands? The evolutionary process is rife with happenstance, contingency, incredible waste, death, pain and horror. Millions of sperm and ova are produced that never unite to form a zygote. Of the millions of zygotes that are produced, only a few ever reach maturity. On current estimates, 95 percent of the DNA that an organism contains has no function. Certain organic systems are marvels of engineering; others are little more than contraptions. When the eggs that cuckoos lay in the nests of other birds hatch, the cuckoo chick proceeds to push the eggs of its foster parents out of the nest. The queens of a particular species of parasitic ant have only one remarkable adaptation, a serrated appendage which they use to saw off the head of the host queen. To quote Darwin, "I cannot persuade myself that a beneficent and omnipotent God would have designedly created the *Ichneumonidae* with the express intention of their feeding within the living bodies of caterpillars."

Whatever the God implied by evolutionary theory and the data of natural history may be like, He is not the Protestant God of waste not want not. He is also not a loving God who cares about His productions. He is not even the awful God portrayed in the book of Job. The God of the Galapagos is careless, wasteful, indifferent, almost diabolical. He is certainly not the sort of God to whom anyone would be inclined to pray.[31]

Although Johnson did not feel it necessary to reply in detail to this argument, he did quote from it and discuss it at times as confirmation of his claim that "Darwinian theory is fundamentally incompatible with a theistic understanding of reality."[32]

From a rhetorical-critical point of view, how effective was this argument picturing Judeo-Christian monotheism as unthinkable in light of biology? Though it may have scored points with nontheist readers of *Nature,* there are three distinct problems with this rhetorical strategy that render Hull's line of argument ineffective and even mark it as a major blunder.

First, to base one's rejection of the possibility of intelligent creation on the moral problems with natural theology is to play into Johnson's hand. It appears to confirm (by the comparative length devoted to it) that the strongest case for embracing Darwinian theory is the distaste with the theistic alternative. Hull's argument thus highlights the very theological functions, beliefs, and values Johnson has claimed are at the core of Darwinism.

Second, Hull seems to imply there are no answers to his questions about evil in nature. Any reader of Hull's review who is conversant with the history of theology stretching back thousands of years knows of a rich flow of ongoing discussion, and a resultant vast literature, focused on these very questions. A modern example among many would be C. S. Lewis's *The Problem of Pain.* This book is just one of hundreds that address such questions and develop various solutions. Thus, Hull's antitheology may impress those who have no knowledge of this ongoing discussion, but to others his comments seem quite puerile or even absurd.

Third, some of Hull's examples seem to reflect not so much "waste and inefficiency" as they do the necessary parameters of a well-functioning system (e.g., the number of sperm or ova produced) or the current state of ignorance of biology (e.g., the purpose of 95 percent of DNA that originally seemed to be superfluous but is increasingly found to have important functions). A verdict of "inefficiency" and thus "bad design" seems premature or questionable.[33]

The Warning in *Science* and Behe's Response

Science, published by the American Association for the Advancement of Science (AAAS), is almost as prestigious in scientific circles as *Nature.* When *Darwin on Trial* appeared in June 1991, the editors of *Science* had to decide among three responses: (1) Ignore the book completely, (2) print a brief "book notice," or (3) publish a full review (a Hull-type attack piece). *Science* chose the middle option, and in an anonymous column entitled "Johnson vs. Darwin," on the "Briefings" news page, *Darwin on Trial* was panned as a potentially dangerous book.[34] In language reminiscent of hurricane warnings posted by Florida newspapers, Eugenie Scott alerted AAAS members and science educators to be braced for the confusion the book would generate. In the ensuing months and years, Scott herself engaged in several radio debates with Johnson over National Public Radio in an effort to neutralize his critique.

Michael Behe was one biologist who noticed the "Johnson vs. Darwin" column in *Science*. As mentioned, Behe had already become skeptical about Darwinism through his reading of Denton's *Evolution: A Theory in Crisis* in 1987. By 1991 he was using Denton along with Dawkins's *The Blind Watchmaker* and Kuhn's *Structure of Scientific Revolutions* in a freshman seminar course at Lehigh entitled "Popular Arguments on Evolution." Behe designed this course to examine the strength of arguments for or against Darwinian evolution advanced in popular books and articles.[35] As soon as *Darwin on Trial* was published, he read it and was impressed with Johnson's handling of the scientific issues. He felt Johnson had made a strong case for his main theses. When he saw *Science*'s book notice about Johnson, he was excited that a leading science journal would take note of *Darwin on Trial*. His enthusiasm dissolved to anger[36] as he realized that *Science*'s coverage amounted to an "anti-intellectual dismissal," laced with a string of subtle inferences (called "enthymemes"[37] by scholars of rhetoric), which tapped into scientists' deep antipathy toward creation science.

For example, *Science* said that although the Berkeley law professor who wrote the "new anti-evolution book"[38] claimed not to be a defender of creation science, nevertheless the book was endorsed by the Institute of Creation Research, and it used many of the same arguments as creationists. Another link to classic creationism is motive: "Johnson admits that *religion fuels his personal beef with evolution*. 'There is no room [in evolutionary theory] for a life force,' he says, 'for something . . . that cannot be perceived through the tools of science'" (emphasis added).[39]

The third paragraph continues the same rhetorical theme of Johnson's close tie with creationism: "But like the creationists he dissociates himself from, Johnson claims to 'examine the scientific evidence [for evolution] on its own terms.'" Johnson is said to have criticized evolutionary theory as "constantly reformulated, in the manner of Marxism, on account of the failure of its predictions to come true." Thus, Darwin's theory should be abandoned; scientists "should admit that the origin of species can't be explained without invoking supernatural processes."

Eugenie Scott's critique and general warning dominate the last two paragraphs:

Johnson's arguments demonstrate his misunderstanding of the scientific process, in which theories are continually tested and refined, says Eugenie Scott of the

Berkeley-based National Center for Science Education. The problem, says Scott, is that Johnson is a lawyer, not a scientist. "Theory, proof, and law are different terms to scientists than to lawyers," she says.

Johnson is busy on the talk-show circuit publicizing his book, and Scott worries that his academic position and his approach will win him a wide following. "I hope scientists find out about this. They really need to know [the book] is out there and is confusing the public."[40]

Scott's worry about *Darwin on Trial* recalls Philip Spieth's dread about potential damage from Denton, which I discussed in chapter 4. Scott here implicitly calls for a vigilant watch in light of the possible spread of a new and dangerous rhetorical virus. Also note Scott's conclusion that Johnson does not understand how science works. In the coming years Scott would repeat these words so often (including during her National Public Radio debates with Johnson) that Johnson began to quip, "When they bury me, they will put on my epitaph, 'He didn't understand how science works.'"[41]

Incensed at the treatment of *Darwin on Trial,* Behe wrote a letter to *Science* that was published 30 August 1991. It not only captured the attention of Johnson and the fledgling Design Movement but also endures as a tiny masterpiece of early Design rhetoric. Behe begins by noting that the briefing on *Darwin on Trial* is

> a good illustration of the failure of the scientific community to follow its own advice about the perennial evolution controversy. Instead of simply addressing the skeptical arguments advanced in the book, the article relies on ad hominem remarks. It is pointed out that Johnson's religious views predispose him against naked materialism (although in his book he states that he finds nothing a priori incredible in God's using Darwinistic evolution to produce life), and a science educator is trotted out to opine that Johnson misunderstands the scientific process. Johnson is also found guilty by association because Creationists like his book.

Then Behe turned the tables, showing how similar inferences or enthymemes could be constructed against Darwinism and adding that they, too, would be irrelevant:

> Well, now. It is also true that fascist governments have embraced Darwinism, that most scientists are not trained logicians, and that many commentators on evolution are predisposed in favor of naked materialism. But all of this is name calling and quite beside the point. In his book Johnson appears to be an interested, open-minded, and very intelligent layman who sees large conclusions drawn from little evidence, notices anomalies in current evolutionary explanations, and will draw his own conclusion, thank you, about the validity of Darwin's theory. A man like that deserves to be argued with, not condescended to.

Behe's third and final paragraph is built on the practical importance of arguing with intelligent skeptics rather than condescending to them. He echoed Johnson's point in *Darwin on Trial* that Darwinian theory, if true, does not require a sophisticated or advanced reasoning or special mathematical training, since Darwin and others wrote for a general audience with the assumption that the intelligent lay reader will be persuaded by the evidence. In Behe's own words: "The theory of evolution by natural selection is not a difficult concept to grasp, and Charles Darwin addressed . . . a general audience. But neither is it self-evident to many people that natural selection can fully account for the world they observe. Thus when questions about the theory arise in public forums, the scientific community would do much better in the long run to patiently list supporting facts and frankly admit where positive evidence is lacking, rather than paternalistically maintaining that an understanding of the theory of evolution is reserved for the priesthood of professional scientists."

By concluding with a stigma-phrase, the "priesthood of professional scientists," Behe chained the same projection theme that started with Denton and continued on with Johnson's fourth (T–4) thesis, namely: *Darwinian scientists are our current cultural high priests who mediate knowledge to the masses. Their paradigm is known to be true* a priori *and is not open to question.* In terms of this projection theme, the Darwinian priests are rigid and dogmatic; they are the villains. The intelligent questioner is the potential victim, but he or she also is the *courageous hero,* for in this fantasy, it is very risky for outsiders to question the authority or specific pronouncements of the priests. Of course, Behe and his Design cohorts would argue that such a theme is grounded in a symbolic reality and a visible social reality behind it—the communication style and strategy seen in the book warning of *Science* and the reviews of Gould and Hull.

This was Behe's debut as a Design rhetor, and his virtuosity was clearly displayed. Equally remarkable is the congruence between his own rhetorical vision and Johnson's, although the two had never met. How could Behe, without contact with Johnson or the Ad Hoc Origins Committee, develop a virtually identical rhetorical vision? One major factor is the role of Denton in fathering the dissenting consciousness of both. In the prototypical rhetorical vision of Denton, the evidence for macroevolution was seen as embarrassingly weak, and the evidence against the Darwinian mutation-selection mechanism was deemed overwhelming. Only the priority of the paradigm kept Darwinism afloat, according to Denton's projection theme that chained out from his final chapter. This rhetorical vision (both as intellectual critique embracing an array of evidences and arguments and as a network of projection themes) went through a process of adaptation to the worldview of robust but nonliteralist theism of Behe and Johnson. Thus, the separate but parallel development of Johnson and Behe is not surprising. They were endowed with Denton's rhetorical vision, and they possessed common metaphysical space for the role of a designer who transcends the cosmos.

After reading Behe's letter, Johnson immediately wrote him to express thanks and invite his collaboration. Before long Behe was integrated into the network of dissenters. When plans took shape in late 1991 for a conference in Dallas to bring together top scientists and philosophers on both sides of the debate to discuss the theses of *Darwin on Trial,* Michael Behe was one of the first invitees.

The Darwinism Symposium: Repositioning and Dialectical Fallout

Michael Ruse, who grew up in a Quaker home in England, is now a Floridian champion of Darwinism. He accepted a distinguished professorship in philosophy at Florida State University in early 2000 after teaching for over twenty years at the University of Guelph in Ontario. He emerged in the latter decades of the twentieth century as one of the world's most productive and prestigious philosophers of biology, and he was especially known as a dedicated and able defender of evolution. Many of his writings focus on the Darwinian revolution in scientific thought, its philosophical justification, and its far-reaching cultural and intellectual implications. We earlier encountered Ruse's vigorous polemic against creationism found in his review of Denton for *New Scientist.*[42] He also wrote *Darwinism Defended* (1982) and edited *But Is It Science?* (1988)—two of many books Ruse wrote or contributed to that have confronted the threat of scientific creationism[43] by exposing it as "pseudoscience."[44]

Ruse gave crucial testimony at the Scopes II trial in Little Rock, Arkansas, in 1981. Both the trial and his testimony are discussed in chapter 9 of *Darwin on Trial.* Ruse's five points that mark "true science" were used in Justice Overton's January 1982 decision, which ruled Arkansas's "Balanced Treatment Law" unconstitutional. Two of Ruse's points about science—"It is guided by natural law," and "It has to be explanatory by reference to natural law"—specify naturalistic explanations as the *only ones that are acceptable* within science.[45] This connects to Johnson's T-2 thesis which holds that the credibility of Darwinism is dependent upon naturalistic assumptions. These assumptions act to preempt or muffle any incipient doubt, even when the empirical evidence would lead one to question the sufficiency of material mechanisms to produce macroevolution. Ruse's testimony at the Arkansas trial played a key role in Johnson's argument.

Johnson's first interaction with Ruse was in the summer of 1988, just prior to the Berkeley faculty seminar. Johnson sent Ruse a copy of the paper he had written in England and requested that it be considered for publication in *Biology and Philosophy,* which Ruse edited. Johnson held out little hope for his paper's publication but sought to provoke thoughtful interaction with his work. As expected, the paper was turned down, and a round of correspondence followed as Johnson questioned the academic integrity reflected in the comments of an anonymous reviewer.[46]

However, when supporters of Johnson approached Ruse three years later, inviting him to be the lead speaker opposite Johnson in an academic symposium based on *Darwin on Trial,* Ruse did not turn the organizers down. In late March 1992, Ruse, Johnson, and nine other scholars[47] converged upon Dallas, Texas, for a three-day academic symposium on the campus of Southern Methodist University. The "Darwinism Symposium" (as it has been commonly called) was formally entitled "Darwinism: Scientific Inference or Philosophical Preference?" At the heart of the symposium were ten major participants: five Darwinists and five Design proponents,[48] each of whom presented a paper reflecting his own analysis and research related to the given field of specialty. Each participant also had prepared a critical response to one of the papers from the other side. As mentioned earlier, this was Behe's first time to participate with the adolescent Design Movement, and it also represented the public debut of two of the "four horsemen"—William Dembski and Steven Meyer.

The chief guideline for all ten papers was that they had to connect somehow to the arguments in *Darwin on Trial.* Specifically, the conference thesis to be debated was *"Darwinism and Neo-Darwinism, as generally held in our society, carry with them an* a priori *commitment to metaphysical naturalism, which is essential to make a convincing case on their behalf"* (emphasis added).[49] The ten papers responded to the thesis at a rather technical level, covering such topics as transitions in the fossil record, protein structure, the biochemistry of the immune system, probability calculations of origins, philosophical analyses of entailment, and the rules of reasoning in evolutionary thought. The three sponsoring organizations also managed to arrange for a Saturday night debate between Johnson and Ruse in a packed ballroom at SMU and secured funding to videotape the entire three-day conference, including all twenty presentations, group discussion, and the Ruse-Johnson debate with the questions and answers that followed.[50] The Darwinism Symposium also yielded a book containing the transcribed debate and the ten papers, each with its response from the other side.[51] Both Johnson and Ruse reported later that they enjoyed the interaction greatly—as did all of the participants.[52]

The most significant result from this encounter is what it led to a year later in February 1993. At the annual meeting of the AAAS in Boston, Michael Ruse was invited to make a presentation at "The New Antievolutionism," a seminar organized by Eugenie Scott. A number of speakers gave talks, each designed to understand and effectively confront the challenge of the fledgling Design Movement.

Ruse was invited to advise the audience on how to deal with the growing "Phillip Johnson problem." When Ruse began to speak, he referred to Johnson's book but spoke primarily of the symposium in Dallas where he had met Johnson face-to-face. After some criticism of Johnson's book, he began to reflect on their personal interaction: "And as I always find when I meet creationists or nonevolutionists or critics or whatever, I find it a lot easier to hate them in print than I do in person."[53]

In terms of the actual give-and-take of debate, Ruse acknowledged the value of the symposium: "I thought we had a really quite constructive interchange. Because basically we didn't talk so much about creationism. . . . But we did talk much more about the whole question of metaphysics, the whole question of philosophical bases. And what Johnson was arguing was that, at a certain level, the kind of position of a person like myself, an evolutionist, is metaphysically based."

Ruse then abruptly startled his audience by saying he had been rethinking the issue of philosophical bases in recent years, and after his participation in the symposium, he had changed his mind on a key point: "I must confess, in the ten years since I performed, or I appeared, in the creationism trial in Arkansas, I must say that I've been coming to this kind of position myself." What position? Ruse explained that those in academia especially "should recognize, both historically and perhaps philosophically, certainly that the science side has certain metaphysical assumptions built into doing science, which—it may not be a good thing to admit in a court of law—but I think that in honesty . . . we should recognize [this]."

Ruse then expanded his agreement with Johnson on another point—the quasi-religious role of Darwinism for some scientists. He reviewed the history of science to show that "for many evolutionists, evolution has functioned as something with elements which are, let us say, akin to being a secular religion." He summarized the case for placing the Huxleys (both T. H. Huxley and grandson Julian) along with the father of sociobiology, E. O. Wilson of Harvard, in that group for whom Darwinism had not only scientific but also religious meaning.

As he wound up his talk, Ruse went back to his central thesis that "at some very basic level, evolution as a scientific theory makes a commitment to a kind of naturalism, namely, that at some level one is going to exclude miracles and these sorts of things, come what may." He anticipated the obvious rejoinder: "Now, you might say, does this mean it's just a religious assumption; does this mean it's irrational to do something like this? I would argue very strongly that it's not."

Ruse then stressed that the *pragmatic results* of Darwinian theory shore up the truth value of evolutionary science in spite of its ultimate basis in philosophical assumptions. Nevertheless, Ruse repeated that "evolution, akin to religion, involves making certain *a priori* or metaphysical assumptions, which at some level cannot be proven empirically. . . . And I think that the way to deal with creationism, but the way to deal with evolution also, is not to deny these facts, but to recognize them, and to see where we can go as we move on from there."

Ruse found it necessary to repeat several times in his talk the disclaimer that he was as much an evolutionist as ever. Nevertheless, Ruse's talk was greeted by an awkward silence—and perhaps gapes of disbelief—from the audience. As Ruse closed his talk, Eugenie Scott stepped to the lectern and invited questions from the audience. During an initial pause of several seconds, Ruse exclaimed, "State of shock!"

News of Ruse's talk spread quickly in the Design camp and was celebrated as a key breakthrough. Copies of the audiotape circulated, and the transcript was studied, savored, and eagerly quoted by Design proponents. In addition, Johnson highlighted Ruse's comments in the epilogue chapter of a revised edition of *Darwin on Trial* (1993). To put the reaction in terms of my narrative theory, this event generated a significant scientific personal narrative, and it was also the springboard for projections or fantasies, chained out in many retellings and extrapolations, suggestions, comparisons, speculations, and dreams of having turned a corner.

In an effort at damage control, zoologist Arthur Shapiro, one of the Darwinian participants in the Dallas symposium, wrote an article entitled "Did Ruse Give Away the Store?" The very title suggests a whimsical (exaggerated, nightmarish) projection theme from the Darwinian side of what happened in Ruse's speech, and yet the article itself takes the form of a sober analysis and reassurance, dispelling this panic theme and instead placing in perspective "what really happened."[54] Shapiro argued that, in fact, Ruse did not "give away the store"; he had "merely upset those scientific practitioners who really believe the self-justifying positivist propaganda about ultimate objectivity." Shapiro then defused the impact of any Darwinist admitting his acceptance of naturalism. He created a disclaimer: "Darwinism *is* a philosophical preference, if by that we mean we choose to discuss the material Universe in terms of material processes accessible by material operations."[55]

This defensive position is very similar to Eugenie Scott's comment following Ruse's speech. She suggested that Johnson confused a "working naturalism," which is unavoidable in the lab, with a "metaphysical naturalism," which makes flat, absolute statements about ultimate reality.[56] Scott and Shapiro were making a distinction between (1) what scientists *must assume* in order to do their job—that there are, and have been, no interventions in nature by a supernatural entity—and (2) what some scientists *proclaim* (as the only rational worldview) about the nonexistence of God. The first is legitimate, Scott suggested; the second is not. This principle has been boiled down into a new maxim: *Darwinism entails only methodological naturalism, not metaphysical naturalism.*

Scholars during the 1990s vigorously debated the Darwinists' new distinction between the two kinds of naturalism. The typical response of design was,

> Methodological naturalism may work well in the lab, where we have direct observation of nature, and where repeatable processes reveal the laws and processes behind nature. Yet what requires that everything in the universe, especially history-of-life singularities in the distant past, must conform to naturalistic (atheistic) assumptions? Why does biology require a "working atheism" in that context? Doesn't that produce circular reasoning? And when some branches of

empirical science have developed methods of *reliably detecting the action of an intelligence within a given system,* why must biologists ban those methods or tools from the arena of biological origins? What about the example of this latter "detection of design," when a biologist fudges data in the lab, and his (unethical) intelligent intervention becomes empirically detectable?[57]

At this point the rhetorician who is taking notes of both sides of this debate raises her hand and comments, "Please note *how* both sides are deciding the issue of what role, if any, materialism should be allowed in science. The question is not being settled by an appeal to 'scientific facts,' nor even to any universally valid philosophical principles. Rather, rhetoric itself—practical reasoning—is being called on to tackle this fundamental question." This is a perfect example of what Steve Fuller, in his book *Thomas Kuhn,* pointed to as the crucial "missing frame" in Kuhn's theory, in which subtle rhetorical-argumentative negotiation and repositioning occur during a sophisticated paradigm challenge.[58]

The rhetorician also points out how the debate is progressing in a classic dialectical pattern through subtle rhetorical transformation in a paradigm challenge. This process can be described as "imperceptible softening, inevitable tactical repositioning, and quiet erosion by rhetorical engagement." These phrases spring from the images and ideas emanating from Steve Fuller's own words in *Thomas Kuhn.* There, Fuller pointed out

> the ways in which partisan positions shift, often unintentionally and imperceptibly, in the course of debate, as the stakes and implications of acceding to one argument over another appear in different contexts. A position that one would never have adopted at the start of a dispute may become easier to accept later, in large part because *the very practice of arguing* will have made one accustomed to the other's position. Moreover, the person may not believe that she has conceded anything "essential" to her position along the way. Only in retrospect can a historian detect that a subtle shift in the burden of proof took place that enabled the acceptance of a previously intolerable point of view.[59]

In the results of the 1992 interactions of Johnson and Ruse, a thesis-versus-antithesis struggle is visible (over the scientific and philosophical bases of Darwinism), being played out rhetorically in the Darwinism Symposium. A year later this clash yielded a startling synthesis—Ruse's qualified admission of the role of philosophical assumptions.[60] Then with a clever rhetorical twist, Ruse's colleagues (Scott, Shapiro, and others) turned this synthesis into a new thesis. Ruse's admission of the role of naturalism became somewhat modified to yield the new thesis: *Darwinism's naturalism is merely methodological; it makes no metaphysical statement as to the ultimate nonexistence of a designer.* This thesis, in turn, has since spawned a vigorous new antithesis from Design: *The distinction between methodological and metaphysical naturalism is functionally meaningless and misleading, since to exclude intelligent causes from consideration in science is really the same as excluding them from reality.*[61]

Since the Ruse lecture, the distinction between methodological naturalism and metaphysical naturalism has become one of the strategic battlegrounds of the debate. Its shadow falls across many discussions and exchanges that have taken place since 1993, and in a sense, the distinction has become the Darwinists' rhetorical firewall after Ruse's statement. In the PBS televised debate (summarized earlier), this distinction was reinforced through a surprisingly religion-friendly tone of argumentation from the Darwinist side. It was as if the Darwinists said, "We freely admit that God might exist, and if he does, Darwinian evolution is clearly his way of creating. What is the problem?"

Design advocates are now employing rhetorical strategies to defeat this distinction. At the end of the decade this battle began to depend to a large extent on whether and how well Johnson's group was able to appeal to "empirical evidence of design" and could propose a principled, empirically rooted method for detecting (in a generic sense) intelligent activity. Thus, Behe has created, as a mode of design detection, his rhetorical universe of irreducibly complex molecular machines.[62] Likewise, Dembski created his "explanatory filter"[63] as a method for empirically detecting any kind of design anywhere in the universe.

These developments will be discussed later, but in retrospect the story of the Darwinism Symposium (and Ruse's AAAS talk) marks a major rhetorical-historical watershed. After 1993 the Design versus Darwinism debate began to break genuinely new ground. In a sense the Johnson-Ruse conversation was fashioning what Symbolic Convergence Theory (SCT) calls a new type of "special communication theory." Special communication theories are sets of rules or principles that guide the smooth functioning of many structured types of communication, types that are rooted in a given cultural setting with a distinctive dynamic activity underway. For example, there is a special communication theory for the writing and peer review of a scientific article in a technical journal. Other theories pertain to more mundane communication types such as funeral orations, "roasts" of celebrities, or even playing bridge. If a person follows the rules for one theory (e.g., a funeral oration) in a totally different setting or venue (a roast), the clash becomes obvious quite quickly and efforts may be made to terminate the inappropriate communication.

In this fascinating perspective of SCT theorists, before the Darwinism Symposium in 1992 there existed both (1) a Darwinist special theory and (2) a Design special theory, regulating scientific discussion within those two communities.[64] However, there was no joint communication theory for close interaction between these two sides. Through the Darwinism Symposium in 1992, a new *joint communication theory* was born. The videotape of the Darwinism Symposium (along with the published papers) reveals this early display of the joint special theory with its several rules. Four of these rules are prominent: (1) The credibility (or plausibility) of knowledge claims of Darwinism or Design may be freely questioned, but issues relating to a literal Genesis (e.g., the age of the earth, or whether rock strata were produced by a global flood) are not welcome. (2) The agenda shall encourage the consideration of two sets of issues—issues of evidence and issues related to metaphysical or philosophical bases that undergird research or shape thinking on origins. (3) The highest level of scholarly integrity shall be expected. (4) Mutual respect shall be maintained. The Darwinism Symposium at SMU was not only the debut of this special joint theory, it was the first occasion where parties in academia fashioned a balanced rhetorical process of metaphysical negotiation on the credibility of knowledge-claims about macroevolution. This event began to give a glimmer of hope that other such models could be developed in those educational flashpoints that flare up frequently now within America's democratic pluralistic culture.

As the implications and reverberations of Ruse's talk were working themselves out (a summary went into the epilogue of the 1993 revised edition of *Darwin on Trial*), the Ad Hoc Origins Committee came together for a three-day meeting in Seattle, held 3–5 August 1993. By this time Michael Behe was already recognized as the preeminent scientist within the Design community, and at the Seattle meeting he presented a talk outlining several of the main ideas about the complexity of molecular machines and his initial literature search, which showed that no researcher had suggested how these systems might have evolved.

Also in 1993 Behe presented a more detailed elaboration of his ideas at a private conference of ten Design theorists, including Johnson, William Dembski, Paul Nelson, Dean Kenyon, and the German biologist Siegfried Scherer from the University of Konstanz. One interested outsider attended—the evolutionary paleontologist David Raup. Held at the Pajaro Dunes resort in California, this meeting served many functions, but for Michael Behe it was the dress rehearsal for his book-length argument for design from "irreducible complexity." Incidentally, the sophisticated documentary *Unlocking the Mystery of Life* (2002) has now made this private meeting at Pajaro Dunes famous, pointing to it as a major turning point. Indeed, the writers of the video's script used this meeting as the narrative frame for the early discussion and development of many Design ideas that were discussed at the beach resort.

At Pajaro Dunes, Behe became confident that the conceptual backbone of his argument was complete. He was convinced that the time was ripe to publish. Through an agent he contacted Free Press, and the publisher quickly agreed to a contract. Behe began in 1995 to tap out the manuscript of *Darwin's Black Box* on his computer, and by early 1996 Johnson's e-mail discussion group was percolating with excited discussion about the potential impact of the new book. The fast-changing narrative of the Great Debate was about to experience its greatest rhetorical shock yet—a new "molecular chapter" in its story was ready to begin.[65]

8

The Dam Breaks

Michael Behe and the Explosion of Design

The Public Debuts of Berlinski and Behe

In the summer of 1996 a pair of rhetorical bombs jarred the world of biological science. The first bomb, dropped in June, was a lengthy essay entitled "The Deniable Darwin" in *Commentary*, a scholarly journal published by the American Jewish Committee. It was written by the Jewish intellectual David Berlinski, whose 1995 work, *A Tour of the Calculus*, had been hailed as a significant achievement in the history of mathematics. Berlinski's *Commentary* article, a full-scale attack on the credibility of Darwinian evolution, was listed in the table of contents with the eye-catching heading: "The fossil record is incomplete, the reasoning is flawed; is the theory of evolution fit to survive?"

The article provoked a tsunami of fierce indignation and warm congratulation. In *Commentary*'s September 1996 issue, thirty-three pages were devoted to the torrent of letters—including denunciations from several of the world's most prestigious evolutionists, most notably Richard Dawkins and Daniel Dennett—along with Berlinski's response to each letter.[1] Editors had expected Berlinski's article to trigger tremors; what they got was a rhetorical earthquake.

In August 1996, as evolutionists were responding to Berlinski's jolt, a second anti-Darwinist bomb exploded. *Darwin's Black Box*, written by Lehigh biologist Michael Behe, was released by Free Press, a subsidiary of Simon and Schuster. American and British journalists and science writers immediately spotted Behe on their radar screens. *Darwin's Black Box* was soon being discussed in *Newsweek,* the *Wall Street Journal, National Review,* the *Chronicle of Higher Education,* and the prestigious British science journal *Nature.*

The *New York Times Book Review* was one of the first to profile Behe's work. The *Times* summarized Behe's argument that many of the biochemical machines inside the cell, such as the tiny outboard motor called the *flagellum,* exhibited an eerie kind of complexity that defied Darwinian explanations. These systems, said Behe, could not function if any one part were removed; they were "irreducibly complex." The *Times* took note of Behe's "heretical" proposition—namely, that such complexity presents us with "an overwhelming argument for intelligent design."

It was at this point that James Shreeve, writing for the *Times,* parted company with Behe. Science, he warned, should not jump to such a radical conclusion but rather should be patient, leaving a few mysteries for the next generation to solve. His "wait-and-see" response was echoed by nearly every other evolutionist who responded to Behe's book.[2]

Both Behe and Berlinski wrote in such a way that most readers would see their criticisms did not fit comfortably in the creationist genre. First, they attacked the credibility of neo-Darwinism on strictly scientific grounds. Second, neither of the two could be construed as fundamentalist (Berlinski, Jewish, has been vague about his belief in God; Behe is Roman Catholic). Third, Behe specifically stated that he is not a creationist, since he accepts the conventional timeline of the geological ages and has no quarrel with Darwin's notion of common ancestry of all animals.[3]

Although Behe and Berlinski shared an anti-Darwinist rhetorical vision, observers could discern in Behe's rhetoric a move beyond negative criticism of the reigning paradigm. He was arguing, much as Thaxton had been doing since 1988, that "intelligent causation" is a robust, fruitful alternative explanation of the anomalies that cannot be explained by Darwinian concepts (which hold strictly to design by nonintelligent or "wholly natural" processes). In effect Behe proposed a new, or at least radically remodeled, biological paradigm. One of his chief arguments was that in other disciplines, such as astronomy and cosmology, notions of design and intelligent causation have now been openly discussed for decades. It is time, Behe said, for the biological sciences to likewise open up to the superior plausibility of intelligent design as an explanation for certain kinds of complexity in the biosphere.[4]

Behe's proposal was met with suspicion by evolutionary theorists, some of whom saw his ideas as a clandestine or stealth version of creation science. Behe and other Intelligent Design scholars resisted this attempt to lump them with the creation scientists. They unfurled a principle that resisted any easy link with the Book of Genesis. This principle states that the scientific data that the scientific method can yield indicate only a general conclusion; the data can only argue to the category of "intelligence." The specific designer or designers may or may not be describable as "supreme being(s)." Behe said, "Inferences to design do not require that we have a candidate for the role of designer. We can determine that a system was designed by examining the system itself, and we can hold the conviction of design much more strongly than a conviction about the identity of the designer."[5]

Among Johnson's colleagues, Behe suddenly emerged in 1996 as the movement's most powerful scientific spokesperson to the scientific establishment and the general public. In comparison with Berlinski, Behe's role as a public critic has proved much more powerful for three reasons. First, Behe is a scientific insider—a tenured biochemist at a major research university—while Berlinski is an outsider.[6] Second, whereas Berlinski's *Commentary* article gained relatively little notice outside the journal's pages, Behe's book garnered over one hundred book reviews and published responses in both professional and popular media—a remarkable number by any publisher's reckoning.[7] Third, the brewing storm of controversy over Behe helped propel the book to much-higher-than-expected sales. Over 45,000 copies were sold in fifteen printings during its first year. The publisher delayed the paperback edition for an additional nine months, until March 1998, because the hardback was selling too well. During the year 2000 the book was still selling at the rate of twenty thousand books a year, and it also was selling well in fifteen foreign translations.[8]

The response to *Darwin's Black Box* at times compounded itself. For example, just two months after the *New York Times* reviewed Behe (4 August 1996), a *Times* editor who had just read the book invited him, as a Catholic in good standing, to write an opinion article on the Pope's puzzling statement on evolution that was publicized around the world in late October. The editor agreed that his article would summarize the main points of his new book in relation to the Pope's remarks about "several theories" of evolutionary development, one of which (Behe argued) entailed intelligent design. When the *Times* published Behe's article "Darwin Under the Microscope" on 29 October 1996, it sparked a new round of letters to the editor and gave the book another national publicity boost.

In the introduction, I deliberately used the description of an "earthquake" to describe the jolts from Berlinski and Behe. This is not just my attempt to picture the power of these rhetorical events. This analogy actually sprang to life at the time of the publication of *Darwin's Black Box* and began to thrive and spread widely. I see the "temblor sending out shock waves" as one of the most potent projection themes that chained out in the history of Design.[9] Phillip Johnson actually helped initiate this projection chain. When word came in late July 1996 that the *New York Times Book Review* was about to publish a review of Behe, he typed a message to his e-mail community, predicting that the *New York Times* coverage would result in a "cultural earthquake."[10]

The remainder of this section on Michael Behe will probe the unique contributions of his rhetoric, tracing why and how his writings played a crucial role in triggering new kinds of scientific debate both in the public sector and at high academic levels. I want to ask three key questions: (1) How did Behe's *personal (individual) narrative,* including his key discoveries, constrain and energize his own rhetoric and contribute to the rhetoric of Design? (2) What were the major rhetorical qualities (in terms of ethos, pathos, and logos) Behe possessed that succeeded in gaining a far wider audience than Johnson or any other Design rhetor? (3) In what broad ways did the entire rhetorical phenomenon of Michael Behe (his persona as well as discourses) modify and extend the rhetorical vision of Design, especially as it worked through movement-narrative and projection theme chaining? To answer these questions, I will be probing *Darwin's Black Box* and the opinion piece "Darwin Under the Microscope" (hereafter referred to as "Microscope") that he wrote for the *New York Times*. To map the contours of Behe's interaction story, I also will turn to three published reviews of the book—in *Nature*, the *New York Times Book Review,* and *New Scientist.*

Behe's Background, Rhetorical Situation, and Key Discoveries

In earlier chapters I stressed the key role of scientists' individual stories—a special type of narrative. Such vignettes possess unique persuasive power, especially as they are linked artfully with other personal or incident stories in a stream of history-of-science narrative. Behe's sudden conversion to skepticism of Darwinism is the most important individual story of all the Design theorists. I will return to this story later in the chapter, but here I merely note that Behe's own recollection as to the *cause* of this change is the key to this story's power. His personal religious viewpoint did not drive this shift of thinking. Obviously, his worldview of traditional Catholicism contained within it the metaphysical space for such a shift, but his religious perspective played a permissive or preparatory role, not a determinative one.[11] Rather than the dogma of religion, it was the data of science, as presented by Denton, that induced Behe to reject Darwinism.

Behe's book did not arise directly from this conversion but was led up to by three personal historical processes:

1. The abrupt awakening of doubt through reading Denton triggered an energetic research phase. There began a steady accumulation in Behe's consciousness of evidence and arguments that eroded the plausibility of the Darwinian scenario of origins.
2. Concurrently with his personal research, he developed a growing conviction that many others in academia had been eased into a deep dogmatic slumber. Thus, Behe's scientific critique of the creative power of natural selection needed to be shared.
3. He was encouraged by the positive results as he raised these issues through an undergraduate course he developed at Lehigh and later through his participation in the 1992 Darwinism Symposium and the Pajaro Dunes meeting in 1993.

In terms of personal stakes and motives, it is clear that Behe was risking several kinds of loss in challenging the reigning scientific orthodoxy. Although he had tenure at Lehigh University, he was aware that his promotion to full professor could be affected by such a radical statement to his biological colleagues.[12] He realized that his ongoing ability to obtain research grants, especially from the National Institutes of Health, could be affected in the future, and his reception in the wider community of biochemists could turn chilly after such a heretical critique of his own field was released.[13]

Behe's motivations were therefore complex. They were deeply rooted at the level of a strong moral indignation—a simmering, settled outrage[14]—about the pervasive deception and misinformation he saw in all of the textbooks' discussions of macroevolution. Behe had come to view as highly deceptive those books and media programming that seek to win over the public, and especially high school and college students, to belief in Darwinian evolution as fact. This sense of profound contempt is critical to the subtle underlying passion—the key to the rhetorical pathos—of Behe's writing and speaking, which will be discussed below.

As stated earlier, these feelings can be traced back to the dramatic change in his biological viewpoint as a professor at Lehigh in 1987. Previous to his biological conversion, Behe had an attitude of relaxed acceptance of Darwinian explanations of the origin of biological complexity. His Darwinian point of view was absorbed in high school biology and further reinforced through undergraduate studies and graduate studies of biochemistry. In his mind there was no conflict with his religious views as a Catholic. God simply had created slowly and gradually using the Darwinian mechanisms that he had authored.

One day in 1987 Denton's book came to Behe's attention through a book club advertisement. He ordered a copy and upon its arrival read the book in one long sitting, starting earlier in the day and staying up well past midnight. He describes it as the greatest intellectual shock of his life.[15] It was an intensely intellectual experience, and yet it immediately generated anger that he had been so greatly misled for so long. A theory that he assumed had rested securely upon compelling biological evidence was intellectually shattered in one day. He immediately began to rethink his own area, especially the origin of biochemical systems such as blood clotting, the cilium, and intracellular transport systems. In 1993 after a period of collaboration with Johnson and others, he coined the phrase "irreducible complexity" to describe systems whose function depended upon the interaction of many parts and wherein the removal of any part effectively shut down the function of the system. These systems, reasoned Behe, could not possibly have been built up, step-by-step, via Darwinian pathways. They appeared to be something that only an engineering mind could put together.

It is one thing to experience persuasion, and it is another to set out to persuade others. As Behe began to strategize how to criticize evolutionary theory on his own, he was aware of the pitfalls of being labeled a creationist. In fact, his own religious background has been used as a weapon against his case for design in spite of his disavowal of religious factors in his change of mind. For example, Coyne's review of Behe in *Nature* began, "The goal of creationists has always been to replace the teaching of evolution with the narrative given in the first eleven chapters of Genesis." After tracing the failure of scientific creationists to make headway in introducing their views in public schools, Coyne adds, "In *Darwin's Black Box,* Michael Behe offers *a new and more sophisticated version of scientific creationism*" (emphasis added).[16]

This rhetorical strategy of placing Behe in the same camp with creation scientists was not used by all Behe's critics; many were careful to distinguish the two kinds of criticism as separate genres. When critics failed to distinguish Behe's position from that of creation science, it threatened to backfire. Behe had stated so many basic differences that anyone who labeled Behe a creation scientist could, in turn, be viewed as one engaged in crude name calling. In any case, the cultural fog of the older creation-evolution debate has proved to be a key to what Bitzer would call the "rhetorical situation" Behe faced.[17]

Part of this rhetorical situation, in terms of resources, is the accumulated data on the recently elucidated structures (with their constituent protein parts) in molecular biology. Relatively little of this information had been used by Denton and Johnson, although Denton did discuss the cilium and the flagellum.[18] As noted earlier, these two writers were criticized for saying nothing new but rather "wheeling out and dusting off" old creationist arguments that were answered long ago.[19] Behe avoided this rhetorical ambush by presenting new data on biochemical systems, most of which had come to light in the course of the molecular biology revolution of the previous three decades (1960s through the early 1990s). Behe devoted a chapter in his book to each of six miniature biochemical systems, describing each in detail and showing the new mystery of origins that attended the elucidation of their mysterious workings.

In Behe's hands, this information led to two fundamental factual-conceptual discoveries.[20] First, as noted above, he realized that most of these systems possessed the quality of irreducible complexity. Second, Behe undertook a literature review in 1993–95 that confirmed his suspicion: His colleagues in biochemistry and evolutionary biology had not figured out plausible hypothetical pathways for the origin of any of these systems. During this review he researched a dozen of the most widely used biochemistry textbooks, as well as many technical journals on biochemical evolution, looking for proposed evolutionary scenarios. He was astonished, yet excited, to find in the literature a "thundering silence."[21] Not one biochemist in the past forty years had even *attempted* a testable explanation for the origin of any of the systems about which he was writing. Behe sensed he had the key discovery that would cap off his argument—he would use the silence of evolutionary biologists on this topic as his clincher.

A Rhetorical Analysis of Behe: Ethos, Pathos, Logos

Aristotle, a father of classical rhetorical theory, suggested a triad of means of persuasion in oratory: ethos, pathos, and logos.[22] I have referred already to ethos (the audience's perception of the character and credibility of the rhetor) in conjunction with my analysis of Denton and Johnson. Behe's ethos is rooted in his academic standing in the field of biochemistry, which is the main focus of his book. In reviews and profiles published on Behe, the fact is commonly emphasized (or admitted) that Behe is a tenured scientist at a well-respected research university. This helped him enormously in winning a hearing for his radical proposal. Coyne, writing in *Nature,* referred to the slightly improved credibility of Behe compared to creation scientists: "Unlike his predecessors, Behe is a genuine scientist, a biochemist from Lehigh University in Pennsylvania. The book jacket asserts that he is not a creationist, but believes in the scientific method."[23]

Sometimes the positive ethos of Behe was emphasized, but at times it was complicated with the hint of "seduction." Lee Cullum, a writer for the *Dallas Morning News,* quoted the description of H. Allen Orr, an evolutionary geneticist at the University of Rochester: "Behe is the real thing: a research scientist, someone who does experiments, gets grants and publishes papers." Cullum added, "Mr. Orr went so far as to concede that Mr. 'Behe's work may well represent the *most sophisticated and the most seductive creationist attack on evolution in a quarter-century'*" (emphasis added).[24] This hint of seduction links with the theme of Design as a *dangerous virus*—a "sophisticated, stealth type creationism"—that we have seen all along in the opponents' rhetoric.

Behe's openness about his Catholic persuasion helped establish a relatively innocuous religious persona that could be described as generic Christian yet nonfundamentalist. This aura of religious harmlessness was reinforced by his explanations that it was *not* his religious views that led him to the radical perspective in *Darwin's Black Box* but his sudden empirical awakening to Darwinism's scientific problems.

His character trait as an approachable and warm human being was enhanced by his writing style. He liberally sprinkled his chapters with homespun illustrations—assembling a bicycle on Christmas eve; making chocolate cake; observing snap-lock beads and a dolly wagon on the rug. Finally, Behe's whimsical humor, one of his most powerful rhetorical trademarks, added an attractive and endearing nuance to both his ethos and his logos. All these touches brought Behe within the range of the nontechnical reader, and they helped soften the possible negative image of a scientist who was "dangerous" or who was discussing matters far too technical for the layperson.

The most unusual aspect about Behe's ethos arose early in the book where he said, "I find the idea of common descent (that all organisms share a common ancestor) fairly convincing, and have no particular reason to doubt it."[25] This point of view was unlike Johnson and most of the Design rhetors. The irony of Behe's position was that while he did not go as far as Johnson in questioning macroevolution, he proceeded much further in calling for an explicit scientific theory of design. How does one reconcile these two positions? Behe maintains that all the genetic information for the molecular systems described in his book were probably front-loaded by a creator into the genomes of our earlier ancestors.

As a result of this unconventional perspective, Behe's denial that he is a "creationist as commonly understood" rang true, and this protected him from a quick rejection by those who despised that perspective. With those who were Darwinian in perspective, he made his ethos more winsome by a respectful attitude toward those studying evolution: "I greatly respect the work of my colleagues who study the development and behavior of organisms within an evolutionary framework, and I think that evolutionary biologists have contributed enormously to our understanding of the world."[26]

To sum up this point, a key opening consideration for any reader of a book critical of Darwin is whether the writer is a creationist. Behe specifically denies that he is, and in saying this he has chosen to accept the traditional media definition of a creationist as a biblical literalist. This is a different strategy from Johnson's, who created a more general definition of a creationist as anyone who believes that (somehow) we were created for a purpose. This denial of the creationist label, coupled with his acceptance of common ancestry, have worked together to shape a relatively palatable ethos for the typical neutral reader of Behe.

In addition to ethos, Aristotle pointed to the key factor of pathos—the feeling, emotion, and passion that is woven into or generated by persuasive discourse. Behe's pathos is not so prominent, as one would expect in scientific discourse. Yet there are some important feelings that can be subtly sensed in his rhetoric. Throughout *Darwin's Black Box* one can detect Behe's underlying feeling of *contempt* for the current inflated Darwinian claims about origins, combined with *excitement* at the drama of the coming revolution in biological sciences. A typical passage, whose strong wording conveys this flow of feeling just below the surface, is the conclusion of his chapter on blood clotting. Here he says that his purpose was "to illustrate the *enormous difficulty* (indeed the *apparent impossibility*) of a problem that has resisted the determined efforts of a top-notch scientist [Dr. Russell Doolittle] for four decades. Blood coagulation is a paradigm of the *staggering complexity* that underlies even apparently simple bodily processes. Faced with such complexity beneath even simple phenomena, *Darwinian theory falls silent*" (emphasis added).[27] Such wording dramatizes the overwhelming crisis for Darwinism, and it conveys personal excitement and eager anticipation of what is coming.

The argumentative content, or logos, of Behe is by far the most important element, and it is one that is the object of scrutiny in all the published reviews. His main argument has been outlined; what we need to appreciate is the clever logic it employs. There are three basic steps. First, he quotes Darwin's own wager (which Johnson and Denton also used). In the *Origin,* Darwin said, "If it could be demonstrated that any complex organ existed which could not possibly have been formed by numerous, successive slight modifications, my theory would absolutely break down."[28] Behe seizes this quote as a tool, a falsification test of Darwin's own gradualistic theory.

Behe's second step is to introduce his unique rhetorical invention—irreducible complexity. He says, "By irreducibly complex, I mean a single system composed of several well-matched, interacting parts that contribute to the basic function, wherein the removal of any one of the parts causes the system to effectively cease functioning. An irreducibly complex system cannot be produced directly . . . by slight, successive modifications of a precursor system, because any precursor to an irreducibly complex system that is missing a part is by definition nonfunctional. An irreducibly complex biological system, if there is such a thing, would be a powerful challenge to Darwinian evolution."[29] He argues that any such system, because it depends on all the parts being present before it can function, fits the parameters of Darwin's own test as a system that could not be evolved. The first half of *Darwin's Black Box* is devoted to several major test cases. In the first chapter and chapters 3 through 7 he details the dizzying complexity of molecular systems that fit his description of irreducible complexity, which by his analysis could not have been evolved.

Behe's use of a mousetrap analogy for irreducible complexity merits special attention. Among Behe's many original contributions to the Design arsenal of images, arguments, and evidences, his mousetrap illustration instantly achieved wide fame. The fact that it has grabbed the attention of the scientific world was shown in the spirited critique and defense of the mousetrap argument that broke out in the televised PBS debate in December 1997, which pitted Behe against a Darwinian debater.[30]

How does the mousetrap illustrate irreducible complexity? Behe explains that it has five parts—a base, a spring, a hammer (the U-shaped piece that snaps down on the mouse), a holding bar to hold the hammer back, and a catch (the sensitive trigger that holds the bait and releases the holding bar). In a typical lecture, after showing a diagram of a mousetrap and naming the parts, Behe simply asks, "How many of these five parts do you need to catch a mouse?" The obvious answer, he points out, is all of them; you need all five parts in order to catch any mice. The mousetrap, therefore, is *irreducibly complex.* It could not have evolved step-by-step because its useful function could not be achieved until all the parts were joined in a carefully coordinated system. Behe then notes that the other more technical systems he describes, such as the flagellum, cilia, and blood-clotting cascade, are just like the mousetrap, only these systems have dozens or even hundreds of parts, all of which must be present in order to function.

In both "Microscope" and Behe's book, the mousetrap plays a central role in both explaining and illustrating irreducible complexity. Not only can a typical adult reader, unversed in science, understand Behe through this simple contraption, but even a schoolchild can grasp the concept. Here Behe's teaching ability—his knack in conveying biochemistry on a simple level—bears fruit in a metaphor whose meaning can be very widely grasped. It has become a symbol of Behe's whole rhetorical project.

The next step in his logos is straightforward. Behe combines the first two notions—Darwin's wager and the irreducible complexity that meets that wager—with a tour of the systems and molecular machines that his colleagues in biochemistry have now discovered. This virtual museum tour focuses on a host of irreducibly complex systems, each of which cannot possibly be formed gradually, step-by-Darwinian-step. Thus, looking at these machines and applying Darwin's own criteria, his theory "absolutely breaks down."

The final step of Behe's logos—accentuating the image of a breakdown—is his literature review, which was mentioned earlier. In Behe's view, his argument is confirmed by the absolute silence of scientists in his own field and within evolutionary biology, who make no attempt to suggest possible pathways by which complex molecular machines might have evolved.

Of the forty reviews of Behe that I read, all commented on the clarity of this three-step argument and the accuracy of his scientific detail, and many reviewers noted the persuasiveness (superficially at least) of his argument. Nevertheless, many pled for Behe to be patient. "We need more time!" was the response of many critics.

One example of this call for patience is an incident that Phillip Johnson related to me. It concerns Richard Dawkins's visit in the fall of 1996 to the San Francisco Bay area in California to do book signings of his new work, *Climbing Mount Improbable*. When he came to a large bookstore in Berkeley, he spoke briefly before the book signing, and Johnson was seated in the front row. After Dawkins's remarks, Johnson (whom Dawkins probably did not know) asked whether he had read Behe's book and could offer a response to it. Dawkins said he had read it and complained that Behe was "lazy." He should "get out there and find those evolutionary pathways" by which the complex machines had arisen. This type of response is rhetorically revealing. Never at issue is *whether* natural mechanisms could produce irreducible complexity in the first place; it is automatically assumed that they can. Rather, the scientist's job is merely to find those pathways and to track down those causal mechanisms. From the Design perspective, this is again a problem of basic philosophical assumption; appropriate paths of research are rooted ultimately not just in science itself but in a "preferred" metaphysical worldview.

The persuasive force of Behe's logos is enhanced by his argumentative style. Nearly all the reviewers, including those who rejected his design thesis, commented on his creative, inventive descriptions. Pomiankowski in *New Scientist* says, "Behe is very good at making biochemistry easy to understand." Coyne in *Nature* admits, "His examples of such pathways" are "described with admirable clarity." Shreeve in the *Times* praised Behe's style extensively: "His best weapon is a talent for lively exposition. . . . His principal explanatory tool, however, is

Perhaps the most popular and rhetorically substantive counterargument mounted against the *irreducible complexity thesis* is the notion of "co-option." This idea builds on biochemical research which shows that many protein parts that comprise some of Behe's molecular machines already serve other unrelated functions in the same cell. Darwinists like Kenneth Miller argue that these preexisting proteins could have been "co-opted"—borrowed and pressed into service in new ways that led, step-by-step, to irreducibly complex machines.

Behe has vigorously attacked the co-option argument in print, but perhaps the most fascinating response seen thus far (and one that began to penetrate widely in the U.S. beginning in 2003) emerged in the video documentary *Unlocking the Mystery of Life*, which included a segment on Behe that emphasized the *flagellum*, a rotary motor mounted on the end of some bacteria. To catch the rhetorical thrust of this three-minute segment, let me quote a piece from the video script that includes the narrator's words as well as those of Scott Minnich, professor of biology at the University of Idaho:

> *Narrator:* Some scientists praised Behe's work, while others dismissed it as unscientific and religiously motivated. Behe's critics also insisted that he had underestimated the power of natural selection. They argued that the flagellar motor could have been constructed from parts used to build simpler molecular machines—like this needle-nosed cellular pump. If components of the pump already existed, they could have been preserved by natural selection even before the bacterial motor arose. This theory is called "Co-option."

> *Minnich:* It's essentially saying that evolution or natural selection, at some point, was able to borrow components of one molecular machine and build a new machine with some of these components.

> Narrator: Scott Minnich has studied the flagellar motor for nearly twenty years. His research has led him to challenge the co-option argument.

> *Minnich:* With a bacterial flagellum, you're talking about a machine that's got forty structural parts. Yes, we find ten of them are involved in another molecular machine, but the other thirty are unique. So where are you going to borrow them from? Eventually you're going to have to account for the function of every single part as if originally having some other purpose. I mean, you can only follow that argument so far, until you run into the problem that you're borrowing from nothing . . . but, even if you concede that you have all the parts necessary to build one of these machines, that's only part of the problem. Maybe even more complex is the assembly instructions. That is never addressed by opponents of the irreducible complexity argument.

As I have said earlier, a fairly simple "point-counterpoint" dialectic, phrased in the very words of a contemporary biologist who is skeptical of Darwinian scenarios, is used to defeat the co-option argument. To this observer, it was the personal (narratively rich) skepticism of Minnich that sealed and energized the basic counterargument against co-option.

analogy. Sometimes the analogies are crisply clear. . . . These expository devices
. . . charmingly convey a sense of biochemistry's hidden beauty."[31]

Walking readers through the nuts and bolts, complete with technical names,
is part of this descriptive strategy. Behe's chapters actually contain sections of
such biochemical description of parts with strange-sounding names like "trans-
retinal" and "antithrombin." Here is a typical sentence, from chapter 1: "When
attached to metarhodopsin II and its entourage, the phosphodiesterase acquires
the chemical ability to 'cut' a molecule called cGMP (a chemical relative of
both GDP and GTP)."[32] Such technical sections (usually two or three pages in
length) are marked off with tiny box symbols, and they explain the step-by-step
working of the Lilliputian systems. This rhetorical technique injects a note of
genuine scientific authority—Behe simply presents the scientific facts from his
field, complete with odd-sounding technical names. He said his purpose was
that the reader may *feel* the complexity, after being immersed in the description
of the actual cogs, wheels, and levers that are interacting in complex machinery.[33] This stylistic strategy seems to have worked quite well.

Behe's Narrative Logos: Creating the Molecular Mystery

A particularly muscular part of Behe's logos, which I shall analyze separately,
is his array of narrative arguments, some quite reminiscent of Denton and
Johnson. Before doing that, let me give a brief disclaimer. When I refer to the
potency of Behe's narrative logos, there is no intent to overlook or downplay
his frequent use of a more conventional kind of argument, for instance, when
a criticism is leveled at the notion of design and Behe begins to point out
flaws in the argument. A beautiful example of Behe's rhetorical skill in such
conventional logos is his point-by-point refutation of the "argument from
imperfection," especially as espoused by Kenneth Miller. Miller was introduced
earlier as the highly adept debater who wielded charts and exuded complete
confidence in the 1997 PBS debate. Miller, who teaches biology at Brown
University, has now risen to legendary status as a champion of Darwinism and
a dogged opponent of Design. However, in the hands of Behe, his published
argument against the design of the eye is outlined and then critiqued merci-
lessly. For example, the "demand for perfection" in such structures is shown
faulty; it "overlooks the possibility that the designer might have multiple
motives, with engineering excellence oftentimes relegated to a secondary
role" (223). We don't have the space to cover each of Behe's counterattacks
on Miller—much of which is in the form of conventional logos.

Yet at the end of his conventional-type refutation, Behe then employs a narrative point that is clever: The fact that Miller's argument, which he published
in *Technology Review,* a journal for scientists, would be an "argument that is
based on psychology and emotion, instead of hard science," gives the opposite
message from the one Miller intended to give. Behe's implied message here in

the narrative situation is that it is Darwinism, not Design, that suffers a lack of logical rigor and empirical soundness (225).

One of the most powerful sections of narrative logos, a set of stories gathered into a single long argument, is found in the opening chapter (3–25). Behe traces major steps in the development of biology, from Aristotle all the way through Darwin and Haeckel in the late 1800s. The chapter's main point is simple. Even in Darwin's day, the vast complexity of a single cell was not only unknown—*it was not even glimpsed.* The cell, for Darwin and his contemporaries, was a "black box." Behe says a black box is a "whimsical term for a device that does something, about whose inner workings are mysterious—sometimes because the workings can't be seen, and sometimes because they just aren't comprehensible."[34] Haeckel, Darwin's disciple and popularizer in Germany, even described the cell as a "simple little lump of albuminous combination of carbon"—in Behe's description, "not much different from a piece of microscopic Jell-O."[35]

As a result of such ignorance, theories could be spun about the evolution of animals or organs without a thought of the molecular-level complexities. Neo-Darwinism was hammered out, says Behe, before this unimagined world of machines and intricate systems was revealed: "The scientific disciplines that were part of the evolutionary synthesis are all nonmolecular. Yet for the Darwinian theory of evolution to be true, it has to account for the molecular structure of life. It is the purpose of this book to show that it does not" (25).

Here Behe has effectively heightened the drama of the molecular evidence, because only at this level does the plausibility of Darwinian theory reach its ultimate test. In terms of practical function, you cannot get tinier than the molecular parts from which everything in the cell is built. This is a microcosm-analogy of the persuasive notion of "dramatic supremacy" (ultimate explanatory importance) that I began to develop in chapter 1.

Behe opens his second chapter with a section that is similar to the strings of controversy stories in Denton or Johnson. It is a powerful and intense barrage, including a dozen cases in which a scientist (most of them contemporary) puzzled over the inability of Darwinism to account for the patterns of biological evidence. This is a miniature history-of-science narrative, gathering the scattered dissent or puzzlement into a sizzling collection, erecting Behe's own tree.[36] This leads into a different kind of narrative discussion—an analysis of the debate-in-print between Dawkins and Hitching over the evolution of eyes and the weird apparatus on the bombardier beetle that periodically squirts a blast of scalding water on a would-be predator. Behe concludes that both writers misunderstand the issues because neither is facing the crucial reality of the molecular explanations. No one is descending to the crucial level—the molecular world. At this point, Behe begins his full development of the core theses about irreducible complexity.

Narratives of Revolution and Darwinian Projection Themes

At the end of his book, when he is summing up his argument and outlining the implications for future science, Behe weaves a different kind of double narrative. At this point, many readers would be convinced of the validity of his arguments and might wonder why scientists do not recognize the case for design that emerges. To answer this, he creates a humorous fantasy scenario about the "might-be-expected" result or appropriate outcome of this sort of dawning of design in the scientific community[37] versus the actual situation that has developed. This trio of paragraphs is some of his most often-quoted writing, and I have placed it here to allow the humor-fantasy to be felt:

> The result of these cumulative efforts to investigate the cell—to investigate life at the molecular level—is a loud, clear, piercing cry of "design!" The result is so unambiguous and so significant that it must be ranked as one of the greatest achievements in the history of science. The discovery rivals those of Newton and Einstein, Lavoisier and Schrodinger, Pasteur and Darwin. The observation of the intelligent design of life is as momentous as the observation that the earth goes around the sun or that disease is caused by bacteria or that radiation is emitted in quanta. The magnitude of the victory, gained at such great cost through sustained effort over the course of decades, would be expected to send champagne corks flying in labs around the world. This triumph of science should evoke cries of "Eureka!" from ten thousand throats, should occasion much hand-slapping and high-fiving, and perhaps even be an excuse to take a day off.
>
> But no bottles have been uncorked, no hands slapped. Instead, a curious, embarrassed silence surrounds the stark complexity of the cell. When the subject comes up in public, feet start to shuffle, and breathing gets a bit labored. In private people are a bit more relaxed; many explicitly admit the obvious but then stare at the ground, shake their heads, and let it go at that.
>
> Why does the scientific community not greedily embrace its startling discovery? Why is the observation of design handled with intellectual gloves? The dilemma is that while one side of the elephant is labeled intelligent design, the other side might be labeled God. [232–33]

The reference to an "elephant" is from an earlier humor-scenario in which a dead man lies crushed in a room, "flat as a pancake," but since the detectives assume this to be a murder case, they scour the room for fibers, fingerprints, or any other clues that can point them to the perpetrator. All along, they quietly ignore the huge elephant that stands near the dead body. Because they *assume* what sort of cause was operative, the elephant—standing in this case for the intelligent designer—is completely ignored.[38]

Behe implies that one compelling factor prevents the imagined victory celebration from erupting in the labs of the world—scientists' dread of entering the uncharted, fearful intellectual territory of "God acting in nature." In his concluding chapter, which these paragraphs launch, he tackles the scientific and practical difficulties associated with intelligent design since it is a notion closely connected to God. His key introductory question is this: Since about 90 percent of Americans believe in God anyway, and God's name is constantly invoked by presidents and sessions of Congress, especially during crises, why the difficulty?

Behe suggests four reasons. The first two—which take up only a tiny fraction of the chapter—are scientific chauvinism and the historical patterns of tension and clash between religion and science.[39] The last two reasons occupy the bulk of this chapter, and they correspond to the two kinds of naturalism that emerged from Ruse's speech in 1993—first, methodological, and second, metaphysical. The first kind, methodological naturalism, is like a functional atheism within the lab, which is said to be simply necessary for doing science. The second, much stronger type, metaphysical naturalism, is a personal conviction about ultimate reality, and it works at the level of individual worldviews of scientists.

I explained that methodological naturalism was defended by Arthur Shapiro and Eugenie Scott as merely a rule that *had to be observed in science* because of practical considerations. I also sketched how David Hull insisted on this same point through a *reductio ad absurdum* in his review of Johnson in *Nature*. To illustrate the practically "mandated banishment" of design, Behe quotes a dictum from a prominent biochemist, Richard Dickerson, who happens to be an evangelical Christian.[40] His "Rule No. 1," which might practically debar the consideration of design in all cases, is "Let us see how far and to what extent we can explain the behavior of the physical and material universe in terms of purely physical and material causes, without invoking the supernatural." Behe makes many points about Dickerson's rule. One chimes in with Hull's image—the fear about where the supernatural explanations would stop:

> Behind Dickerson's rule are vague images of Vikings attributing thunder and lightning to the work of the gods and of witch doctors trying to drive out evil spirits from sick people. Closer to modern science are memories of Isaac Newton himself proposing that God occasionally intervened to stabilize the solar system. The anxiety is that if the supernatural were allowed as an explanation, then there would be no stopping it—it would be invoked frequently to explain many things that in reality have natural explanations. [241]

In my view, Behe is seeking to identify and deflect an important, widespread, and fairly robust Darwinian projection theme—*the inevitable ruination of the purity of science if design were admitted as a possible cause of biological complexity.* To deflect this fear, he uses two *narrative tools:* First, he spins some "absurd fantasy" to show how unlikely it is that science will dive into spiritual themes; then he drains fear of belief in God by invoking the historical role of theistic religion in nurturing science's investigation of the universe. Behe's skills of humor and compression help to drive these two narrative arguments home:

If my graduate student came into my office and said that the angel of death killed her bacterial culture, I would be disinclined to believe her. The *Journal of Biological Chemistry* is unlikely to start a new section on the spiritual regulation of enzyme activity. Science has learned over the past half millennium that the universe operates with great regularity the great majority of the time, and that simple laws and predictable behavior explain most physical phenomena. Historians of science have emphasized that science was born from a religious culture—Europe in the Middle Ages—whose religious traditions included a rational God who made a rational, understandable, law-bound universe. Both science and religion expect that the world will almost always spin according to the fixed law of gravity. [241]

Having argued that design would never undercut science's task of unraveling of regularities of nature, Behe turns to the issue of odd, special events—called "singularities"—for which unusual causes are proposed on a case-by-case basis. This is the explanatory situation in which intelligent design fits. Behe points out that the proposal of design will have to be established with data that are overwhelming; that is, a scientist will have to "support that assertion with observable evidence. The scientific community is not so frail that its healthy skepticism will turn into gullibility" (242).

What about the issue of possible design, when the designer is unobserved? Behe asks, "But how can an intelligent designer be tested? Can a designer be put in a test tube?" Of course not, says Behe, but neither can we observe or test the forever-invisible "common ancestors" that are invoked by Darwinists.

The problem is that whenever science tries to explain a unique historical event, careful testing and replicability are by definition impossible. Science may be able to study the motion of modern comets, and test Newton's laws of motion that describe how the comets move. But science will never be able to study the comet that putatively struck the earth many millions of years ago. Science can, however, observe the comet's lingering effects on the modern earth. Similarly, science can see the effect that a designer has had on life. [242–43]

Behe's fourth point about scientists' aversion to design is that of metaphysical belief: "Many people, including many important and well-respected scientists, just don't *want* there to be anything beyond nature. They don't want a supernatural being to affect nature, no matter how brief or constructive the interaction may have been. In other words, like young-earth creationists, they bring an a priori philosophical commitment to their science that restricts what kinds of explanations they will accept about the physical world" (243). What Behe says here should not surprise us, since this point is an echo of one of Johnson's main theses. Note also that, like Johnson, Behe positions himself in intellectual space between and above the two *a priori* positions—the metaphysical materialists and the young-earth creationists.

What is rather surprising for the rhetorician is how Behe argues this point. One might expect that Behe, never having mentioned Denton or Johnson in the course of his book, would now at last connect his own work with theirs and with other scholars in the Design Movement who had argued the point of metaphysical preference, but he does not. Behe's purpose in not bringing his colleagues into the picture was simple—he did not want to muddy his rhetorical water (here or anywhere else in his book) by negative preconceptions about others in the Design Movement. He decided not to invoke the work of Denton or Johnson; he wanted his work to stand on its own.[41] Instead of connecting with the story and insights of his own movement, Behe uses a different kind of narrative strategy. He draws upon two test cases from mainstream science in the twentieth century that demonstrate the appropriateness of design inferences in biology. The first test case is a history-of-science narrative, the development of the big bang theory. The second test case is an individual narrative—the development of the panspermia theory by Francis Crick, codiscoverer of the structure of DNA. Let us see the unique kind of rhetorical electricity that flows through each of these narratives.

In the case history of the big bang, profound religious implications were apparent to many scientists, including Einstein, but these religious implications did not deter the scientists from embracing the theory as soon as compelling evidence arose. In his story of the theory's development over time, Behe details the several subhypotheses that arose (imaginary time, countless millions of "bubble universes," quantum fluctuation) by which scientists have been able to deflate the religious implications of the big bang. The bottom line is clear: If we managed just fine with the religious implications of the big bang, we will do fine with similar implications from design.

Behe has a somewhat different point in retelling the story of Crick's notion of life being sent here by an intelligent race in another part of space (panspermia). He discusses this strange hypothesis, showing how it lends itself to comparison with design ideas. For example, since a creature with the intelligence to send life could also conceivably play a role in designing the structures of life, Crick could have posited the design of life just as naturally as the "shipping of life to earth by spaceship." Behe points out that such a possible hypothesis would automatically raise the question of where these beings came from, and (just as in the case of the big bang) all kinds of non-God or naturalistic hypotheses are possible. In the hands of Behe the storyteller, the two vignettes of the triumph of the big bang and Crick's proposal of panspermia illustrate how current theories, which imply theological *possibilities,* are discussed and debated openly but are certainly not rejected out of hand. Behe asks why his own field of biology must now invoke an ad hoc "no design" rule or balk at considering a theory with theological implications—design—in its own search for answers to puzzling phenomena.

In his narrative with which he ends the book, Behe reviews the train of disturbing, strange discoveries, which, beginning with Copernicus and Galileo, extended step-by-step through the bizarre pictures of the universe that emerged from Einstein and modern physicists. Then he places the discovery of the design of life into this train:

> Now it's the turn of the fundamental science of life, modern biochemistry, to disturb. The simplicity that was once expected to be the foundation of life has proven to be a phantom. Instead, systems of horrendous irreducible complexity inhabit the cell. The resulting realization that life was designed by an intelligence is a shock to us in the twentieth century who have gotten used to thinking of life as the result of simple natural laws. But other centuries have had their shocks, and there is no reason to suppose that we should escape them. Humanity has endured as the center of the heavens moved from the earth to beyond the sun, as the history of life expanded to encompass long-dead reptiles, and as the eternal universe proved mortal. We will endure the opening of Darwin's black box. [252–53]

The power of narrative contextualization (a subtle form of narrative reasoning) is again pressed into service. As Behe crafted his closing argument, calculated to defend design from attack as nonscience, he found no greater rhetorical weapon than a miniature history-of-science of previous revolutions. In this march through strange conceptual lands, often whipping with theological winds, the scientists who led the way in unraveling the truth of nature were the ones who had the courage to confront the strangeness of each new shock. Behe's lesson of this history-parable is clear. We too need the intellectual courage to be braced for biochemistry's shocking new turn in science. It is a narrative that is bolted to an unabashed projection theme in its conclusion: Intelligent Design will be the next paradigm shift, the successor to the big bang and the quantum shocks of twentieth-century physics. Design, like these recent discoveries, is painted in heroic hues—revolutionary red, yet ultimately fruitful green and triumphant gold.

9

Mere Creation and Beyond

William Dembski and the Explanatory Filter

The summer and fall of 1996 constituted a momentous and enormously action-packed season for Intelligent Design—or "ID," as it quickly came to be known. Recall that in July Berlinski's article "The Deniable Darwin" appeared in *Commentary,* followed in September by the maelstrom of published reaction. Berlinski, a longtime friend of Murray Eden and the French mathematician Schutzenberger (key skeptics at the Wistar Symposium), had been previously unknown to the Design community. After his article appeared, he was warmly invited to join the movement. Berlinski responded positively to the invitation and quickly became a major theorist and spokesperson.

In early August James Shreeve's mildly complimentary review of Behe appeared in the *New York Times Book Review.* After seeing an advance copy of the review, Behe remarked to Johnson's e-mail discussion group, "On a scale of one to ten, it's an eight."[1] As mentioned earlier, by late October the *Times* had even printed on its editorial pages Behe's own summary of the biochemical argument for design, "Darwin Under the Microscope," in connection with the Pope's statement on evolution to the Pontifical Academy of Science.[2]

It was a fitting climax to this string of events when in early November about 180 members of the Design community converged upon Biola University in Los Angeles for Mere Creation, the first major international conference on design theory. Three stated goals were to "unite on common ground," to "build a community of thought," and to "share ideas and knowledge."[3] In the volume of papers that came out of this conference, the writer of the foreword, chemist Henry F. Schaeffer III, called it an "unprecedented intellectual event" because it brought together "scientists and scholars who reject naturalism as an adequate framework for doing science and who seek a common vision of creation united under the rubric of intelligent design." Schaeffer added,

> The . . . participants, primarily academics, formed a nonhomogeneous group. Most had never met each other. Yet virtually all of the participants questioned the reigning paradigm of biology—namely, that natural selection and mutation can account for the origin and diversity of all living things. At the same time virtually none of the conference participants were creationists of the sort one frequently reads about in the popular press. In particular a very large majority of the participants had no stake in treating Genesis as a scientific text.[4]

Of the eighteen presenters of plenary talks, the most famous were Johnson, Behe, and Berlinski (substituting for Charles Thaxton who fell ill). Yet for all the importance of their presence and talks, the most novel and strategic scientific arguments and ideas came from the core of young scholars known as the "four horsemen"—Stephen Meyer, Paul Nelson, Jonathan Wells, and William Dembski. These four had recently completed Ph.D.s or were in the final stages of their dissertations at prestigious universities and had been working together since 1993 on conceptualizing new lines of criticism. The first target of this criticism was Darwin's doctrine of "descent with modification." For example, Nelson and Wells had researched new types of evidence (especially in embryology) that showed the problems with the notion of common ancestry. Nelson's dissertation in the philosophy of science at the University of Chicago, entitled "On Common Descent," argued that this key notion consistently escapes testing in Darwinian theory. Rather, common descent functions as a given—*a foundational axiom*—which is taken as a starting point of rational discussion on the problem of origins.[5]

The second area, pioneered by Stephen Meyer (building on the work of Thaxton), dealt with the notion of "specified complexity" of DNA and other information-bearing molecules as indicators of design. *Specified complexity* is any complex sequential pattern of symbols that is "aperiodic" or nonrepeating but requires that each symbol be specified. Since the specified complexity of DNA (in this argument) is "inferred" to have arisen from design, it was necessary for Meyer to address the legitimacy of the design inference. His core argument was that the "inference to design" is the methodological equivalent of the "inference to descent." This topic was the subject of Meyer's Ph.D. dissertation at Cambridge University, and he argued both at the 1992 Darwinism Symposium and in his chapter in *The Creation Hypothesis* that there is no good reason, from the philosophy of science at least, for any ad hoc rule that forbids the consideration of intelligent causation. Meyer's paper at Mere

Creation, published as "The Explanatory Power of Design: DNA and the Origin of Information," was a compact summary of his argument based on the information content of DNA.

In terms of conceptual innovation, the most important presenter at Mere Creation was William Dembski, who had earned a doctorate in mathematics and was nearing completion of his second doctorate in the philosophy of science. Dembski's importance was rooted in the theoretical and rhetorical centrality, both at the conference and in the book *Mere Creation,* of his detection system known as the "explanatory filter." Whereas the arguments of Meyer, Nelson, and Wells were either attacks on current theory or specific inferences to design in one area (such as DNA and the "specified complexity" of its information), Dembski's idea was much broader and more fundamental to design theory. Developed during the early 1990s, the filter took the form of a three-tier system of conceptual sieves, linked with a step-by-step procedure that Dembski claimed could positively identify the action of intelligent design in any system or phenomenon in nature. Dembski argued that the filter is analytically "robust" (one of Dembski's favorite words).[6] That is, it can be shown as highly precise and reliable and can be applied to many different fields of research. He supported its legitimacy in science by noting that certain specialized branches of scientific research—such as forensics, SETI, archaeology, patent review, and data falsification analysis—already use reasoning akin to his filter in order to distinguish cases of intelligent action from cases produced naturally. Thus, the filter formalizes and makes precise the processes of design detection that are going on daily in many fields.[7]

The filter, although based somewhat on statistical and mathematical concepts, is not very difficult to comprehend. Dembski begins by specifying an event "E" that is to be submitted to the filter to see if it may have resulted from intelligent design. It is examined at the first level of the filter to see if E is a high probability event. If it *is* a high probability event (called an HP event), it is concluded right away that E was produced through the operation of natural law. The conclusion of "design" is thereby ruled out immediately. If E is not an HP event, then it passes to the second level of the filter. At this level E is examined again to see if it is a medium probability event (MP event). For example, if E is the dealing out in a poker game of a royal flush in spades (which, using statistical terminology, has a medium probability, or one chance in 2,598,960), then the event is attributed to chance and design is again ruled out. E has been adequately explained as an MP event.

However, if E is neither high probability nor medium probability, then by definition it is a small probability event, and it passes to the third level of the filter. This is called the specification level, and it is the final step in determining if E is the result of design. When E arrives at this level, it is not automatically assumed that it is an intelligently caused event. To return to the poker example, any exact hand dealt in a given round has an extremely remote likelihood—as small as the tiny "one chance in 2,598,960" mentioned earlier. So every possible hand, in a sense, has a certain vanishingly low probability of appearing in a given deal of the cards. Therefore, in Dembski's third filter, his specification level, in order for E to be clearly indicated as "designed," it must first be judged to be of very low probability, and secondly it must conform to an independently given pattern or an "ideal specification." In the case of the poker hand, if five royal flushes in a row were dealt to the same person, one would definitely suspect cheating. This is because the five hands are now clearly an event in the realm of very low probability events (which can be quantified mathematically), and this event conforms to an independently given (ideal) pattern. Therefore, the fellow poker players can solidly infer design—someone has cheated. It would be foolish and illogical to attribute the five consecutive royal flushes to mere chance. In any such situation, E can be tagged as "designed" if and only if it simultaneously is a low probability event and also conforms to a specification—an independently given pattern or ideal specification.

I have presented the explanatory filter in a way that distils the thrust of Dembski's arguments.[8] The key point for the application of the filter, according to Dembski, is how universally this method of "detecting design" may be fruitfully extended. He argues that any system or physical state of affairs, in any branch of science, will be detected as "designed" if it successfully passes through all three levels (or nodes) of his filter.

I see the rhetorical power of Dembski's filter in establishing and defending design theory as having four dimensions. First, *the filter places design theorists in the context of currently accepted science.* They are portrayed as merely proposing to apply to biology what other scientists, such as SETI astronomers who sift radio static from space looking for messages, are already doing. Second, *the filter is a regulative procedure that projects caution.* Note that design theorists don't just go out attributing all kinds of phenomena to design (or God) but rather have strict criteria through which any phenomenon must pass before it is considered to be "designed." Thus, Hull's fear-fantasy of design theorists attributing earth's magnetic field to God's will is hammered flat.

Third, the assignment of "design" to any event E is made more impressive by its *use of a principled system of statistical analysis.* By this the inference is apparently solidified. The tests that establish design are said to rest on mathematical criteria rather than relativistic notions such as intuition, personal incredulity, or alleged or perceived implausibility. Fourth, Dembski's filter does not specify God or any transcendent designer; it merely *specifies some species of intelligence as the cause.* Thus, the filter buttresses a basic plank of intelligent design theory: Science by itself cannot indicate that some deity produced the design. Just as a forensic pathologist can only tell that a person was murdered, not who did it, so also the design theorist can only indicate that a designer was responsible. The issue of the identity of the designer is made separable from the bare conclusion of design itself.

Just as it is hard to exaggerate the role of Johnson and Behe in the historical development of design theory, it is equally hard in this connection to overstate the importance of Dembski and his filter. The filter has become the most important procedural plank of a putative new paradigm that is "geared to the detection of design" in biological systems. Viewed from one angle, both Behe's notion of "irreducible complexity of molecular machines" and Meyer's idea of "specified complexity of DNA" are specialized cases of the application of Dembski's filter. That is, Behe's and Meyer's specific inferences of design can be subsumed under the big tent of the explanatory filter.

Dembski's Stories and the Narrative Persuasion of Design

Dembski's rhetorical importance for the Intelligent Design Movement is connected primarily to the explanatory filter's perceived pragmatic excellence—its inherent logic, its broad explanatory power, and its mathematical precision. Yet this conceptual apparatus should not be separated from the profoundly human story of Dembski himself, who is often visualized in projection themes of Design as something of a youthful prodigy, a mathematician-storyteller whose life itself may become a heroic parable. It is at this level that narrative sparks fly, that diverse stories proliferate. Some stories have become famous as Dembski's own tools of persuasion, and others serve as the larger frame or context, enriching the meaning of Dembski's ideas and his role in Design. As a result, important narratives and projections are conceived and wielded both inside and outside Dembski's rhetoric.

For example, Dembski's career is somewhat of a "roller-coaster" contextual story line. It begins with a bright doctoral candidate in mathematics, and later in the philosophy of science, who duly impresses his Ph.D. committees. One of his dissertations is eventually published by Cambridge University Press after passing through a rigorous peer-review process. As he begins to engage the "scientific priesthood," he is ultimately hired to assemble the first U.S. academic center for the study of design theory at Baylor University. Continuing the ascent, he manages to organize an extraordinary and unprecedented conference on design in the spring of 2000, involving some very prestigious defenders of naturalism and Darwinian macroevolution. Yet, in the midst of steady progress, he is harshly pummeled and publicly vilified by his own hostile colleagues at Baylor. Ultimately, their strenuous attacks provoke an investigation, and—ironically—at the moment of vindication by an academic review board, a bit of incautious rhetoric in his victory statement immediately and tragically swallows the string of victories into the swamp of sudden defeat. This main story line is the most important out of the many narratives that are linked to the suasory career of Dembski.

Dembski's Academic Persona

First, as I unfold this story, I must focus on the prelude to Dembski's rhetoric—his gathering of impressive academic credentials. With two earned doctorates in mathematics and philosophy of science as well as a master's degree in theology from Princeton Seminary, his achievements rival that of any other Design scholar. His academic degrees confer an aura of erudition and mathematical authority as he presents his filter and details its application.

Lynn Vincent beautifully captured this persona in her article "Science vs. Science": "It's easy to imagine what William Dembski's wife finds in the dryer lint trap after washing her husband's pants: equations. Long, elegant equations replete with tangents, vectors, and permutations tangled unceremoniously with tissue shreds in the lint trap. When Mr. Dembski speaks, equations come out. When he writes, equations come out. Surely he must keep a few spare equations in his pockets."[9]

His aura of scholarly competence was further boosted when Cambridge University Press published his highly technical book on the explanatory filter, *The Design Inference*. Johnson, in an apt description, calls it "remorselessly rigorous."[10] The middle chapters of the book (2 through 6) are humorously indecipherable to the average reader, filled with esoteric and incomprehensible mathematical discussions and statistical equations. For example, in chapter 3, "Probability Theory," the reader is greeted by these two opening paragraphs:

> Our aim throughout the next four chapters is to explicate and justify the Law of Small Probability (LSP). To accomplish this aim, let us start by identifying the conception of probability we shall be using. Conceptions of probability abound. Typically they begin with a full theoretical apparatus determining the range of applicability as well as the interpretation of probabilities. In developing our conception of probability, I want to reverse this usual order, and instead of starting with a full theoretical apparatus, begin by asking what minimally we need in a conception of probability to make the design inference work.
>
> One thing that becomes clear immediately is that we do not need a full-blown Bayesian conception of probability. Within the Bayesian conception propositions are assigned probabilities according to the degree of belief attached to them. Given propositions E and H, it makes sense within the Bayesian conception to assign probabilities to E and H individually (i.e. $P(E)$ and $P(H)$) as well as to assign conditional probabilities to E given H and to H given E (i.e., $P(E|H)$ and $P(H|E)$). If E denotes evidence and H denotes a hypothesis, then of particular interest for the Bayesian probabilist is how believing E affects belief in the hypothesis H. Bayes's theorem is said to answer this question, relating the probability of H given E (i.e., $P(H|E)$, known as the posterior probability) to the probability E given H (i.e., $P(E|H)$, known as the likelihood) and probability of H by itself (i.e., $P(H)$, known as the prior probability):
>
> $$P(H|E) = \frac{P(E|H)P(H)}{P(E|H)P(H) + \Sigma_j P(E|H_j)P(H_j)}$$
>
> Here the H_j are alternate hypotheses that together with H are mutually exclusive and exhaustive. Bayes's theorem is therefore of particular interest if one

wants to understand the degree to which evidence confirms one hypothesis over another.[11]

If the reader's eyes are not sufficiently glazed by these explanations, note what appears several chapters later on, when Dembski presents the final, polished formula for his "Law of Small Probability." This is precisely how it appears:

Suppose a subject S has identified both a requisite precondition $\Sigma = (H, P, I, \Phi = (\phi, \lambda))$ *and probabilistic resources* Ω. *Then the following formula defines the Law of Small Probability:*

$$\forall X\{oc(X) \ \& \ \exists D[sp(D, X) \ \& \ SP(D^*)] \rightarrow \sim ch(X)\}.$$

Here X ranges over events and D over patterns.[12]

At this point in the text, Dembski presents his final six-step formulation of applying the LSP (essentially his explanatory filter) to the specific task of the design inference. Why have I occupied this space in quoting the paragraphs and his final formula? I simply wanted to convey the acute sense of Dembski's *expertise* that would strike any ordinary reader not acquainted with statistics in just thumbing through the book. Any notion of Dembski as a "Bible-toting creationist" seems ludicrous in the light of these symbols of refined scholarship.

The dust jacket of *The Design Inference* confirms this sense by means of kudos from Dembski's mentors in his doctoral studies. They clearly hint that this work is indeed groundbreaking. From William Wimsatt comes such a glittering endorsement:

Dembski has written a sparklingly original book. Not since David Hume's *Dialogues Concerning Natural Religion* has someone taken such a close look at the design argument, but it is done now in a much broader post-Darwinian context. Now we proceed with modern characterizations of probability and complexity, and the results bear fundamentally on notions of randomness and on strategies for dealing with the explanation of radically improbable events. We almost forget that design arguments are implicit in criminal arguments "beyond reasonable doubt," plagiarism, phylogenetic inference, cryptography, and a host of other modern contexts. Dembski's analysis of randomness is the most sophisticated to be found in the literature, and his discussions are an important contribution to the theory of explanation and a timely discussion of a neglected and unanticipatedly important topic.[13]

Behe's foreword for Dembski's recent book *Intelligent Design* (1999) has the same effect of projection theme construction around the persona of the new rising theorist. There, Behe briefly explains the explanatory filter and compares it favorably with the greatest intellectual advances in the history of humankind:

But until Dembski, thinking about how we detect design was like writing before the alphabet or calculating before Arabic numerals. Indeed, in the past, ascribing something to design often seemed a matter of mere taste. . . . Socrates, for example, thought that the progression of the seasons, of day and night, of rain and dry weather, pointed toward design. But what else could follow day except night? What could come after drought but rain? How then can we reliably detect design? Dembski's insight, first elaborated in his scholarly monograph *The Design Inference* and explained here for a broader readership, is that we recognize design in what he calls "specified complexity" or equivalently "specified small probability." In other words we apprehend design in highly improbable (complex) events that also fit some independently identifiable pattern (specification).[14]

On that book's jacket, Dembski's individual story and ethos were boosted into the zone of exhilarating projection chaining by a blurb from Rob Koons, a philosopher at the University of Texas. Koons described Dembski as the "Isaac Newton of information theory, and since this is the Age of Information, that makes Dembski one of the most important thinkers of our time."

Dembski's Narrative-Type Logos

Another important species of narrative is discovered as one descends into Dembski's rhetoric, in which he builds his conceptual arguments (logos) with the help of vivid anecdotes. Some are true stories that illustrate the function of the filter. Dembski has recycled in many of his discourses the story of Nicolas Caputo, a clerk of elections in New Jersey who placed his party first on the ballot forty out of forty-one times. After an election-fraud suit was filed, the case came to trial and the prosecution's lawyers established, based on statistical proof of *small probability* in conjunction with *specification,* that Caputo's procedure could not have been based on random choice, as Caputo claimed. Dembski tells the story of Caputo with a touch of humor that tends to endear the reader to the line of argumentation. By sheer repetition, this story has begun to function as a key rhetorical commonplace.[15]

At times, Dembski's stories are fanciful yet vivid thought experiments that show how design by an unidentified intelligence can be clearly and easily stipulated by mathematical calculation of the statistical resources of the physical universe. The most famous of these fable stories, the "Incredible Talking Pulsar," has been belittled by design critics.[16] Other stories are not invented but rather adapted by Dembski. Typical is the story told in the movie *Contact,* starring Jodie Foster, of the SETI detection of intelligence through the reception of a radio signal containing prime numbers from 1 through 101. In one article, Dembski folds this fictional story into the heart of his explanation of the filter.[17] One reason these stories are important to the rhetoric of design is that some of his examples or stories (especially that of Caputo and the movie

Contact) have begun to circulate as rhetorical commonplaces within other texts when the explanatory filter is being explained.

Dembski also uses true stories in his explanation and testing of the filter. A prime example is his discussion of the comet that crashed into Jupiter in 1994, in comparison with the plagiarism theory of the founder of Christian Science: "The fact that the Shoemaker-Levy comet crashed into Jupiter exactly twenty-five years to the day after the Apollo 11 moon landing is a coincidence best referred to chance. But the fact that Mary Baker Eddy's writings on Christian Science bear a remarkable resemblance to Phineas Parkhurst Quimby's writings on mental healing is a coincidence that cannot be explained by chance and is properly explained by positing Quimby as a source for Eddy." Dembski in the context then explains why the filter screens out the comet crash as not designed but rather a chance event:

> The explanatory filter is robust and easily resists counterexamples of the Shoemaker-Levy variety. Assuming, for instance, that the Apollo 11 moon landing serves as a specification for the crash of Shoemaker-Levy into Jupiter (a generous concession at that), and that the comet could have crashed at any time within a period of a year, and that the comet crashed to the very second precisely twenty-five years after the moon landing, a straightforward probability calculation indicates that the probability of this coincidence is no smaller than $[10^{-8}]$. This is not all that small a probability . . . when considered in relation to all the events astronomers are observing in the solar system. Certainly this probability is nowhere near the universal probability bound of $[10^{-50}]$ proposed by Borel.[18]

It is worth noting that such exemplary stories do not fit into any of the three categories of my taxonomy of factual narrative commonly found in Design's discourses. These may constitute a fourth category of narrative, of much less importance in genesis-type argumentation. Perhaps this type of narrative might be called "test case stories and scenarios."

Dembski's Personal Narrative: The Nadir and the Zenith

Finally, and most important for the movement story, Dembski's own recent personal narrative has become something of a living parable on both the potential and the pitfalls of Design's goal to transform the current evolutionary paradigm in biology. This is especially true of Dembski's career after being hired by Baylor University in 1998. Though I sketched this tale above, I need to flesh it out with details at this point. On the one hand, the pitfalls were symbolized by the recent controversy story (in 1998–2000) when the faculty at Baylor University struggled to shut down the Michael Polanyi Center, which President Sloan had hired Dembski to create and operate as a center for the development of intelligent design theory. When the faculty complained strenuously about the president's failure to consult with them and later voted overwhelmingly to shut the center down, President Sloan deferred a decision on the future of the center to an outside panel of experts from other universities.

Their finding, delivered in October 2000, was generally supportive of Dembski, stating that "the committee wishes to make it clear that it considers research on the logical structure of mathematical arguments for intelligent design to have a legitimate claim to a place in current discussions of the relations of religion and the sciences." On the other hand, it made some administrative recommendations, such as changing the name of the center and creating an advisory committee of Baylor faculty to "assist in planning and reviewing its activities."[19] In an e-mail statement, Dembski thanked the panel for its support but seemed to gloat in a victory over his Baylor opponents, who he claimed had tried to throttle free speech:

> The report marks the triumph of intelligent design as a legitimate form of academic inquiry. This is a great day for academic freedom. I'm deeply grateful to President Sloan and Baylor University for making this possible, as well as to the peer review committee for its unqualified affirmation of my own work on intelligent design.
>   The scope of the Center will be expanded to embrace a broader set of conceptual issues at the intersection of science and religion, and the Center will therefore receive a new name to reflect this expanded vision. My work on intelligent design will continue unabated. Dogmatic opponents of design who demanded the Center be shut down have met their Waterloo. Baylor University is to be commended for remaining strong in the face of intolerant assaults on freedom of thought and expression.[20]

Dembski's confrontational language about "intolerant assaults" and "[d]ogmatic opponents" who "have met their Waterloo" was too much additional aggravation to an already strained network of relationships. President Sloan immediately rebuked Dembski for his lack of collegiality and ordered him to apologize and recant the e-mail. When Dembski refused, he was fired from his directorship of the center, which was turned over to Bruce Gordon, the assistant director. Dembski was allowed to stay on in the Baylor Institute for Faith and Learning in a research position. Two years later, at the writing of this chapter, Dembski was even hesitant to discuss the details of this painful incident. After this personal nadir, he became busy healing the relationships with fellow faculty who had become hostile to his presence and his work at Baylor.[21]

If the October 2000 brouhaha represents a key setback in Dembski's young career, then it is somewhat balanced by the zenith that had occurred just six months earlier. Dembski's carefully cultivated efforts at engaging critics in dialogue enabled him to host successfully a conference of considerable intellectual reach. This conference, entitled "The Nature of Nature," asked whether science had progressed in recent decades to a place where evidence suggested that "nature was pointing beyond itself to something that transcended nature."

Held at Baylor University in April 2000, the three-day conference placed Design scholars in an intense exchange with a dozen leading Darwinists, including two Nobel laureates.[22] In addition to the many papers that were presented, a stream of valuable dialogue took place in the fields of biology, chemistry, physics, mathematics, philosophy, and the history of science.

Not surprisingly, the purpose statement for the conference steered a cluster of scientific issues into a head-on confrontation with metaphysical (worldview) issues that were portrayed as undergirding all scientific investigation:

> Is the universe self-contained or does it require something beyond itself to explain its existence and internal function? Philosophical naturalism takes the universe to be self-contained, and it is widely presupposed throughout science. Even so, the idea that nature points beyond itself has recently been reformulated with respect to a number of issues. Consciousness, the origin of life, the unreasonable effectiveness of mathematics at modeling the physical world, and the fine-tuning of universal constants are just a few of the problems that critics have claimed are incapable of purely naturalistic explanation. Do such assertions constitute arguments from incredulity—an unwarranted appeal to ignorance? If not, is the explanation of such phenomena beyond the pale of science? Is it, perhaps, possible to offer cogent philosophical and even scientific arguments that nature does point beyond itself? The aim of this conference is to examine such questions.

For a week, both during and after the conference, Johnson's e-mail discussion group swarmed with reports and commentary about "The Nature of Nature." A few complained that the Darwinian speakers, by their slight numerical advantage and ceaseless hammering on design theory, seemed to drown out the relatively cautious and low-key ID presentations. However, many participants on the Design side indicated that the rhetorical exchange was nicely balanced and that many ID talks, especially Behe's, had a considerable impact on the audience.[23] Paul Nelson, one of the horsemen who was also a Design speaker at the conference, started his own new projection chain of progress by reporting the remarks made at a closing dinner by Darwinian biochemist and Nobel laureate Christian de Duve: "For me, the conference was captured by Christian de Duve's wine-glass-lifted toast to the conference at the private Saturday night dinner for the speakers. He said—and what a pity a tiny video camera hadn't been present, so the tape could be replayed dozens of times for the oppositional Baylor science faculty—that although the conference included widely-divergent viewpoints, the discussions were conducted with remarkable patience and good humor, even—and here the room burst into delighted laughter—with 'intelligent design!'"[24]

Of course, in analyzing the rhetoric of Design, the most important point about the conference rests not in a closing toast or in projections of a dramatic turning point that may have chained out from it. Rather, the significance lay in the new, more powerful phase of engagement that was completed and a new set of social and intellectual relationships that had been restructured. This recalls a similar sort of deepening and irreversible legitimizing that had flowed out of the five Darwinists' participation in the Darwinism Symposium in Dallas eight years earlier. This inexorable transformation through engagement, a process noted earlier as the "Fuller-Kuhn frame," was here not just a softening on one side or a tactical repositioning of the various defenders. It was the creation of a crucial dialogue, a forceful, point-by-point comparison of two paradigms. A heuristic paradigm (Design) was expressed by its brightest, most knowledgeable, and most articulate proponents in a direct encounter with the opposition's strongest criticisms brought by intellectual rhetorical personas, complete with their assumptions and motivations. In Design's own reckoning of its dialectical progress and hope for the future, the Baylor encounter ranked as the most substantive academic milestone to date.

Continuing Criticism: Provoking the Paradigm Crisis

The Baylor conference is one side—the academic side—of the engagement with secular American culture. The other side of this engagement is the public side, and I said at the outset of this book that some very important coverage of Design by major media took place in the spring of 2001. Front-page articles appeared both in the *Los Angeles Times* and in the *New York Times*. An analysis of the texts of these journalistic awakenings reveal a subtext: Design's new significance in the eyes of the media grows out of a crucial perception that a significant number of credentialed scientists have legitimate empirical reasons for claiming Darwinism is entering a paradigm crisis.

The two articles reveal that Design's strategy of attesting a paradigm crisis had succeeded in establishing in the consciousness of the media[25] both of its conceptual edges—*critical* of the old paradigm and *constructive* of a new paradigm. The constructive edge somewhat dominated the newspaper articles, with their emphasis on Behe and Dembski, and especially the claim that mathematics can be used as a tool in the detection of design. Yet the critical edge is also quite visible—especially in the questioning of natural selection's ability to create novel biological complexity.

This balance between constructive and critical approaches was perfectly symbolized by a pair of polished videotaped documentaries on Design that appeared in April 2002. These tapes should prove to be highly influential in the years to come because of their extraordinary educational and video quality, reminiscent of a special on the Discovery Channel. One of them, *Unlocking the Mystery of Life,* produced by Illustra Media, is a compact sixty-five-minute overview of the rise of Design, introducing the listener to all the major figures in the Design Movement. It strongly presents the positive or constructive side—the reasoning process that led to the design inference, specifically including the molecular complexity of the flagellum and the information embedded in DNA. The other video, *Icons of Evolution,* by Cold Water Media, is primarily critical in approach. It highlights the empirical problems embedded in the teaching of Darwinian evolution in high school classrooms and the campaign of censorship waged against a Burlington, Washington, biology teacher—Roger DeHart—who sought to expose his students to these problems.

Such empirical problems (as covered in the *Icons of Evolution* video) have not been the focus of this chapter—for good reasons. I have been emphasizing Dembski's new contributions to the "constructive" methodology of ID, and have focused on the development and rhetorical deployment of his explanatory filter. As a result, I have not traced the heavy flow of rhetoric in the late 1990s and beyond that continued to drive home the "critical edge" of design theory. It is important to note that almost all ID discourses leading up to the *New York Times* and *Los Angeles Times* stories argued along the lines of Denton's "crisis"—*that evolutionary biology, forced now to confront its anomalies, has arrived at the brink of empirical collapse of its truth claims.* This plot has remained absolutely essential to the success of the movement.[26]

In order to understand and appreciate this side of the rhetorical onslaught in the late 1990s, I shall give four examples in the Design rhetoric that sought to buttress the impression of a paradigm crisis by focusing on fundamental anomalies. First, the newer writings frequently describe the latest puzzles and problems in the origin of life field, which are said to render textbook theories null and void, thereby aggravating the stalemate in this field.[27] Second, closely related to the origin of life problem is the "information problem," which in a sense has become the new heart of the design critique of macroevolution. Using a model of information as "specified complexity," developed by Thaxton, Meyer, and Dembski, Design rhetors feel they have a tool for demonstrating the inadequacy of naturalistic explanations of the rise of new genetic information.[28]

Third, Design writings now stress the dramatic fossil discoveries at Chengjiang in southern China since the late 1980s. These fossils reveal dozens of bizarre creatures hitherto unknown in the Cambrian explosion, making the event seem greater and more perplexing than ever. In this connection, Design authors have begun telling a new controversy story that grew out of the criticism of Darwinian theory during a 1999 tour of the United States by Jun-Yuan Chen, the principal Chengjiang paleontologist. He became puzzled as he received hardly any response to his criticism of Darwinian theory, which was woven into his public lectures, and he finally asked one of his hosts about the silence. He was told that criticizing Darwinism is unpopular in U.S. academia; scientists tend to become embarrassed about such talk. "At that he laughed and said: 'In China we can criticize Darwin, but not the government; in America, you can criticize the government, but not Darwin.'"[29] This amusing story is now being replicated widely in Design texts and was used by Johnson as his opening illustration in his 1999 column in the *Wall Street Journal.*

Fourth, and most important, Jonathan Wells's major book that appeared in 2000, *The Icons of Evolution,* has taken on its own life as a withering Denton-style attack on ten major textbook symbols of Darwinian macroevolution—often visually enshrined in the form of pictures or diagrams. Each chapter of his book is devoted to one of these icons, first explaining the evidentiary problems surrounding that topic and then showing how this qualifying information is never mentioned in the ten high school and college introductory biology texts that cite these classic proofs of evolutionary cosmology. In the appendix, Wells assigns individual grades based on detailed criteria for these textbooks. Three of the books receive an overall grade of D, and the rest receive an F.[30]

What made Wells's attack potentially explosive was his allegation not only of widespread distortion and misinformation in the textbook discussions of these icons but far worse—*of known and tolerated fraud.* One fraudulent icon—Haeckel's faked drawings of vertebrate embryos—apparently has been known for many decades, but authors of textbooks left the famous drawings, without comment, in their discussions of evidence for evolution. Wells charges that such fraudulent "proof" of evolution was sometimes knowingly printed as a tool to convert generations of unsuspecting schoolchildren.[31] Wells's closing comment on fraud is very important and is sufficiently nuanced and well-connected with my narrative thesis to make it worth quoting at length:

> Fraud is a dirty word. In their 1982 book, *Betrayers of the Truth . . .* , William Broad and Nicholas Wade distinguish between deliberate fraud and unwitting self-deception. Conscious faking of data is an example of the former, but is relatively rare. Unconscious manipulation of data by researchers convinced that they already know the truth is an example of the latter, and is much more common. There is a continuum between fraud and self-deception, and most cases of misrepresentation fall somewhere between them.
>
> Some textbook-writers, such as Douglas Futuyma, may not even know that one or more of the icons of evolution are failures. Futuyma might reasonably be criticized for his ignorance—especially since he is supposed to be an expert on this subject—but his ignorance is not conscious misrepresentation.
>
> What about Stephen Jay Gould, a historian of science who has known for decades about Haeckel's faked embryo drawings? All that time, students passing through Gould's classes were learning biology from textbooks that probably used Haeckel's embryos as evidence for evolution. Yet Gould did nothing to correct the situation until another biologist complained about it in 1999. Even then, Gould blamed textbook-writers for the mistake, and dismissed the whistle-blower (a Lehigh University biochemist) as a "creationist." Who bears the greatest responsibility here—textbook-writers who mindlessly recycle faked drawings, people who complain about them, or the world-famous expert who watches smugly from the sidelines while his colleagues unwittingly become accessories to what he himself calls the "academic equivalent of murder"?[32]

When it is shown that fraudulent textbook "evidence" has been long tolerated and concealed, it severely compounds the rhetorical challenge for Darwinian rhetors and risks an angry public backlash.[33] It leads inevitably to deeper narrative questions that expose the background of this scandal, such as "How was this allowed to persist?" and "Why did those who knew better never come forth?"

Notice that Wells has turned Gould's treatment of such matters (a micro-narrative) into a relevant *social fact,* useful in persuasion. I view all such empirical falsification discussed above (whether the evidence is physical or social-narrative in nature) as pure extensions of Dentonian rhetoric. They imply that Darwinism is to be rejected not for its religious incompatibility but for its scientific inaccuracy and cumulative implausibility. From my narrative perspective, Wells's book and other similar Design discourses tend to shred the plausibility of the tale of macroevolution (in Wells's case, leaving behind a mystery). Then they put in its place an intriguing story line—Darwinian science is now found to be resting on problematic foundations. In Wells's presentation, the situation is actually far worse. The publishers of biology texts are revealed to have been complicit in deception, fraud, and even cover-up.

What makes these scientific criticisms rhetorically momentous is their convergence upon an increasingly explicit Kuhnian plot line. Whether or not the rhetoric of Intelligent Design issues in anything like a scientific revolution, such persuasion is one of the most important examples since Kuhn of a scientific-rhetorical movement taking full advantage of the new self-consciousness of paradigmatic change and invoking his terminology (and narrative vision) as part of the strategy of conquest. Kuhn's vision, itself a broad projection theme archetype, has become, for good or ill, an available tool for creating a *rhetorical crisis* regardless of whether future (or neutral) observers eventually (in retrospect) affix the tag of a genuine *paradigm crisis.*

Nelson and Wells, for example, recently theorized: "A *design paradigm* would resolve the anomalies listed above. . . . Unlike Dobzhansky's 'anti-evolutionist' view, a *design paradigm* does not entail the immutability of species or a recent origin of life; unlike neo-Darwinian evolution, however, it does liberate biology from having to insist on common descent and the sufficiency of natural selection and adaptive genetic changes."[34] In the eyes of Design rhetors, the emotional severity of the criticism they have received from respondents like Jukes, Provine, and Gould only underscores the revolutionary nature of the situation. In a recent literature review of design texts and rebuttals, rhetorician Thomas Lessl concluded that the "leaders of the ID Movement would insist that the [Darwinian] hyperorthodoxy and heavy-handedness . . . are merely signs that the evolutionary paradigm is indeed in crisis."[35]

Emotional Reaction vs. Intellectual Rebellion

Kenneth Miller says that Johnson, Behe, Dembski, and their cohorts possess and are driven by a powerful emotional opposition to Darwinism, rooted in the theory's inherent tendency to portray a meaningless universe that is filled with contingency, happenstance, and horror.[36] This attribution of crucial emotional substrata that control the motivation of Design creates a certain distinct explanatory account of the whole Design Movement. This is an important part of the dominant plot line of Darwinian projection themes that seek to make sense of the new challengers.

The recognition of this emotional factor as key to the Darwinian conceptualization is crucial to understanding the rhetorical situation in which Design arguments are deployed. However, the emphasis on *fear* is one of Darwinism's clearest and most important distortions of the crucial motives at work.[37] Rather than emotions of fear driving Design's crusade, the psychosocial reality is very much more *a profound intellectual dissatisfaction, leading to academic rebellion.* If there is emotion at work, it tends to be one of anger—righteous indignation—flowing out into a network of defiant attitudes and acts. Furthermore, the goals that loom on Design's horizon are limited not just to the correcting of the misinformation about origins, or even to the pursuing of a paradigm shift in the one field of biology, but rather include something more profound—a remodeling of scientific and academic epistemology and freedom of criticism. Central to this project is the prying of long-sequestered issues of *metaphysics* out of the intellectual closet. Design rhetors ask, "What is the effect of materialistic assumptions on the freedom of inquiry? On the kinds of questions that can be asked? On the cogency and credibility of scientific claims outside of the professional guilds that approve those claims for publication?" This highlighting of the agency of metaphysics, so clearly visible in the writings of Johnson, is equally central to the rhetoric of Behe and Dembski. In the latter half of the 1990s, Design's rhetorical vision matured around this central point of revealing the role of metaphysical assumptions.

What then is Design's key driving motivation? It is reflected in the individual narratives of its rhetors—a deep revulsion to perceived academic scandal and an angry rebellion against a pervasive state of intellectual oblivion within academia about the empirical problems with Darwinism. To borrow Kant's fantasy theme about Hume, the motives coalesce in a battle to "arouse" the academy—and society at large—from the "dogmatic slumber" of evolutionary naturalism. The question is not just the plausibility of Darwinism but even more the assumed truth of naturalism.

This Kantian image, "aroused from dogmatic slumber," is an apt and compact symbol of the most powerful analytical metanarrative of Design's rhetorical vision. The *passion to awaken* is vastly more of a powerful driving force than any *fear* or even moral revulsion. I submit that the individual stories of Johnson and Behe coincide with many other such awakening stories of those active in Design. I solicited and recorded these stories in the course of my research and have preserved these personal anecdotes that were shared by a variety of academicians who are affiliated with Design.[38]

Of course, there *are* concerns for moral implications of Darwinism among Design rhetors, but they are secondary. The chief *moral* concern is to defend the universal values associated with truthfulness, which are vital to the integrity of the universities: the truthful reporting of all the facts relevant to a paradigm's puzzles, the willingness to test or falsify a paradigm's most cherished axioms or truths, and an honest assessment of all possible inferences that can be drawn from the layers of evidence. Restoring these values in biology and in other spheres is a key concern that inhabits Design's rhetorical vision. Ultimately, the reformative urgency of Design (according to its own rhetorical vision) will be satisfied when scientists and their discourses reach a critical point—a clear realization of the role of metaphysics in scientific work (and in all academic research), coupled with a willingness to submit to questioning the all-important role of naturalism. Behe and Dembski argue that both functions of science—the *transmission of knowledge* as well as its research function of the *discovery of new knowledge*—can be reformed fairly easily and that the changes can be made relatively quickly.

In summary, Behe and Dembski were the chief innovators of Design's rhetorical vision after 1996. They contributed to the clarity of goals and gave rise to an attitude of optimism, even spurring projection themes of success in the first half of the twenty-first century. According to this vision, the movement's goals will be achieved by two simultaneous resurrections—the resurrection and exposition of the *unsolved biological mystery* (the question of how life and humanity arose from lifeless chemicals), and the resurrection of Paley's idea, the *inference to design,* that has lain buried in the academic graveyard for over a century. Adapting Berlinski's famous projection theme, this Paleyan idea and the mystery it claims to solve are being disinterred before our eyes and brought "clattering back to life, dragging their winding sheets behind them."[39] In Design's consciousness, the rhetorical artifacts of Behe and Dembski have already set in motion these symbolic resurrections.

10

A Revolution Built on Recalcitrance

Intelligent Design
in Historical Perspective

I have now finished tracing the birth and rapid development of the Intelligent Design Movement, focusing on the dynamic role of a complex matrix of narrative that organizes and energizes Design's scientific persuasion. This rhetorical history began with the heady triumphalism of the 1959 Darwin Centennial and traced the subsequent lines of evidentiary trouble that arose for Darwinism. These empirical threats to "evolution as fact" became the threads Michael Denton wove together in *Evolution: A Theory in Crisis,* his radical call for a rejection of the Darwinian paradigm. Denton's argumentative style and his array of objections to Darwinism were directly absorbed and adapted by the earliest Design rhetors. Essentially, in 1986 Denton galvanized the preexisting sophisticated skeptics of Darwinism and fathered the Intelligent Design Movement as a rhetorical enterprise.

Through *Evolution,* Denton also converted several American scholars who were to play a leading role in the Design Movement. One of these, Phillip Johnson, quickly rose to prominence as he developed his own critique of Darwinism and placed it before his colleagues at Berkeley and before the American academy in general. Responses to his criticism ranged from savage opposition (Gould) to quiet support (Raup). Johnson's early critique, developed in London, grew into his book *Darwin on Trial,* and I have focused on its theses, rhetorical strategy, and crucial role in shaping the thought of Design as a fledgling revolutionary movement. Johnson's symbolic act of "outing" naturalism as a major player in evolution's drama of origins was foundational to his scientific argument. It became the cornerstone of Design's criticism. After 1991 Intelligent Design was targeting not only Darwinism but also the naturalistic set of rules and norms that define the academic notion of "knowledge" and that constrain the diverse methodologies of research in modern universities.

Finally, biologist Michael Behe—another Denton convert—and mathematician William Dembski became persuasive advocates in the late 1990s and beyond. Behe's argument, built around *irreducible complexity,* proved effective in attacking the naturalistic evolution of life's Lilliputian molecular machinery. Behe and Dembski's work drastically accelerated the expansion of Design in the universities, and Dembski's explanatory filter seemed to provide the needed foundational structure of an Intelligent Design paradigm. In this paradigm-building period, empirical attacks on Darwinism continued undiminished, as seen in Jonathan Wells's *The Icons of Evolution* and its charges of textbook misinformation and Darwinists' toleration and even propagation of long-known fraud.

The narrative of Intelligent Design was greatly enriched by the steady stream of criticism aimed at the movement and its products. Interactions with key Darwinists, especially through symposia on university campuses, produced their own persuasive effects, including a subtle but irreversible repositioning and softening of Design's opponents. Such persuasion hinged partly on the presentation of key evidentiary arguments, but equally important were the skilled use of factual scientific narratives and the creation of captivating projection themes. These combined to "arouse from dogmatic slumber" many in academia and beyond who assumed the firm status of Darwinism as fact.

This book was first written as my doctoral dissertation at the University of South Florida. That version transitioned at this point into a ten-page essay tracing four major stages in the development of the rhetoric of science and its growing scrutiny of Design (see appendix 4).[1]

One eventful sliver of that survey of the rhetoric of science is preserved here. I am framing it as a vivid personal memory from college days, which connects with Thomas Kuhn and the academic fallout that precipitated in the 1960s. It is a fascinating story leading up to a key debate that is still simmering. This debate is vital because it lays a foundation for my major conclusions. Let's turn back the clock and set the stage with campus scenes from the noisy, radicalized days of the Vietnam era.

The Recalcitrance Issue: Is Science "Rhetoric without Remainder"?

In the fall of 1968, when I entered Princeton University as a freshman, one could hear in campuses across the United States the furious gushing and splashing of strange revolutionary currents. One of these was the leftist campus movement "Students for a Democratic Society" (SDS), which was busy staging protests across the country against the Vietnam War. Angry shouts echoed through ivy-covered college courtyards, and Princeton was no exception. Nassau Hall, the stately granite building that two centuries earlier shook with the boom of George Washington's cannon battery firing on the British troops holed up inside, on this occasion echoed with a different kind of angry confrontation. Nassau Hall's elm-shaded front green was the favorite spot for students and young professors who staged SDS "teach-ins"—impromptu open-air seminars on the evils of Vietnam, the military-industrial complex, and the U.S. exploitation of third world countries.

Princeton radicals also seized two buildings[2] to communicate their protest and force their demands, and this pattern of building seizure was replicated on dozens of universities across the country. But the revolutionary currents were not just surging in courtyard speeches and occasional building takeovers. From campus to campus inside those same classroom buildings one could hear the ruthless, poetic atheism of Nietzsche, the Eastern-tinged ideas of Hermann Hesse and Alan Watts, openly neo-Marxist socioeconomic analysis, and B. F. Skinner's radical psychological theory of behaviorism. In dorm rooms, under the musical blast of Jimi Hendrix and Janice Joplin, there were quiet babbles of drug experimentation mixed with a dozen other rivulets of the rebellious counterculture.[3]

In these revolutionary days, I took a curious dive into one such current, signing up for an eye-opening course on Nietzsche and existentialism, taught by a world authority and a Princeton philosophy superstar, Walter Kaufmann. I enjoyed the course, and in retrospect I wish I had double majored in philosophy along with history, because I might have met and studied under another superstar and revolutionary in that same department, Thomas Kuhn.[4] Although you would not have known it at the time, Kuhn's own creeping revolution in thought was every bit as momentous as the other campus currents, probably even *more* momentous.

By 1968 virtually everyone in academia was beginning to hear about Thomas Kuhn's 1962 academic bombshell, *The Structure of Scientific Revolutions*. His ideas—scientific *paradigm*, suffering an *anomaly overload*, leading eventually to *crisis* and then to *revolution*—were being discussed and applied in departments far removed from the history of science, even in the field of rhetoric. In the wake of Kuhn, many rhetoricians realized that science itself, which before had been nearly invisible to deep rhetorical analysis, now loomed as a vast, juicy target.

Kuhn himself was quickly described as an "implicit rhetorician of science" because of the scattered references in his book to the role of "persuasion" in paradigm crises. Deirdre McCloskey implies that during these years Kuhn had keen interest in this new genre of rhetorical analysis of science. She pointed to a conference held in 1984 by the Project on the Rhetoric of Inquiry, which brought together British sociologists of science and American rhetoricians of science for the first time. Of this event, she said, "Kuhn himself participated enthusiastically in the 1984 conference that initiated the rhetoric of inquiry," and added (somewhat controversially) that Kuhn and the British sociologists of science "are rhetoricians, if sometimes ignorant of their own tradition."[5]

This bold baptizing of Kuhn himself as a pioneer of the rhetoric of science was not an isolated incident. Others in fields outside communication theory, whose work was nevertheless identified as contributing to the rhetoric of science, were hailed as *implicit* (or unconscious) rhetoricians of science. Work began to mushroom across disciplinary boundaries in the late 1970s and early 1980s, and the *explicit* rhetoric of science field (working under that name) was born. In its fast-growing youth—the 1980s and 1990s—the field was dominated by those in communication[6] and English departments, but it included philosophers, historians, sociologists, and even those in the natural and social sciences. Early publications laid down the theoretical principles and perspectives of the rhetoric of science and performed countless case studies. A classic example of such case studies was the work of John Angus Campbell, a scholar who devoted his career to the rhetorical analysis of Charles Darwin and especially his *Origin,* thereby establishing himself as a leading figure in the new field.[7]

One early task was to establish the jarring reality of scientific rhetoric: that *scientists are rhetors in disguise.* This thesis was the title of Herbert Simons's (1980) early essay in the rhetoric of science. He staged a courtroom drama—a presentation of both pro and con responses to his question: Are scientists rhetors in disguise? As presiding judge, he handed down a balanced verdict. *Science is indeed rhetorical through and through, but this does not invalidate the unique values of science as a knowledge-building practice.*

Simons said this tension—science *is* rhetorical yet uniquely powerful— makes sense when rhetoric is viewed not as a "devil term" but in its positive sense of the scientific community's giving of *good reasons* on matters of judgment. He suggested that the new rhetoricians of science play an advisory role, pointing out when scientific rhetoric violates its own canons and time-honored norms. Simons does not explore the potentially questionable nature of those canons and norms, especially in a paradigm challenge such as Intelligent Design's, which questions key norms.[8]

Simons's prudent moderation contrasts with the intoxicating radicalism of Alan Gross. When Harvard University Press published his *The Rhetoric of Science* in 1990, it was a major milestone. With Gross's adaptation of ancient principles of rhetoric from Aristotle,[9] his book probably seemed conservative to some, but this was somewhat misleading. His foundational ideas are radical to the core, as shown when he extends the reach of rhetorical analysis to all kinds of science, including experiments. In his view, the rhetoric of science "does not deny the brute facts of nature; it merely affirms that these facts, whatever they are, are not science itself, knowledge itself. Scientific knowledge consists of the answers to three questions: (1) What range of brute facts is worth investigating? (2) How is this range to be investigated? (3) What do the results of these investigations mean? Whatever they are, the brute facts themselves mean nothing; only statements have meaning, and of the truth of statements we must be persuaded."[10]

On the other hand, this summary need not be seen as radical. It stands as a remarkably elegant and helpful taxonomy of the research targets that rhetoricians of science can study. One of the most oft-quoted summaries of the field's agenda, its insights seem eminently sensible. Yet, on the other hand, Gross's system of thought cuts much deeper here, descending to the scientific bedrock of "factuality." The controversy pivots on two key questions: (1) Is Gross saying that scientific claims are not, at their base, *factual*—that is, they are purely human constructions that forfeit factuality as they move on to express the results from the "investigations of brute facts"? (2) When rhetorical analysis is completed on a scientific publication, is there no solid *factual remainder?* Gross answers these questions with a boldly radical "yes." This startling claim—*the discourses of science are wholly rhetorical without remainder*—is the heart of his radicalism. Gross wants to "entertain a possibility Aristotle could never countenance: the possibility that claims of science are *solely* the products of persuasion" (emphasis added).[11] Here we have moved well beyond the polite, cooperative vision of Herbert Simons!

The radical position, represented by Gross and others, prompted a critique by J. E. McGuire and Trevor Melia through two important articles in *Rhetorica*.[12] McGuire and Melia argued that researchers in the rhetoric of science should listen carefully to Kenneth Burke, a towering genius whose rich tomes cast a shadow across communication theory in the latter half of the twentieth century. Burke's key insight in this connection is a neglected doctrine of *recalcitrance*.

What does this term mean? *Recalcitrance,* "the state of being recalcitrant," says *Webster's Dictionary,* is to be "obstinately defiant of authority or restraint," "difficult to manage," or "resistant."[13] My mental picture for *recalcitrant* comes from a recent visit to our good friends John and Ruth in New Hampshire who had arduously cleared their backyard of the hundreds of granite stones embedded in the soil to plant a new lawn and garden. With great effort the boulders were dug up, one by one, and placed along the back of the property to create a crude stone wall. However, one rock was "recalcitrant"—*stubbornly resistant* to extensive efforts to dig it up. Soon the excavators realized that it was like the tip of an iceberg. The little recalcitrant rock was actually an incredibly gigantic boulder, perhaps thirty feet or more in diameter. In this analogy, the stubborn, unyielding rock is a symbol of a brick wall that scientific rhetoric hits—the recalcitrance of nature. McGuire and Melia use the phrase "the states-of-affairs in nature" to describe this recalcitrant network of *objective factual realities.* Thus, they conclude, "scientific texts encounter a special 'recalcitrance' from the world they hope to describe"—a stubborn facticity that does not permit a promiscuous or unlimited range of textual construction, as scientists routinely intervene in nature.

Because this recalcitrance is central to my conclusions about Design's unique rhetorical power, Burke's timeless caution is worth repeating:

> But the discoveries which flow . . . are nothing other than revisions made necessary by the nature of the world itself. *They thus have an objective validity.*
>
> [T]he point of view, in seeking its corroboration or externalization, also discloses many significant respects in which the *material of externalization is recalcitrant.* And our "opportunistic" shifts of strategy, *as shaped to take this recalcitrance into account,* are *objective.*[14] [emphasis added]

McGuire and Melia sum up their own position by suggesting that the rhetoric of science cannot necessarily proceed on the same assumptions that guide rhetorical analysis in other disciplines:

> Science is the result not only of textual representation, but also of extra-textual interventions with nature. That is, scientific texts, unlike other texts, are not only the product of libraries, but also and notably of laboratories.
>
> It is one thing to stress the social, the artifactual, and the conventional aspects that characterize the purposive production of scientific knowledge. It is quite another to claim that warranted scientific knowledge and talk about the "factual quality" of consensual statements are at bottom about the activities of scientific knowers and not about states-of-affairs in nature. . . . From methodological premises about how people behave we can scarcely derive ontological conclusions about what is the case in nature. This is to confound *how* knowledge is produced with what knowledge is *about.*[15]

In response to this criticism, Gross left open the door for *some role* for "recalcitrance" of nature. But the critics "cannot be right before the fact; they must turn out to be right."[16]

Although I appreciate Gross's major contributions and his brilliant adaptation of neo-Aristotelian rhetoric, I consider his radicalism faulty and agree with McGuire and Melia. The next section will show how the "recalcitrance of nature" began to play a role as a leading actor on the stage of Design's narratives. But I also will point out a second and generally overlooked dimension of scientific recalcitrance that empowers and constrains the rhetoric on both sides of the debate.

The Narrative Themes of This Study

In writing a rhetorical history of an entire movement, I have veered somewhat off the beaten path; most case studies in the rhetoric of science focus on much smaller targets.[17] As I reviewed the breadth of the drama I was preparing to tell, with its swirl of strong personalities, its prodigious output of writing and speaking, and its astonishing variety of personal interaction, I pondered how to frame such a sprawling, bustling story. Very early, as I thought about Design's challenge to Darwinian domination *as its own fascinating drama,* I was struck by the pivotal role that is played throughout by narrative—a matrix of stories, with several distinct kinds toiling away at different levels and time frames. I noted that narrative was not just a pragmatic tool used in specific discourses (although there was plenty of that). It was also playing a grander role as a complex substratum of reasoning and persuasion. In fact, on both sides of the debate, narrative was unavoidably at work, building the main frame of understanding. What's more, besides the factual type of scientific narrative there was a plethora of imaginative projection themes—stories that are dramatizations or mixtures of fact and faith. It dawned on me that all of these stories were steadily filling and molding Design's rich rhetorical consciousness.

I was hesitant at first to rush in with such a narrative approach, being unaware of any previous attempt to combine such varied perspectives in one analytical project. Yet, encouraged by my dissertation committee and scholars who pioneered the various narrative approaches, I threw caution to the wind and set off to tell the "drama of Design" as a *system of stories.*[18] How well did this approach work, and what did we learn?

"The Design Movement Story"

Clearly, one of the most uniquely effective tools of persuasion in the toolkit of Design is its own "noble tale." It is absolutely central to the matrix of evidences and stories deployed by Design theorists. Telling this story does have a partisan effect, of course, but in addition, telling the story fully and accurately is nurturing to the health of science itself and even to the civic health of American society. This is because there persists today among many in our better-educated population a severe and malignant distortion of Intelligent Design. This blind spot is evidenced in hostile caricatures in the media and vitriolic attacks in the universities, which lump Design into biblical creationism and dismiss it as nothing more than "creation science in a cheap tuxedo."[19] Such dismissive rhetoric is understandable given the scientific and cultural stakes, but speaking charitably,[20] it betrays a failure to grasp the historical meaning and rhetorical motives in the rise of Intelligent Design *as a novel story.*

What I mean here as a "novel story" is the entire plot line of the movement. Essentially, respected professors at prestigious secular universities are rising up and arguing that (1) Darwinism is woefully lacking factual support and is rather based on philosophical assumptions, and (2) empirical evidence, especially in molecular biology, now points compellingly to some sort of creative intelligence behind life. In very fundamental ways, I am arguing, this story veers away from the usual *theistic evolution story* ("based on the evidence, theistic scientists are now concluding that God worked through evolution") and from the classic *creation science tale* ("scientists are recognizing that Genesis is literally true after all").

One fact about the story of Intelligent Design is undeniable—it is a significant and growing movement, rooted in the universities but possessing tendrils that multiply and grow ever deeper into the soil of American society. It has now penetrated the consciousness of major U.S. media, and the tale of the emergence of Design is beginning to play its own role as a powerful factor in persuasion. For example, James Glanz's April 2001 front-page story on Design in the *New York Times* was fundamentally a movement story. Glanz highlights Design as an emerging band of scientists—one that possibly poses a cogent or legitimate challenge to evolutionary theory. Glanz's various descriptions and his quotes of approval and disapproval dramatized and caused only slight problems for the plausibility of Design's own plot line. They did nothing to squelch it. On the contrary, the article was so fair and well balanced, it had the effect of *encouraging the consideration of the story.* Such textual representations in major media are growing more frequent, and they have legitimized the need of my deeper exploration of that movement story, sketching key characters, texts, interactions, and its emerging rhetorical consciousness.

At this point, the reader is acquainted with Design's major texts and interactions through my retelling of that story. In a sense, the historical watershed of this story came in the late 1990s, after the addition of Behe's molecular argument and Dembski's mathematical filter for the detection of design in nature. Here a turning point was reached—a dawning of the perception that a new genre had been born. Design was finally being seen as a separate intellectual and rhetorical phenomenon from creation science; it was being taken seriously. The precipitating event was Behe's *Darwin's Black Box,* with its explosion of excitement and brisk sales.[21] This led to Behe's extensive speaking tours in universities and the cascade of media coverage, punctuated by the string of five Design articles in the *New York Times* that emphasized Behe, including two invited columns written by Behe himself.[22]

This perception of legitimacy was reinforced and rendered nearly irresistible by the inherent attractiveness of the positive type of plot line of the rise of Design. A Darwinian (negative) telling of the Design story is a flat, boring, and increasingly implausible tale of grim siege by emotional, ignorant villains. Therein, Design scholars are construed as professors driven by "religious motivations," who are "fearful" of evolution, who aren't "patient enough" to wait for the answers to be found, who don't understand how science works, or who want to shut down science with a vision of a universe bursting with miracles.[23] By contrast, the tale of Design told from a friendly point of view is a story that is fresh, profoundly interesting, and fascinating—dramatic, in the most basic sense of the word. It seems to have what Walter Fisher described as "narrative fidelity," in ringing true to the experiences of our lives.

One does not have to hold already to some sort of divine intervention in nature to grasp this point. By thought experiment, anyone can imagine this story—that scientists might be found profoundly misguided in their decades-old pronouncement of "overwhelming evidence" for evolution—and in this imagining, one realizes that such a drama contains powerful human-interest themes. It is a moral tale of self-deluded blindness, long overlooked, but finally—at great sacrifice—brought to light. Also, it is a story with broad cultural impact.

I am not contending that merely the tale itself, however fascinating, is enough to convince. The arguments and evidence brought forth are absolutely critical—they must go on to clinch it. That is why, for example, Denton's book proved overwhelming in its rhetorical power. It linked the "new story" with a withering blast of scientific argumentation. For someone who is "pre-Denton" but inclined to be *open* to the movement story of Design, the tale is just plausible and fascinating enough to encourage a hearing of the supporting evidence.

Of course, to convinced evolutionists who are still unfamiliar with the specific arguments and evidences employed by Design, this imagined plot line will be virtually unthinkable, extremely unpalatable, or both. Others who already know the story and arguments of Design and who see them as horrendously seductive and potentially disastrous will view this movement with deep antipathy. Gould's pair of debates with Johnson—in person and later in print—are two out of scores of examples that could be given of this latter reaction. Design critiques will succeed in penetrating such hardened personal and institutional strongholds only with a drastic sociological and historical reordering of the current situation.[24]

While the defenders of the Darwinian paradigm—especially those who have spent their entire career in the field of evolutionary biology—are predictably the hardest to influence, even in prying out limited concessions to Design, the educated public is another matter entirely. This is where the sociological percentages come into consideration. The percentage of the hard core of evolutionary sentiment in the United States—those who believe in macroevolution *with no intelligent guidance at all*—has been measured by Gallup polls as relatively small. Over the past fifteen years Gallup has repeatedly listed the percentage of those who hold to a "recent creation" view (vaguely described by Gallup) to be about 40 to 45 percent, while the "God-guided evolution" view garnered another 40 to 45 percent.[25] The third category, holding to a strictly naturalistic evolution, in which there was no participation by a preexisting intelligence, has consistently hovered at or slightly under 10 percent.

Note that in the Gallup analysis, American adults have been split into a three-segment cross section, tilted decisively (eight to one or better) toward potential interest in Design's story and its scientific case. Phillip Johnson frequently quotes the Gallup figures, arguing that the "view of the nine percent" is enshrined as textbook orthodoxy.[26] However, the most significant figure for the rhetorical landscape of Design is not the nine percent but the larger figures combined. Already, nearly half of the American people hold a recent-creation position (they are implicitly friendly to Design), and nearly another half—holding a God-guided view of evolution—are potentially open to the story. They are generally "pre-Denton," and many of this group will begin to listen to the story of Design in the coming years.

Let me explain what I mean by "pre-Denton." Those who believe in "God-guided evolution" are theists who assume the scientific accuracy of the story of macroevolution presented in biology textbooks, in *National Geographic,* and on nature programs on TV. This 45 percent of America who *assume God worked through evolution* are candidates for conversion to Design-type skepticism. Any such conversion would be driven through scientific rhetoric—that is, through exposure to Design critiques by reading a book or article, hearing a video or speech, or encountering a combination of such texts over time. In this group, religious belief is not the driving factor for change but rather is a permissive factor in the sense that such people have the metaphysical space to consider Design as a possibility.[27]

In this group that holds to God-guided evolution, conversions to a Design perspective would appear (ironically) as the opposite of a stereotypical religious-type process of persuasion, which supposedly involves mystical or intuitive experiences or the acceptance of the authority of ancient writings or a religious figure. Rather, it would be more accurate to describe this Design persuasion (such as that which Behe himself experienced) as "empirical-intellectual," engaging a range of practical reasons, focusing on (1) logic and empirical evidence, (2) the reality and role of philosophical assumptions, and (3) the history of science with its proliferation of narratives. Of these three types of persuasive reasons, it is the narrative-historical that frames the decision and makes it understandable and plausible; the other two clinch it. All three types of argument are woven into Design texts, but it is narrative—both factual stories and projections—that is a crucial and yet highly underappreciated dimension, begging for analysis.

The Foundational Level: Factual Scientific Narrative

At the outset I proposed that there are two levels of narrative at work in the controversy between Darwinism and Design: factual stories and projection themes ("fantasy themes" in rhetorical theory).[28] Later, I further subdivided the level of factual stories (which strive for factual fidelity) into three types of narrative that are being created or destroyed in any branch of cosmological science that is under heavy fire:

1. *Genesis or cosmological narrative,* which is the ancient cosmic-history narrative that is in dispute. This is always a fierce battleground in such rhetoric.

2. Human narrative tracing the *history of science*—how we have arrived at our present state of knowledge on a given topic by progressive discovery and rethinking of old views. The story of Intelligent Design and the HEB or "History of Evolutionary Biology" are both special versions of this category of narrative.

3. A variety of micronarratives called *individual or personal stories*. They often focus on a single scientist's life story or even one single statement, an experiment, or a dialogical encounter (Patterson's embarrassing question comes to mind). Other personal narratives, which I called *incident stories,* involve several persons as protagonists (e.g., the British Museum story that dominates chapter eleven in *Darwin on Trial*).

These categories of stories have been shown to play distinct and important roles throughout the history of Design—roles that changed hardly at all over the various stages that have been surveyed. For example, the Darwinian cosmological story is subjected to an equally thorough shredding at each stage—by Denton, Johnson, Behe, and now Wells. A rare variation in this shredding is Behe's acceptance of common ancestry. The fact that he provisionally accepts common ancestry and yet remains a star in good standing shows Design's flexibility in tolerating members' evolutionary beliefs on certain topics.[29]

How is the attack on Darwinian cosmology being carried out? Again, there is a remarkable consistency in strategy that has been shared by Denton, Johnson, and Behe. All three separate microevolution (which they grant as plausible) from macroevolution of new complex systems (which they radically question). The questioning process proceeds down two avenues: First, Denton and Johnson seek to show that there is no empirical (fossil) evidence that the morphological chasms *were* bridged, step-by-tiny-step, moving along the putative macroevolutionary lineages. Second, all three argue that the chasms between two different structures in the same lineage (or the path to Behe's irreducibly complex structures) cannot possibly be conceived in terms of hypothetical intermediates; the attempted thought experiments go nowhere. In connection with this second line of reasoning, all three also invoked in their books the crucial test of Darwin, saying that if one can find an organ that could not possibly have developed step-by-step, "my theory would absolutely break down." At this point, a clever narrative argument is drawn out: Darwin proposed this falsification test in 1859, and now, applying that test, we should accept the results and conclude that Darwin's theory is falsified.

I have emphasized the many human factors in the creation and reception of cosmologies, whether evolutionary or Design.[30] Now I need to counterbalance these human factors with the decisive importance of what Gross called the "brute facts" of nature, or what Burke called the "recalcitrance of nature." With the moderate rhetoricians, I argue that Design discourses, intent on presenting true claims, *are not rhetorical without remainder* but are responding to a network of stubborn realities that drive and constrain the products of the movement.

Nature's recalcitrance is the ultimately determinative factor (or limiting factor) in the shredding, shaping, or vindicating of any cosmological narrative.

The reader will have noted that certain stubborn realities of nature keep coming up time after time, and they serve as the main fuel in the evidentiary debate: (1) the Cambrian explosion, now underscored and heightened in the recent discoveries in China, (2) the general absence of transitional fossils between the higher taxonomic categories outside of the Cambrian, (3) the cell's molecular systems of breathtaking complexity, recently elucidated, and (4) the quiet experiment-driven collapse of confidence in "chemical soup" scenarios for the origin of life.[31] These four sets of brute facts, of course, mean nothing by themselves, as Gross has reminded us; they must be interpreted and woven by rhetorical skill into some explanatory scheme. Yet, such facts are ultimately stubborn; they surprise and confound; they often trigger a heightened sense of mystery. They are the stuff of anomalies, which of course in the Kuhnian vision of science may lead eventually to a genuine paradigm crisis.

The four scientific realities cited above (a list that could easily be expanded) cannot be ignored in their foundational role as rhetorical weapons in the hands of Design. They and their range of possible interpretations have become the turf on which some of the fiercest battles are now fought. So the recalcitrance of nature, I propose, is the foundation of the Design assault on the Darwinian paradigm. In dramatic terms, it is cast as the ultimate protagonist in all of the major scenes. Phillip Johnson refers to the power of recalcitrant nature (without using that name) in his "sinking ship" projection theme, found in the "How to Sink a Battleship" essay in *Mere Creation*. The key reference to recalcitrance is in the final line—"reality." To appreciate the context, I will quote the entire projection theme:

> When I finished the epilogue to *Darwin on Trial* in 1993, I compared evolution-ary naturalism to a great battleship afloat on the ocean of reality. The ship's sides are heavily armored with philosophical and legal barriers to criticism, and its decks are stacked with 16-inch rhetorical guns to intimidate would-be attackers. In appearance it is as impregnable as the Soviet Union seemed a few years ago. But the ship has sprung a metaphysical leak, and that leak widens as more and more people understand it and draw attention to the conflict between *empirical science* and materialist philosophy. The more perceptive of the ship's officers know that the ship is doomed if the leak cannot be plugged. The struggle to save the ship will go on for a while, and meanwhile there will even be academic wine-and-cheese parties on the deck. In the end the ship's great firepower and ponderous armor will only drag it to the bottom. Reality will win.[32]

The deciding factor in the defeat of the battleship—in Johnson's view—is the power of "reality," which is, at root, recalcitrant nature itself. This is also what he has in mind in recent writings when he says that materialist philosophy and the empirical evidence (i.e., recalcitrant nature) are going in opposite directions.[33]

Conversely, recalcitrant nature will prove to be the key to any effective defense of Darwinism in the years to come. This is exemplified in Kenneth Miller's energetic attack on Design in the PBS debate, using charts showing recently discovered fossil lineages. This same power of a recalcitrant nature goes a long way to explain why, for example, Johnson would receive support from one of the world's leading evolutionary paleontologists.[34] In summary, in the Design controversy, all of the myriad social and narrative realities are indelibly stamped with the marks of nature's recalcitrance.

What Design cosmology takes the place of the shredded Darwinian one? Only a sketch is put in the place of the evolutionary narrative. There is little consensus on the details of the origins of different biological kinds, although most Design advocates hold to various forms of a "progressive creation" view, which implies that a creator periodically introduced new creatures into the earth along the conventional timeline. The common denominator in virtually all sketches of the Intelligent Design cosmology is the *introduction of personal agency*—the activity of a preexisting intelligence, which most privately identify with the "God" of their own worldview.[35] At this point a Design caveat is generally voiced: Science can only determine if a system is designed or not; it is not well equipped to ferret out the *who* of this designer on its own. Other research disciplines need to be consulted to move further into this question.

This lack of a coherent, detailed cosmology has proved to be both a boon and a burden for Design. It is a boon in that it accommodates such a wide range of views, and the very mysterious nature of the story projects an aura of excitement in the quest of knowledge. The burden, which is acknowledged by Denton, Johnson, and others, is that science (at least in the Kuhnian vision) abhors a vacuum, and this lack of a detailed story is construed by Darwinian critics as a lack of a coherent alternative paradigm.

I explained earlier that the second and third types of scientific narrative are closely related, as limbs jutting out from the trunk of a tree. The trunk is the history-of-science (HOS) story, and the limbs are the individual stories that flesh out the details of the trunk. This is where Design's primary act of *narrative creation* is taking place. Design theorists concentrate much of their effort in undermining the old textbook history of evolutionary biology (Johnson's anti-HEB argument in chapter 5). Simultaneously, they are reconstructing a new HOS, and we saw Denton, Johnson, and Behe all excelling as storytellers, stringing together a variety of scientific debates, shocking admissions, scandals, conversions, heresies, and other kinds of scientific narratives that contribute to a new Design HOS. This HOS building, which reaches back into Darwin's day and earlier, is progressive, cumulative, and integrated not only with the debates and texts of Design but also with the responses they have received.[36] One surprise in the HOS building after 1996 was a noticeable shift in Design discourses—from a predominance of controversy stories that cause problems for Darwinism toward a greater prevalence of positive-HOS discussions of Paley's design arguments of the early 1800s. Paley's watchmaker argument, which I discussed in chapter 4 in connection with Richard Dawkins, is widely seen in modern science as a thoroughly discredited argument that was defeated in the 1800s. Since the current Design theory deploys arguments similar to Paley's, Design theorists starting with Behe decided to fully vindicate the story of Paley.[37]

Before leaving factual narratives, I need to point out that we have stumbled upon another vast rock in the soil of scientific rhetoric—the second, unrecognized dimension of recalcitrance that I alluded to earlier. Recalcitrant nature is not the only scientific reality that is exerting a powerful influence on Design rhetoric. Another order of reality, equally recalcitrant and unyielding, is the *set of personal or incident narratives* that either undermine confidence in Darwinism or vindicate the legitimacy of Design. In a Kuhnian frame, one can class these stories (many of which are controversy stories) as social-scientific anomalies. I am referring to published texts such as that of Pierre Grassé or the Wistar Symposium, or speeches such as Colin Patterson's in 1981, or spoken words of support such as David Raup's in 1989, or public modifications of prior positions such as Michael Ruse's in 1993. All of these (and much more) fit into this class of recalcitrant stories or snippets of narrative.[38]

Such stories, behaving like the stubborn brute facts in nature, are not moldable like clay; they are hard like chunks of marble. Again, this is not to deny the role of rhetoric in describing and making inferences from these stories. Clearly, there is a necessary place for the interpretation of these stories and for artistic invention in dealing with them. With dogged effort, creativity, and ingenuity, these narratives can be harmonized with a Darwinian HEB (or at least their damage can be minimized). However, the stubborn narratives referred to here, as they are fitted together in the flow of Design's new HOS, have a unique and virtually impenetrable firmness, especially when they are blended with the stubborn rocks of nature's own recalcitrance that seem to falsify macroevolution.[39]

The Superstructure: Projection-Chaining

Projection themes are dramas that "project out" from the facts, mixing fact with faith. As a social movement with ambitious goals, Design would be expected to produce, cherish, and proliferate such projection themes.[40] This is exactly what I found, in great profusion, scattered through published texts and oral presentations. I highlighted many of these, which generally connect with visions of ultimate paradigm victory sometime in the first decades of the twenty-first century.[41]

Projection themes, as stories, seem to build on a prior foundation of factual stories. For example, Johnson's sinking ship story and Behe's eureka fantasy of the biologists celebrating the discovery of design in nature are dependent for their power on the strength of the prior work done on the other side of the spectrum—work in factual scientific narrative (including the arguments and evidences embedded in those scientific narratives). In this picture of the two levels of narrative, factual stories are the foundation; projection themes are the superstructure.[42]

The most permanent and formative aspect of projections is their chaining out and convergence into a rhetorical vision—a movement's collective consciousness as it presses toward common goals. I set out to explore the intellectual content, attitudes, and values embedded in Design's rhetorical vision. Before I turn to that vision, let me tweak the theory and say that the rhetorical consciousness of Design was formed just as much (if not more) by its factual scientific narratives, laced with empirical evidence and arguments, as it was by fantasy themes. In other words, the rhetorical vision derives from both levels of narrative. Why is this so?

Projection themes have their own powerful, socially binding, and motivating effects. Yet these themes tend to be highly vulnerable in a scientific movement, which places such a supreme value on compelling evidence. As a result, the strength of these projections in a movement of scientific reform is heavily dependent on and derivative from the empirical rhetorical resources. If there exists within Design's rhetorical vision—as I claim—a deep conviction about the bankruptcy of Darwinian truth claims, then the core of that conviction will be composed first of cogent arguments and patterns of compelling evidence, all set within a verifiable framework of scientific narrative. This conviction will not spring simply from projections, with their imaginative hopes for the future. Projection themes are secondary here, not primary.

The Rhetorical Vision of Crisis and Paradigm Shift

That being said, there has emerged from factual narrative (including embedded arguments) and from projection themes a robust "rhetorical vision of Design." To a certain extent this vision reflects the movement's own self-image. But the movement story is not just about a string of debates, books, conferences, and ongoing discussions. In other words, the movement's consciousness is not composed of *what people are engaged in now,* as they carry out acts of coordinated persuasion. Rather, the vision is rooted in an *informed, reformist passion,* culminating in a self-consecration that is both intellectual and visceral.

The rhetorical consciousness is rooted in a sense of profound disgust about widespread gross malpractice in teaching about biological origins. Evolutionary biology, from Johnson's perspective, is applied metaphysics.[43] Design's rhetorical vision is therefore rooted in a perception of evolutionary biology as a field whose scientific content, at the level of macroevolution, is deeply interwoven with unprovable philosophical assumptions and whose integrity is plagued by unrepented patterns of deception and even fraud. This attitude of intellectual contempt (directed toward pronouncements and practices, not people)[44] leads to a reformative urgency dedicated to changing the way the subject of origins is taught and researched.

This drive to reform leads to clearly envisioned goals and to agreed upon strategies for attaining those goals. In fact, the main vision of Design goes well beyond biology. It is an alliance with all who are prepared to question both the claims of Darwinian evolution and also the philosophical doctrine of naturalism that is said to undergird those claims. After Johnson's second book, *Reason in the Balance* (1994), Design became committed to broader goals than just scientific reform. Its vision expanded to take in a second major objective—to subvert the assumptive base of all fields of research where naturalism reigns.

Here we see Denton's anti-Darwinian radicalism raised to a higher level, to a crusade against the domination of a worldview that assumes that God, in the production of new knowledge, can be assumed to be "effectively imaginary or at least uninvolved and undetectable." Academic work, in this naturalistic picture of sound rationality, need not take into consideration even the *possibility* of a mind that transcends the cosmos. The goal of Design was not to rewrite the script in academic life so as to insert a role for deity at any given point but rather to secure the freedom within the academy to range across all possible explanatory perspectives, including the one that entertains the possibility of God's existence.[45]

Consider this claim of Design: "Evolutionary science, in its Darwinian incarnation, is devoted to explaining how complex biological realities came into being without the participation of a preexisting intelligence." Amazingly (and revealingly), this is widely taken as a proven historical fact on both sides of the debate. It is a point on which William Provine, Richard Dawkins, Richard Lewontin, and Daniel Dennett would agree with Johnson and the leaders of Design.[46] In light of this crucial agreement between these Darwinists and Design theorists, one noticeable change that Design has already achieved is to establish among evolutionists a realization that a tactical blunder is being committed whenever Darwinists use their evolutionary findings to promote atheism. Even Eugenie Scott now frequently corrects and warns against all such metaphysical preaching by scientists.

In working for the replacement of the Darwinian paradigm, Design activists have now for a decade been waging intellectual war in the universities. There is no truce in view, and the level of rhetorical conflict is mounting steadily. Clearly we have entered a *rhetorical crisis* for the evolutionary community. This rhetorical crisis could conceivably grow within the next few decades into a generally perceived *paradigm crisis*. (But can such a crisis, if Kuhn is right, even be known clearly as such when it is in the process of building up?) If this happens, such a Kuhnian crisis would eventuate as an exceedingly unusual crisis. The term "paradigm crisis" in this case (and the subsequent revolution) would turn out to be a weird understatement, almost a euphemism, if science—and university research in general—began to permit researchers to take into account the genuine *possibility* of a "creator" who is relevant to the cosmos and to humanity.

Such a change would be more than a change of paradigms in natural science. It would effect a new global research consciousness, a new ethos, which would reverberate throughout the social sciences, arts, and humanities. It would be, in short, a *new kind of life of the mind.*

Design prides itself in having a few nontheists among its ranks, yet its rhetorical vision is centered around the reinstatement of the *possibility* that some intelligent designer of the biosphere exists and can be detected in some basic but rigorous way. Accordingly, its discursive defiance is directed against a basic epistemological rule in universities. This unwritten rule says: "Don't worry about the possible existence of a creator. We must, after all, act as if he or she does not exist if we are to maintain firm analytical footing and clear-headed rationality." The verbal sabotage against this rule is perhaps the most revolutionary and interesting aspect of Design as a rhetorical movement. It aspires to restructure some of the rules of rationality in the world's universities, in the guilds that are dedicated to the production of new knowledge and the resifting of old knowledge. Here the Design debate connects with the theistic implications that have emerged from the big bang theory and the "fine-tuned universe." Indeed, Design's theistic-friendly orientation invites the connection with any field that is discussing the reasons for believing and disbelieving in a creator or a deity of any kind, whether philosophy, psychology, theology, or anthropology.

Johnson and his cohorts are not, they claim, seeking to *impose* a theistic worldview in these arenas; indeed, they charge that such an *imposed religious worldview* is the very situation that is in place now, which they seek to overthrow. Rather, they promise—if successful—to end the domination that forbids the serious consideration of the plausibility of a theistic perspective.

The Rhetoric of Metaphysical Negotiation

I have made it clear that the chief villain in Design's metanarrative is not personal but philosophical: *metaphysical naturalism.*[47] Design theorists accordingly were startled by Michael Ruse's admitting the role played by "unprovable philosophical assumptions." This qualified him as a strange antagonist—one with laudable intellectual honesty. While not rejecting either Darwinism or naturalism, Ruse admitted the vital connection between them. In recent published writings, Ruse has maintained his criticism of Design in a strong but very respectful mode. Quite typical of his engaged stance is his blurb on the cover of William Dembski's *No Free Lunch,* published in early 2002: "I disagree strongly with the position taken by William Dembski. But I do think that he argues strongly and that those of us who do not accept his conclusions should read his book and form our own opinion and counterarguments. He should not be ignored."[48]

On the other hand, Stephen Jay Gould strenuously denied Darwinism's strong commitment to naturalism, both in his review of Johnson and in other texts, such as those dealing with Gould's "NOMA" (Non-Overlapping Magisteria) proposal. Johnson's initial reply to Gould, an essay entitled "The Religion of the Blind Watchmaker," sought to show where Gould was contradicting himself on this point. Johnson quoted a string of Gould's previously published statements, saying how Darwin showed modern man that God's hand cannot be seen any longer in the works of nature.

Two key observations are appropriate here: (1) Metaphysical assumptions of scientists about the existence or nonexistence of God play a crucial role in their thinking about what sorts of explanations can be considered to explain natural phenomena and which ones cannot. (2) In spite of the power of these metaphysical or worldview beliefs, scientists who enter into genuine dialogue over these issues have shown their ability to understand and appreciate the other side and even to enter into a careful process of change, entailing crucial repositionings.

Let me expand on the first observation. Just as recent polls showed that strict Darwinism is held by only 9 percent of the general public (and most of this percentage of respondents would probably view themselves as nontheists of one sort or another), a special poll of biologists in the National Academy of Science showed nearly an inversion of this percentage—about 95 percent say that they do not believe in God.[49] Just as one might expect a receptive audience for Denton-type scientific rhetoric in the general public, one would predict the opposite—an extremely chilly attitude or even a rapid hostile rejection—among NAS biologists. This makes the narrative and projection theme of "paradigm crisis" much more tense and dramatically unpredictable, given the unprecedented situation of a proposed paradigm shift being clearly entangled with the religious worldview of the cognitive elites. Even ordinary scientists, of whom about 40 percent believe in God, would be expected to be somewhat more receptive to Design theory compared to the top biologists in the NAS.[50]

Nevertheless, as I said in the second observation, there is a remarkable potential of persons to change their minds through a close encounter with an articulate defender of another worldview. I am not speaking of persons changing paradigms but rather revising the negative image of their opponents. As I traced the story of Design, I was keenly alert to any changes that took place in the defenders of the status quo as they engaged the arguments and ideas of the skeptics. I have described several occasions when such encounters fostered subtle, inexorable, and even remarkable change in points of view. One of the best examples, of course, is Ruse's announced change of position on the role of Darwinism's philosophical assumptions.[51]

This kind of change was rooted in several communication phenomena taking place at different levels simultaneously. In the case of Ruse, there was a very congenial set of personal interactions at the Darwinism Symposium in 1992.

In the private setting, this involved small talk, storytelling, and laughter over meals, along with correspondence and e-mails with Johnson and others. In the public arena there were scientific papers and responses and open dialogue among the ten invited speakers. The AAAS talk by Ruse a year later, which shocked so many, and the dialectical change this interaction produced (see chapter 7) were not a sudden development but a slow rethinking of old issues and questions. This is the essence of such subtle repositioning. It typically proceeds in many steps: first, exposure to the ideas (initially producing a flat rejection); second, personal interaction with the proponents of those ideas; third, substantive exchange of mutual criticism; fourth, an honest attempt at reassessing one's own position in the light of that criticism; fifth, the making public of one's repositioning—to the extent that it is feasible, given one's personal and professional risks in doing so. I have sketched these steps with Ruse's own repositioning process in mind.

In the rhetoric of science field, I recently observed a similar striking process of close encounter between evolutionists and Design advocates followed by profound rethinking and tactical repositioning. This took place in the "Special Issue on Intelligent Design," published in late 1998 in *Rhetoric and Public Affairs*. (For my analysis of this issue, see appendix 4.) This issue of a scholarly journal became a dramatic microcosm of what is starting to happen more broadly in the academy at large as Design is taken seriously. This superb collection of articles, a "symposium in print," was launched with an introductory article by John Campbell on the formative role of Paley's design theory in Darwin's own thinking. This is followed by four articles by leading members of the Design community (with their criticisms and proposals), and the heart of the issue is a string of seven responses and rejoinders—five from rhetoricians and two others from Phillip Johnson and Bruce Weber, an evolutionary biochemist.

The two most weighty and powerful contributions in the string of responses are the pair of articles by rhetoricians David Depew and John Lyne. Their comments show a careful, respectful engagement with the substance of Design's critique, a shrewd consideration of the rhetorical ramifications, and a balanced analysis of proposals for allowing biologists to entertain intelligent causation. Their essays are models of responsible engagement with Design in the academy. Neither Depew nor Lyne was "converted," but both have already—in thought experiment at least—begun to think through the various advantages and disadvantages of the revamped sort of curriculum in which Design's criticisms and perspectives would be entertained. This issue of *Rhetoric and Public Affairs* is a sneak preview of what could and should happen, and what I suggest eventually will happen, as the academy engages Design theory in the coming years. Its articles model for us the available modes of social thought and communication; they help create novel and better ways to manage this delicate borderland between metaphysical advocacy (on either side) and the ideal of metaphysical neutrality or pluralism.

"Yes, It's Important, But (Gasp) Never in Public!"

In one of Johnson's earlier articles, before he took up the topic of evolution, he drew a comparison between the Victorians' skittishness about sex and the modern academy. The Victorians knew that sex was important but were embarrassed to talk about it in public. In a similar way, said Johnson, academicians in our day know that the nonempirical issue of "which values will dominate" is important, but they are rather hesitant to discuss such issues publicly.[52] What Johnson wrote earlier about a Victorian attitude toward values now applies equally well to the role of metaphysics in science. Johnson's primary project (and the project of Intelligent Design) is to pry such embarrassing "god talk" out of the private sphere into public discourse. (The twin "god terms" of *metaphysical materialism* and a *monotheistic God* are functionally equivalent.) Design seeks to "out" from the academic closet all the relevant gods and god terms in the religious and metaphysical context of all genesis debate.

The rhetorical struggle of Darwinists (especially those like Dawkins and Eugenie Scott who are self-described atheists) is to persuade the public and their scientific colleagues that *Design is really a subtle move to sneak a religious worldview through the back door of biology.* The rhetorical struggle of Johnson, Behe, and Dembski (all theists) is to persuade the public and institutional science that *Darwinism is already implicitly theological, grounded ultimately upon a commitment to naturalism rather than to empirical evidence.*

Eugenie Scott expressed this metaphysical tension rather eloquently in an NCSE fundraising letter that I cited in the first section of chapter 1: "But when a topic has religious implications, such as the big bang, the origin of life, or the origin of human beings, they don't want us to look for natural explanations. They claim these phenomena are 'too complex'; they claim that they couldn't have happened 'by chance' . . . ; they claim that 'intelligence' must have been involved. 'Intelligence,' of course, means divine creation, a subject outside of science."

At this point, the reader will easily spot the error of fact when Scott says that "they don't want us to look for natural explanations." Design builds on this very foundation—the negative results of diligent search for such explanations! In fact, the explanatory filter of Dembski *demands* that a natural explanation be sought first. Yet for our purposes, the key words here are Scott's last ones, "'Intelligence,' of course, means divine creation." Through rhetorical engagement, the origins debate has evolved to the place where metaphysics is irrevocably out on the table, along with its constraining influence on scientific conclusions and truth claims.

Both sides propose radically different ways to manage these issues. Behe and Dembski propose that science itself has been developed, with the assistance of mathematics and statistical reasoning, to the place where the action of any intelligent agent[53] can be reliably detected, and that the historical sciences (such as biological origins) can function far better, and with no political, social, or intellectual sin committed, if they permit intelligent causes to be considered alongside natural causes. Darwinists, on the other hand, are determined that this change will never happen on their watch. They argue that the current naturalistic epistemology of science must be firmly maintained, or science will cease to be *science* and will be ravaged and polluted by the encroachments of religion.

We have arrived at the very crossroads of a demarcation question—*what is science?* The emerging consensus in academia now is that this question is much more difficult, subtle, and vexing than we ever imagined. It is dawning on many in the sciences and its related fields that this crucial question is not just a pristine scientific or philosophical problem. (See my essay in appendix 4, in which I probe recent work by rhetoricians and philosophers of science on this question.) As a result, this demarcation question is growing into an intense sociopolitical and rhetorical struggle over the appropriate role of metaphysics in guiding the interpretation of empirical evidence.

Of course, this discussion and ongoing negotiation—the role of the research-er's philosophical perspective in predetermining his conclusions—is not just an isolated rhetorical or social squabble. It constitutes a new chapter in an even broader and more encompassing level of global narrative. From this larger framework, the Design versus Darwinism debate is a key chapter in, for example, "The Narrative of Post-Renaissance Worldview Conflicts," or "The Narrative of Progressive Enlightenment in Western History." Or, to place this Design chapter in a larger frame akin to my history-of-science narrative, it is part of "The History of Rationality."[54]

Here at the battlefront, where rationality is being defined and worldviews are assessed in accordance with the prevailing criteria and norms of rational-ity, is where the most practical conclusion of this book emerges. The Design discourse genre—the genre that combines the Design genre and all thoughtful responses it has evoked from scientists, philosophers, and rhetoricians—serves as a provocative and exploratory model for fruitful conversation and negotiation within a wide range of metaphysically driven conflicts that have been tearing the American social fabric in the postmodern age.[55]

The post-Victorians eventually learned to talk about sex—*how well* is a separate issue. The contemporary academic culture, through these ongoing rhetorical innovations forced by the campaign of Design, is learning about new motives, new data and conceptualizations, and new modes of open and morally responsible discussion about cosmology and metaphysics. More than that, nearly all Darwinists who participated in these discussions surrounding the issue of the possible intelligent design of the universe found them *decidedly enjoyable,* from both a personal and an intellectual point of view.[56] It appears that the participants who found such discussions bitter and intolerable are those who, like Gould, sought to illegitimize the discussion itself, and—failing in that attempt—later withdrew into a world of their own rhetoric, constructed out of the raw material of crucial distortions, in which (to quote Gould) the only dissenters from macroevolution are "Protestant fundamentalists who believe that every word of the Bible must be literally true."[57] In playing a role of denial and withdrawal, Gould succeeded only in rewriting his own place (to his own detriment, tragically) in the unfolding drama. He was not able, by the sheer force of his symbolic assertions, to transform the drama to his liking. He could not stem the tide of Design's engagement with educated Americans and with the scientific elite.

To be jarred awake to the powerful role that metaphysical commitments play in cosmology—especially their buttressing and protecting the Darwinian knowledge-claims—can be upsetting and unpleasant, but the cognitive health benefits are enormous. This awakening to the power of hidden metaphysical foundations is the most inevitable and immediately beneficial sort of "arousing from dogmatic slumber" that Design could achieve. Through such a rude awakening, the power of reason is newly energized. Indeed, such basic notions as rationality, intelligence, and academic freedom are given new impetus, a new breath of life. The rhetorical history of Design, which I have described and analyzed, shows that this momentous awakening has already begun.

Afterword

Speculation about "intelligent design" behind the world around us dates from the first time our ancestors noticed that their surroundings were comprehensible. The regularity of the seasons and its relationship with the growth of plants and the migration of animals seemed to provide clear evidence for intention behind it. Indeed, the *appearance* of design was never an issue through the times of Charles Darwin, or even to the present. Rather, the challenge to Darwin and his contemporaries was to explain apparent design without resorting to a "designer."

The brilliant concept of natural selection seemed to fit the bill. Those best suited to their environment would have the greatest chance for survival, and with time their progeny would predominate. There is no "designer" here, only the inexorable and inevitable process of response to environment. Such a mechanism requires vast amounts of time, and the geological record supported this. The great antiquity of the earth was established by a variety of analyses involving physics and chemistry.

With this foundation in the most basic of the sciences, the intimate relationship between Darwinism and science seemed assured. A challenge to Darwinism was thus a challenge to science. Such challenges could be made on religious grounds, for example, but they had to be recognized as such and not based on science. Young-earth creationism conveniently fell into this mold. It was clearly related to a literal reading of the Bible, a religious document. It also appeared to confront the very core of conventional physics and chemistry, the validity of which is proved beyond question with all the technological conveniences and inconveniences of the modern world.

What then of the new "Intelligent Design Movement?" The conundrum lies in the fact that its adherents are not questioning science in the way of young-earth creationists. The advocates of Intelligent Design, or ID, generally accept the evidence for the antiquity of the earth and, indeed, agree with the conventional wisdom common to nearly everyone in the natural sciences. Within this framework, however, are questions which are still unanswered. How did life originate? What gives rise to the complexity we observe in living systems?

The conflict lies not so much in the questions themselves but in the approach permissible for answering them within the context of science. On the one hand, the adherents of Darwinism or, in a broader sense, a naturalistic worldview would claim that only answers relying on "natural causes" are permissible. In contrast, the proponents of ID would assert that the very process of looking for natural causes could lead to the conclusion that naturalistic explanations are the least probable. To draw from the story line of countless detective novels, perhaps we can find out if what happened to our "victim" was due *either* to natural causes *or* to the work of an "intelligent agent."

As with the crime investigation, it clearly does matter whether or not there is "design," and by implication a designer, behind what we observe. For those who wish to exclude *a priori* the possibility of an "intelligent agent," the motive is based not on "just the facts" but on philosophical prejudice. This is a difficult admission for many Darwinists. They feel that it is those who question Darwinism, and they alone, who show philosophical or, equivalently, religious preference.

The confrontation has thus been turned by Darwinists into one between science and religion. This explains their recurring tendency, for example, to make all challenges to Darwinism equal to young-earth creationism. They can then invoke physics and chemistry to discredit a young earth and, by implication, anything that questions Darwinism. Indeed, this justifies ignoring *any* challenges to Darwinism. To do otherwise is a waste of time, since opposing Darwinism is then equivalent to opposing physics and chemistry, which are unmistakably science. The error of any challenge to Darwinism is thus a foregone conclusion. At best, it might offer the kind of solace religions do, but it certainly has no scientific value.

Such ways of thinking can be endlessly frustrating to those critics of Darwinism who are not young-earth creationists. Woodward helps us understand these attitudes by explaining their importance to the *rhetoric* of the controversy. Rhetoric is the study of the principles and rules of expression. It has been appreciated from antiquity to the present, especially in the legal profession, but many of us in the sciences have a prejudice against it. We believe in the primacy of "just the facts," and in the physical sciences we are accustomed to verifying them by checking mathematical predictions or repeating experiments. We are at a loss, however, when there is real controversy but there are no calculations or data that are in dispute.

The debate between the adherents and critics of Darwinism is not in the mathematical language of the physical sciences but in the framework of two competing narratives, as Woodward reminds us. On this basis, Darwinism and ID are both in the same arena. If Darwinism wishes to have the better claim for being science, it would have to be based on a restricted definition that mandates quantitative mathematical arguments as in the physical sciences. Because Darwinism fails this criterion, it would have to join the young-earth creationists and everybody else Darwinists would cast outside the pale of science.

On the other hand, if science permits the narrative approach, then both Darwinism and ID fall once again in the same category. Woodward's focus on the rhetoric of science permits this bigger "tent" for science and, interestingly, provides a justification for classifying Darwinism as well as ID under science. The "price for admission," however, is the acknowledgment that the only basis for excluding one or the other from science is philosophical choice.

Andrew Bocarsly, professor of chemistry, Princeton University

Robert Kaita, principal research physicist,
Plasma Physics Laboratory, Princeton University

Appendix 1

Notes on the Berkeley Faculty Colloquium of 23 September 1988

DICTATED BY PHILLIP E. JOHNSON ON 26 SEPTEMBER 1988

I am dictating my recollections of the faculty colloquium that was held on Friday, 23 September 1988, and at which I defended the August 1988 version of my paper on "Science and Scientific Naturalism in the Evolution Controversy." There were about twenty persons present, the most active of whom were the nonlawyers. These included two first-class philosophers, who defended scientific naturalism on philosophical grounds. They were David Lyons of Cornell and Jeremy Waldron of Oxford and our own faculty. Also present at my invitation were Professor Sherwood Washburn of Anthropology, an emeritus professor of human evolution with an enormous reputation, and two professors from the Biological Sciences Division. They were Thomas Duncan (Botany) and Montgomery Slatkin (Zoology, Evolutionary Biology). I invited Duncan because he serves with one of my law colleagues on the Graduate Council and is a leader in designing the biology curriculum here. I invited Slatkin at the suggestion of William Provine of Cornell, who said that he would be a particularly effective critic of my paper. The turnout among the law faculty itself was relatively light because of conflicting commitments and also because I insisted that all participants read the entire paper and notes as preparation. Most of the law faculty who did come participated very little in the discussion. Two of my colleagues who have told me they agree with my position were present but generally silent; three of four other colleagues who have said they were persuaded, did not come. Professor Paul Carrington, formerly dean

214

at Duke University, was present as a visitor; he is also favorable to my thesis and participated briefly in the discussion. Paul Ruud, a newly tenured associate professor of economics, was also present as a visitor; he also takes my side, and his economics scholarship is heavily involved in assessment of empirical research and its relationship to concrete reality. The questions from members of the law faculty who did participate mainly dealt with the legal-political question of just what adjustments in present law would satisfy me.

I began the discussion by saying that I would assume everyone present had read the paper and therefore would only state its thesis briefly. I started with the title, which focuses attention on the difference between scientific evidence or scientific method on the one hand and the philosophy of scientific naturalism on the other. My argument was that although most people believe that an enormous amount of empirical evidence supports the general theory of evolution, this is in fact an illusion. Most people in the intellectual world are certain that evolution must be true because it is the only tenable naturalistic explanation for the development of complex life, or life in general, and it therefore must be true if nonnaturalistic explanations such as creation are ruled ineligible for consideration. The evidence is then built up upon this preexisting theoretical certainty based on philosophical presupposition. Nonevolutionary explanations of the evidence are not considered, and therefore the evidentiary support that seems to exist is the product of the cultural certainty rather than its cause or support.

I explained that my thesis could be attacked in three ways. First, someone could deny that the official doctrine of evolution is in fact based on a conclusive presumption in favor of scientific naturalism. I doubted that anyone would want to deny so obvious a fact, however. Second, one could defend scientific naturalism on philosophical grounds, arguing that it is a very good philosophy and we are therefore justified in assuming it to be true. The problem with taking this line is that it removes the question from the area of biology or scientific expertise and puts it firmly in the camp of philosophy, where scientists cannot claim to be experts. Third, science could offer evidence that is capable of proving the truth of the doctrine of evolution without support from the philosophical presupposition that only naturalistic explanation can be considered. I doubted any such evidence exists but expressed myself as eager to hear of any.

Most of the ensuing discussion took the second path, arguing that the presumption of naturalism is philosophically valid and that any other approach was tantamount to irrationalism or mysticism. Professors Lyons and Waldron were particularly aggressive in taking this line, but the biologists also spent more time arguing philosophy than scientific evidence. The philosophers insisted on labeling the alternative to naturalism as "magic," in what I considered a transparent effort to prejudice the inquiry from the start. I protested this briefly but then let the matter pass. Lyons argued that all science and not just evolution is based on naturalism and that the commitment to naturalism is what makes science possible. Basically the strategy here was to insist that by asserting "magic" as a legitimate possibility for study I was making scientific inquiry impossible, since such inquiry must presuppose uniform physical laws. How do we know that an experiment performed yesterday will have the same result today, that a ball will not start falling upward tomorrow, etc.? Thus my position was characterized as an attack upon science generally and a form of irrationalism. Of course I denied this, claiming that my attachment to the scientific method was in fact greater than that of my critics, since I refused to rule out certain hypotheses as out of bounds from the start. On the other hand, I insisted that it may well be the case that science cannot answer all the questions that we would like to have it answer (such as the origin of life and so on). That is a real possibility, and every bit as plausible and eligible for serious intellectual consideration as the contrary assertion that there is a naturalistic explanation for everything.

On the question of scientific evidence, Washburn invoked the hominid fossils, stressing that human paleontologists were willing to change their mind when faced with contrary evidence and that they sometimes found confirming evidence that they had not expected to find (the Laetoli footprints). He did not seem to have comprehended the argument in part three of my paper in the slightest. Slatkin argued that the fruitfly experiments show that every characteristic of a fruitfly can be modified by inducing mutations. He conceded that the resulting populations were infertile and composed of creatures that could not survive in the wild but argued that all this was a problem only in the laboratory and would not impede evolution in the wild where populations are larger. In short, there was nothing new on the evidentiary line. I did not reply to these arguments particularly vigorously, being content to use this session for exploring the issues.

My law colleagues were particularly eager to press questions about what I want the schools to do—teach "creation science," teach religious doctrines generally, and so on. There was a bit of suspicion that I might have some kind of hidden agenda, and there was some bafflement when I insisted that I would be content with a more modest science class of the kind outlined at the end of my paper, in which there was candid teaching about the many problem areas that could lead to doubt about whether evolution is true. Some of my law colleagues then insisted that there could be no possible objection to reforming the curriculum along these lines, a point I labeled as naive. Professor Carrington from Duke broke in with the statement that he is a veteran of education politics and school board issues, and he knows for a fact that there is a powerful lobby that simply will not stand for a curriculum that expresses any doubt about the truth of evolution. I noticed that the scientists present were staring stone-faced at this suggestion from the lawyers that they would be willing to agree to a science class of the kind I proposed.

All told, the session was remarkably calm considering the controversial nature of my argument. I think the scientists and Professor Washburn were baffled to see such a topic discussed in all seriousness in a law faculty, and I imagine they are passing the word now that some people over at our school have gone out of their minds. I urged one and all to provide written as well as oral critiques of my paper. In all I consider this session a success from my point of view: A previously unthinkable claim—that the general theory of evolution is not true—was put out on the table for serious consideration and debated just as if it were any other academic topic.

The professor who presided over the colloquium, a nonreligious Israeli who is very sophisticated in philosophical issues, told me privately afterwards that he was very impressed by the paper and even more impressed by my performance in the debate, particularly in view of the intellectual courage he thought it required to advance such an argument in front of hostile experts from the sciences.

Appendix 2

Phillip E. Johnson's
Position Paper on Darwinism

November 30, 1989
To: Campion Center participants
From: Phillip E. Johnson

The August 1988 draft of my paper, which was distributed to you only a few days ago, is a bit lengthy and dense, and it has also been superseded by later drafts in its progress toward publication in book form. I therefore thought it might be helpful to prepare this informal summary of my views for our discussion. I have numbered the paragraphs to facilitate identification of specific controversial points.

1.0 The important issue is not the relationship of science and creationism but the relationship of science and materialist philosophy.

1.1 The provisional agenda defines our topic as "Science and Creationism in Public Schools." "Creationism" to most people means biblical literalism, and specifically the doctrine that all "basic kinds" of living organisms were separately created by God within the space of a single week about six thousand years ago. I have no interest in promoting or even discussing creationism in that sense.

1.2 If I could define our topic I would call it "Science and Scientific Materialism in the Schools, the Universities, and the public Broadcasting System." The question I raise is not whether science should be forced to share the stage with some biblically based rival known as creationism, but whether we ought to be distinguishing between the doctrines of scientific materialist philosophy and the conclusions that can legitimately be drawn from the empirical research methods employed in the natural sciences.

1.3 Scientific materialism (or naturalism, as in my 1988 draft) is the philosophical doctrine that everything real has a material basis, that the path to objective knowledge (as distinguished from subject belief) is exclusively through the methods of investigation accepted by the natural sciences, and that teleological conceptions of nature ("we are here for a purpose") are invalid. To a scientific materialist there can be no "ghost in the machine," no nonmaterial intelligence who created the first life or guided its development into complex form, and no reality that is in principle inaccessible to scientific investigation, i.e., supernatural.

1.4 The metaphysical assumptions of scientific materialism are not themselves established by scientific investigation, but rather are held *a priori* as unchallengeable and usually unexamined components of the "scientific" worldview. Materialist science therefore does not investigate whether the first living organisms evolved from nonliving chemicals without the intervention of any preexisting intelligence; likewise, it does not investigate whether the emergence of complex plants and animals, human consciousness, and so on was the product of purely natural (i.e., mindless, nonteleological) processes. The naturalistic evolution of life from prebiotic chemicals and its subsequent naturalistic evolution into complexity and humanity is assumed as a matter of first principle, and the only question open to investigation is how this naturalistic process occurred.

1.5 The question is whether this refusal to consider any but naturalistic explanations has led to distortions in the interpretation of empirical evidence and especially to claims of knowledge with respect to matters about which natural science is in fact profoundly ignorant.

2.0 The continued dominance of neo-Darwinism is the most important example of distortion and overconfidence resulting from the influence of scientific materialist philosophy upon the interpretation of the empirical evidence.

2.1 Claims that natural selection is a force of stupendous creative power, which is capable of crafting the immensely complex biological structures that living creatures possess in such abundance, are not supported by experimental evidence or observation. The analogy to artificial selection (in which conscious intelligence strives to produce greater variety) is faulty. In any case artificial selection does not continue to produce change in a particular direction indefinitely. Observational evidence (e.g., the famous peppered moth study) shows mainly cyclical changes in the relative frequency of characteristics already present in the population. There is circumstantial evidence pointing to somewhat more impressive changes (e.g., circumpolar gulls, Hawaiian fruitfly species), but the empirical evidence gives no reason for confidence that natural selection has the creative power, regardless of the amount of time available, to build up complex organs from scratch or to change one body plan into another.

2.2 Darwinists deny that natural selection is a tautology when the issue surfaces explicitly. The concept is in fact not inherently tautological, and it is capable of being stated in testable form. It is also capable of being formulated and used as a tautology, however, and in practice Darwinists continually employ it as the invisible cause of whatever change or lack of change seems to have occurred. If new life forms appear it is due to creative natural selection; if old forms fail to change the credit goes to stabilizing selection; and the survival of some groups during mass extinctions is explained by their greater "resistance to extinction."

2.3 The claim that selection in combination with random micromutations can craft new forms and complex organs by gradual steps is disconfirmed by the impossibility of proposing plausible advantageous intermediate forms in many cases. This difficulty can be met with various ad hoc speculations, such as hypothetical mutations in rate genes affecting embryonic development, but experimental confirmation that such processes can create complex organs and new body plans is unavailable. Whether the Darwinistic evolution of wings, eyes, and so on is conceivably possible is not the question here. The question is whether it is more than a speculative possibility.

2.4 The decisive disconfirmation of neo-Darwinism comes from the fossil record. Even if we generously grant the assumption that neo-Darwinist macroevolution is capable of producing basic changes, it does not appear to have done so.

2.4.1 Darwin's hypothesis requires the existence of an immense quantity of transitional forms that became extinct as they were gradually replaced by better adapted descendants. That the fossil record shows a consistent pattern of sudden appearance of new forms followed by stasis—i.e., the pervasive absence of the indispensable transitional intermediates—was a problem in 1859, but it is a far more serious problem when the condition persists after 130 years of determined search for the missing transitional intermediates. [See David Raup's paper contrasting Darwinist expectations for the discovery of intermediates with what has actually occurred.]

2.4.2 The "Cambrian explosion" of the animal phyla in a world previously composed of algae and bacteria (excepting the Ediacara fauna, which do not fill the gap as transitional intermediates) and the failure of life to diversify into new phyla thereafter, can be reconciled with Darwinism only by the strenuous application of ad hoc hypotheses. [See S. J. Gould's critique of the "artifact theory" of the Precambrian fossil record—and its distorting effect in inducing the "Burgess shoehorn"—in *Wonderful Life,* 271–77.]

2.4.3 The importance of mass extinctions—practically invisible for many decades due to the influence of Darwinist prejudice—further disconfirms Darwinist claims that continuous natural selection gradually weeds out the unfit and accounts for the absence of surviving transitional intermediates.

2.4.4 Again, the question is not whether ad hoc hypotheses can be invented to save Darwinism in the teeth of all this unfavorable evidence. The question is whether the evidence, fairly considered in its entirety, gives us cause for confidence that Darwinist evolution took life all the way from the hypothetical first living microorganism to where we find it today.

3.0 The refusal (or inability) of the scientific establishment to acknowledge that Darwinism is in serious evidential difficulties and probably false as a general theory is due to the influence of scientific materialist philosophy and certain arbitrary modes of thought that have become associated with the scientific method.

3.1 Science requires a paradigm or organizing set of principles, and Darwinism has fulfilled this function for more than a century. It is the grand organizing theoretical principle for biology—a statement that does not imply that it is true.

3.2 Once established as orthodox, a paradigm customarily is not discarded until it can be replaced with a new and better paradigm that is acceptable to the scientific community. Disconfirming evidence (anomalies) can always be classified as "unsolved problems," and the situation remains satisfactory for researchers because even an inadequate paradigm can generate an agenda for research.

3.3 To be acceptable a paradigm must conform to the philosophical tenets of scientific materialism. For example, the hypothesis that biological complexity is the product of some preexisting creative intelligence or vital force is not acceptable to scientific materialists. They do not fairly consider this hypothesis and then reject it as contrary to the evidence; rather they disregard it as inherently ineligible for consideration.

3.4 Given the above premises, something very much like Darwinism simply must be accepted as a matter of logical deduction, regardless of the state of the evidence. Random mutation and natural selection must be credited with shaping biological complexity, because nothing else could have been available to do the job. Hence even those evolutionary biologists who are most frank in acknowledging that Darwinism is in trouble frequently end up saying in the next sentence that no reputable biologist seriously doubts the importance of (creative) natural selection in evolution. Because the escape from Darwinism seems to lead nowhere, Darwinism for scientific materialists is inescapable.

3.5 What makes this situation particularly misleading is a confusion (at times convenient for Darwinists) about how the category "science" relates to the category "truth." Outside critics who point out the disconfirming evidence are frequently put off with the rejoinder that they do not understand "how science works" or with the disclaimer that Darwinists are noticing that theistic or teleological interpretations are false but merely that they "should not be taught in science class." Having put aside certain important possibilities as inherently unworthy of consideration, however, Darwinists do not hesitate to assert that their conclusions are objectively true: e.g., evolution (i.e., naturalistic evolution) is a fact; natural selection really has the powers claimed for it. These statements carry the implication that the philosophical premises on which they are based are also objectively true and therefore that competing philosophical premises are false. To put this somewhat abstract point in the vernacular: If God was around and capable of creating, there is no reason to assume that those Cambrian phyla must have evolved through purely naturalistic mechanisms.

4.0 The difficulties of Darwinism cannot be avoided by retreating to some supposedly unchallengeable "fact" of evolution, or by proposing "alternatives to phyletic gradualism" that attempt to occupy a middle ground between Darwinism and saltationism.

4.1 Some Darwinists have attempted to distinguish a purportedly indisputable "fact" of evolution from the concededly debatable theory of evolution by gradualistic natural selection. But what, precisely, is the fact to which they refer? The occurrence of microevolution is a fact, and that life has a common biochemical basis (the DNA code, etc.) is also a fact. The existence of natural relationships of greater and lesser similarity (classification) is also a fact. If "evolution" is merely a shorthand expression for "microevolution" and "relationship," then its use tells us nothing about how those relationships came into being.

4.2 In practice, the "fact of evolution" turns out to be Darwinism. Thus Gould's essay "Evolution as Fact and Theory," distinguishes the fact of evolution from a hyper-Darwinism (pan-selectionism) that Darwin himself repudiated, not from Darwinism properly understood. This reservation is unavoidable because the important claim of "evolution" is not that fish and man have certain common features but that it is possible for a fish ancestor to produce a descendant human being given sufficient time and the right conditions, without intelligent intervention. Absent the supposed creative power of natural selection, the transformation of fish to man would be nearly as miraculous as the instantaneous creation of man from the dust of the earth.

4.3 Nonmaterialistic theories of evolution (e.g., that God or a life force takes an active supervisory role) are nearly as unacceptable to scientific materialists as outright creationism. Saltationism likewise is not really evolution at all but a meaningless halfway point between creation and evolution. As Dawkins put it, you can call the creation of man from the dust of the earth a saltation.

4.4 Semisaltationist "alternatives to phyletic gradualism" are not genuine alternatives to Darwinism. To the extent that "punctuationalists" are merely saying that Darwinist evolution occurs rapidly (in geological terms), and in populations so small as to escape preservation in the fossil record, they are playing a variation on the usual theme that the fossil record is incomplete. As such they assume the whole point at issue, which is whether Darwinist macroevolution actually happened. To the extent that the punctuationalists are saying that the missing intermediates never existed as living creatures, they incur the disadvantages that led to the discrediting of Goldschmidt. The former option seems to be the lesser evil, and so we are now being assured that punctuated equilibrium is within the Darwinist framework and really just an elaboration of the implications for paleontology of Ernst Mayr's theory of speciation.

4.5 Retreat to the fact of evolution therefore inevitably brings Darwinism in again through the back door, without the need for defending its vulnerable aspects. The person who accepts the fact of evolution, thinking that he is acknowledging only that living things are related, will quickly learn that he is deemed to have accepted also that the relationships are based upon common ancestry.* If fish and man are descended from a common ancestor, then the immense differences that distinguish fish from men must be the product of an accumulation of the minor variations that differentiate offspring from their parents. From this acceptance of gradualism it is only a tiny step to full-fledged Darwinism, since something (i.e., natural selection) must guide the process of descent with modification. The fact of evolution turns out to be gradual change through decent with modification from common ancestors, guided where necessary by natural selection, and this is Darwinism.

5.0 The important debate is not between "evolutionists" and "creationists," but between Darwinists (i.e., scientific materialists) and persons who believe that purely naturalistic or materialistic processes may not be adequate to account for the origin and development of life.

5.1 Once separated from its materialistic-mechanistic basis in Darwinism, "evolution" is too vague a concept to be either true or false. If I am told that the phyla of the Cambrian explosion evolved in some non-Darwinian sense from preexisting bacteria or algae, I do not know what the claim adds to the simple factual statement that the prokaryotes came first. It conveys no information about how the new forms came into existence, and the "evolution" in question could be something as metaphysical as the evolution of an idea in the mind of God.

5.2 Similarly, whether "creation" occurred over a greater or lesser period of time, or whether new forms were developed from older ones rather than from scratch, is not fundamental. The truly fundamental question is whether the natural world is the product of a preexisting intelligence and whether we exist for a purpose that we did not invent ourselves. If Darwinists have not been overstating their case, they have disproved the theistic alternative or at least made consideration of it superfluous.

6.0 Whatever may be its utility as a paradigm within the restrictive conventions of scientific materialism, Darwinism has continually been presented to the public as the factual basis for a comprehensive worldview that excludes theism as a possibility. A few representative quotations will suffice to make the point:

*The logical jump described in this sentence is essentially a play upon the term "relationship." Human family relationships (siblings, cousins, etc.) are based on more and less recent common biological ancestry, and it is understandable but illogical to assume that the same must be true of the "relationship" between bats and whales or mammals and birds. Common ancestry as an explanation for natural classification is entirely reasonable as a hypothesis, but when enshrined as an irrebuttable presumption it is a projection of common-sense prejudice. As the physicists have been telling us, materialist common sense is not necessarily a reliable guide to scientific truth.

6.1 George Gaylord Simpson: "Although many details remain to be worked out, it is already evident that all the objective phenomena of the history of life can be explained by purely naturalistic or, in a proper sense of the sometimes abused word, materialistic factors. They are readily explicable on the basis of differential reproduction in populations (the main factor in the modern conception of natural selection) and of the mainly random interplay of the known processes of heredity. . . . Man is the result of a purposeless and natural process that did not have him in mind" *(The Meaning of Evolution)*.

6.2 Douglas Futuyma: "By coupling undirected, purposeless variation to the blind, uncaring process of natural selection, Darwin made theological or spiritual explanations of the life processes superfluous. Together with Marx's materialistic theory of history and society and Freud's attribution of human behavior to influences over which we have little control, Darwin's theory of evolution was a crucial plank in the platform of mechanism and materialism—of much of science, in short—that has since been the stage of most Western thought." (This is not from a popular polemic, but from Futuyma's college textbook.)

6.3 Richard Dawkins: "Darwin made it possible to be an intellectually fulfilled atheist" *(The Blind Watchmaker)*.

6.4 William Provine: "The destructive assumptions of evolutionary biology extend far beyond the assumptions of organized religion to a much deeper and more pervasive belief, held by the vast majority of people, that nonmechanistic organizing designs or forces are somehow responsible for the visible order of the physical universe, biological organisms, and human moral order" ("Evolution and the Foundation of Ethics").

7.0 Whether the materialist-mechanist program has succeeded as the Darwinists have so vehemently claimed is a legitimate subject for intellectual exploration. Scientists rightly fight to protect their freedom from dogmas that others would impose upon them. They should also be willing to consider fairly the possibility that they have been seduced by a dogma they found too attractive to resist.

7.1 With respect to the public schools, the providing of information about the evidence pertaining to Darwinism should be distinguished from efforts to indoctrinate students in "what scientists believe." Specifically, textbooks should be candid in acknowledging the origin of life problem, the fossil record problems, the limited results of selective breeding, and the inability to confirm experimentally the hypothesis that natural selection has creative power.

7.2 More importantly, the universities should be opened up to genuine intellectual inquiry into the fundamental assumptions of Darwinism and scientific materialism. The possibility that Darwinism is false, and that no replacement theory is currently available, should be put on the table for serious consideration.

Appendix 3

Letter from the Ad Hoc
Origins Committee

CRAFTED BY STEVE MEYER
WITH THE HELP OF OTHER SIGNATORIES

Dear Colleague:

We would like to alert you to a stimulating controversy that has been ongoing within American academia over the issue of biological origins.

This controversy has developed as a result of a provocative book by UC Berkeley law professor Phillip E. Johnson. Johnson's book *Darwin on Trial* (Regnery 1991) provides what we regard as a penetrating and fundamental critique of modern Darwinism.

In it Johnson questions the adequacy of what he (following Oxford Zoologist Richard Dawkins) calls the "Blind Watchmaker" thesis—i.e., the idea that nature possesses the power to create new organisms and complex organs without the assistance of a preexistent intelligence.

Johnson finds this to be the one and only permissible idea, in spite of its inadequacies, in modern biological literature. Like a number of biological scientists, he finds much lacking in the mutation/selection mechanism that neo-Darwinism credits with producing large scale structural (macroevolutionary) transformations in living things.

Unlike many critics of neo-Darwinism, however, who seek to retain the Blind Watchmaker idea in a revised evolutionary synthesis, Johnson calls macroevolution itself into question.

He suggests that the Blind Watchmaker is rendered ultimately credible only as a position of philosophy. He argues that many scientists have confused a putative methodological principle (i.e., "acceptable scientific theories must be wholly materialistic") with what must actually have happened (i.e., "therefore, life must have originated and diversified solely as a result of materialistic forces").

As you might imagine, Johnson's analysis has met with criticism from those who disagree with the implication that they do not recognize the role that philosophical assumptions play in their arguments. Some have accused Johnson of biological ignorance; others of misunderstanding the nature of science.

Johnson's vulnerability to criticism as a biological outsider has not, however, prevented his critique from receiving serious attention within scientific circles. His book has already been adopted as a text in courses on biology or origins in a number of major universities (such as Northwestern and Cornell). His argument provided the basis for an interdisciplinary scientific conference held on the campus of S.M.U. This conference featured both skeptics and defenders of evolutionary orthodoxy, including the notable Darwinist Michael Ruse. Johnson's book has also been widely reviewed in scientific journals (including *Nature* and *The Journal of Molecular Evolution*) and other academic publications.

Last July, the well-known Harvard paleontologist Stephen Jay Gould wrote a lengthy and negative review of Johnson's book in *Scientific American.*

Johnson has written a reply to this review which we believe will (in conjunction with Gould's review) open a window for you on this very stimulating debate.

Scientific American has declined to print Johnson's response. In the interest of open academic discourse, we have provided Johnson's response to Gould here. We were denied permission to enclose a copy of the Gould review, so we must ask you to track down the original review. We think that if you read the two pieces in tandem, you'll find them quite engaging.

You should, of course, know something about us. We are a group of fellow professors or academic scientists who are generally sympathetic to Johnson and believe that he warrants a hearing—thus this mailing. Most of us are also Christian Theists who like Johnson are unhappy with the polarized debate between biblical literalism and scientific materialism.

We think a critical re-evaluation of Darwinism is both necessary and possible without embracing young-earth creationism. It is in service of this re-evaluation that we commend the Johnson/Gould discourse to you.

Enjoy.

Appendix 4

The Rhetoric of Science and Intelligent Design

Historical and Conceptual Linkages

In this book I have traced the history of Intelligent Design and its campaign of persuasion from a unique vantage point—the academic crow's nest known as the "rhetoric of science" (hereafter referred to as "RS"). This relatively new subfield of communication theory was introduced in chapter 1, and in my concluding chapter I sketched the Kuhnian influence on its inception and profiled the debate over the recalcitrance of nature. Also, the careful reader has picked up along the way an idea of how RS analysts approach a topic like Intelligent Design, since I used a number of its approaches and instruments. For example, I discussed and used the following tools of rhetorical analysis: the twin perspectives of the "Narrative Paradigm" and "Symbolic Convergence Theory" (formerly "fantasy theme analysis"); the importance of "genre" and "rhetorical obstacles"; neo-Aristotelian notions, such as "enthymemes" and the triad of "ethos, pathos, logos"; Lloyd Bitzer's emphasis on the "rhetorical situation"; and Kenneth Burke's ideas of "perspective by incongruity," "god terms," the "dramatistic pentad," and "recalcitrance." I also developed some of my own RS tools, such as the function of "stigma-words," and the two-level narrative dichotomy, classifying stories into "factual stories" and "projection themes." Finally, I discerned within factual narrative the further subdivision into three types—genesis stories, HOS stories, and individual stories—and I showed the close relationship between the latter two.

I have reviewed my usage of RS ideas and tools to make a simple point: The reader of this history of Design, just by following my arguments and discussions, is to some extent automatically introduced to this field of rhetoric, but that may only whet the appetite to know more of its history and to search out what RS scholars have already said about Intelligent Design. Since my concluding chapter could accommodate only a tiny sliver of RS history, I need to fill in that "bigger pie" of rich recent history of this field, focusing on its links with Design, both historical and analytical, as seen in its budding interest in Intelligent Design. Each of the "acts" I progress through is its own historical-biographical sketch, placed roughly in chronological order, and moving more and more explicitly toward a focus on Intelligent Design. I have placed the key questions for each "act" in an opening italicized summary.

Act 1: Kuhn as Father of RS

How was Thomas Kuhn responsible for unwittingly fathering RS, and what evidence is there for us to make this historical assessment?

In my concluding chapter, I pointed out that it is no coincidence that RS arose as a recognized academic field in the 1960s and 1970s, at the same time that groundwork was being laid for the early assaults on Darwinism by Michael Denton and others. The historical connecting link between these parallel processes is Kuhn's *The Structure of Scientific Revolutions*. We have seen how Kuhn's ideas gave Denton an explanatory framework for his manifesto. Denton argued that the "priority of the paradigm" was the functional reality that kept Darwinism in its ruling position and smothered the peeps of dissent, despite the increasingly massive problems with macroevolution. We also saw how Kuhn's idea of a paradigm crisis and Denton's clear sketch of a Darwinian paradigm crisis were both absorbed into the core of Design's rhetoric.

In a similar vein of influence, according to many RS researchers, Kuhn's ideas were a massive impetus that helped launch their rhetorical assault on science.[1] This stemmed not only from Kuhn's conceptual scheme, implying the human factor in paradigm choice, but also from his scattered comments in *Structure* on the *distinctly rhetorical processes* at work within a paradigm shift. (In the next section I will summarize the most salient points from Kuhn's *Structure* that deal with rhetorical action in a paradigm shift.)

I pointed out in my RS discussion in the concluding chapter that Deirdre McCloskey claimed that Kuhn himself had keen interest in this vein of research. Speaking of a conference held by the Project on the Rhetoric of Inquiry, which opened a link between British sociologists of science and American rhetoricians of science, she says, "Kuhn himself participated enthusiastically in the 1984 conference that initiated the rhetoric of inquiry."[2]

Kuhn's influence, at least as an *unwitting* father of RS, seems clear. Rhetorician Gaonkar says RS scholars generally subscribe to a two-stage narrative of their field, with Kuhn as the turning point: *Stage One*—Before 1960, when philosophers uniformly portrayed science as intrinsically endowed with unique, unproblematic knowledge-guaranteeing norms, "it was difficult to conceive of how science could have anything to do with rhetoric." *Stage Two*—"Then came Kuhn, and everything changed. A new set of descriptions of science (variously identified as postempiricist, Weltanschauungen, and constructivist) became available and made it possible to articulate a connection between rhetoric and science." Both the explicit (self-described) and implicit rhetoricians of science have been called "the children of Thomas Kuhn."[3]

Steve Fuller himself, whom I briefly introduced in chapter 6 and who will figure prominently in this essay, described Kuhn's work as academically "liberating,"[4] although he later made it clear that he is not in fundamental agreement with significant portions of Kuhn's historical thesis, at least as an accurate picture of how science has operated in the modern era. To forestall any misunderstanding of Fuller here, it would be wise to let him speak, in an excerpt from his *Thomas Kuhn: A Philosophical History.*

> Ironically, what Kuhn presented as the "real" history of science in *Structure* itself turned out to be a myth, not only because its own empirical basis was suspect, but more importantly its narrative was used uncritically by social scientists and other inquirers to legitimate their activities as paradigms on the same footing as those of the physical sciences. . . . [*Structure*'s support for an intensified division of academic labor] explains the book's appropriation by a broad church ranging from "normal scientists" to self-avowed "postmodernists." The point of my book has been to explore the background social, philosophical, and historical conditions that have allowed this strange turn of events, in the hope that we may still be in a position to remedy whatever damage has been caused by an unreflective acceptance of the account of science given in *Structure*.[5]

Fuller's position on Kuhn is complex, with layers of analysis and a dense matrix of historical subtleties that need not be explicated here. Amazingly, Fuller holds that Kuhn presents *too comforting and conservative* a picture of the history of science, one that implies modern science is inherently self-correcting by means of the periodic cycles of paradigms and revolutions. Dissenting vigorously, Fuller views today's "big science" as a self-perpetuating, politically insulated monster that has grown arrogant and unaccountable to the public. He foresees a *new role of scientific rhetoric* after a hoped-for revolution that will shatter the power of this monster: Scientists will have to justify their work in the context of a highly democratic assessment by the currently disenfranchised electorate of intelligent, well-informed citizens. The notion of "accountability to the public" is the obvious link here between Fuller's vision and the criticism of Darwinism by the well-informed "citizen-skeptic-in-chief," Phillip Johnson. That shared sensibility is undoubtedly one reason Fuller has publicly declared his support for the work of Intelligent Design theorists.[6]

Nevertheless, despite his criticism of the late giant, from a historical point of view, Fuller has validated the role of Kuhn as the "turning point" for rhetorically oriented science studies in the latter half of the twentieth century, thus agreeing with Gaonkar's assessment above. Before Kuhn there existed a double standard whereby sociologists would study societal processes and practices "scientifically" while science itself was not studied with the same rigor. In that "time-before-Kuhn," science was studied with naïve deference, so that sociologists (Merton, for example) "typically drew conclusions about science based on the authoritative testimony of the great philosophers and scientists, or anecdotal evidence from great episodes in the history of science."[7] By the early 1970s, inspired by Kuhn's work, the atmosphere changed dramatically, and sociologists in England and America launched their new schools of social studies of science. In Act 2 I will describe not one but three such schools of sociology that sprang up after Kuhn to pioneer the social study of science.

My presentation of the role of Kuhn in the founding of RS is not intended to overlook the parallel importance of other philosophers and historians of science, such as Lakatos, Feyerabend, Polanyi, Rorty, and many others, whose work helped render even the hard sciences permeable to analysis in terms of sociorhetorical factors. Rather, my emphasis on Kuhn has resulted from my tracking, as carefully as possible, both Kuhn's sheer presence in RS texts and his ubiquity in the self-analysis within the field as the field began to reflect on its conceptual roots. In these regards, Kuhn's influence dwarfed that of the other scholars. His dominance is seen, for example, in Lawrence Prelli's working toward a comprehensive and detailed theory of RS in his *A Rhetoric of Science* (1989). Prelli not only acknowledged the field's great debt to Kuhn but actually reviewed every speck of Kuhn's own RS heuristic, readapting Kuhn's notions and vocabulary to that of rhetoricians. Kuhn's "proto-RS" thus became the foundation upon which Prelli built his elaborate superstructure.

To recap the Kuhnian link between RS and Intelligent Design, both have taken advantage of *Structure*'s new, "messier" view of how science actually works—*a view that stresses human perspectives and social influences.* The Kuhnian vision points to the role of scientists as *arguers* who are attached to one paradigm, potentially rendering them cognitively blind to its collection of anomalies (or at least profoundly uninterested in radical critiques that focus on this collection of anomalies). RS sees embedded in this new perspective a call for a fresh approach in the study of science, one that brings tools and perspectives of rhetorical analysis to bear on all of science. Meanwhile, as I said in my concluding chapter, Intelligent Design proponents apply this Kuhnian vision to the Darwinian paradigm as a means of explaining why evolutionists can cling to a paradigm which, to their thinking, is thoroughly counterfactual.

Act 2: Historical and Social Rhetoricians of Science—Kuhn and Woolgar

How have the implicit rhetoric of science fields developed, especially the new "social study of science" fields, and what insights can we gather from two pioneers in these fields (Kuhn and Woolgar) whose work relates to Intelligent Design?

One of the most complete overviews of RS penned in the field's adolescence was *Philosophy, Rhetoric, and the End of Knowledge* by Steve Fuller, whose democratizing agenda in science was mentioned above.[8] Fuller begins by tracing the rise of the twin fields of the History and Philosophy of Science, and (startlingly) he claims that this field has entered a steep decline since the appearance of Kuhn's theory of paradigms in 1962. (Kuhn, of course, wrote as the ultimate historian of science.) After this point, says Fuller, new fields of sociological scholarship began to proliferate and have virtually eclipsed the work of historians and philosophers of science. There are three crucial fields to which Fuller and his colleagues refer, and they are frequently called by their acronyms. These fields are (1) the big tent, "STS" (Science and Technology Studies) and its overlapping subdisciplines, (2) "SSS" (Social Study of Science) and (3) "SSK" (Sociology of Scientific Knowledge) (see especially Woolgar 11, 41). Such fields as these three social studies (and to a lesser extent the history and philosophy of science) have increasingly looked into the rhetorical principles and processes at work in science, scrutinizing everything from its daily toil in laboratories all the way up to scientific revolutions. Often when scholars (such as Kuhn himself) doing this rhetorical study do not identify themselves as "rhetoricians of science," they are nevertheless identified as such—"implicit rhetoricians of science"—by those in the explicit (self-identified) RS field.

Fuller, who since the early 1990s has been a major figure in these new fields and would not hesitate to take the RS label for his own work, blends insights from philosophy, sociology, and rhetoric in his own approach, which he labels "social epistemology." He is supremely bullish on his master field of STS, although he is not uncritical of its work. As mentioned above, he asserts that since Kuhn, STS has superceded the sterile field of the history and philosophy of science. Fuller says, "It would be hard to exaggerate the blow that Kuhn dealt to philosophers of science."[9] (One wonders what response would come from the Princeton and MIT philosophy departments, where Kuhn served as a distinguished professor from 1964 until his retirement.)

I find Fuller's narrow claim unconvincing, but a detailed response here would be more distracting than relevant. What is highly relevant is that Fuller, along with virtually every other scholar who has studied science since the 1960s, has been powerfully and irrevocably influenced by Kuhn. In my concluding chapter and in Act 1 of this essay, I outlined the pivotal founding role of Kuhn in the eyes of rhetoricians of science. Here I need to descend deeper into *Structure* itself. Any study of the rhetorical skirmishes between Darwinism and Intelligent Design will benefit from a retrospective glance at what Kuhn himself had to say about the role and processes of persuasion in a paradigm crisis. It will be helpful to review once more his cyclical view of the history of science and summarize his main points on rhetoric.

In chapter 1, I described Kuhn's central term—the *paradigm*—and alluded to the typical cycle through which a paradigm shift occurs. The paradigm cycle is now a commonplace in Western intellectual culture. In a given scientific field, during normal science there is a wide social consensus about the field's fundamental scientific ontology and proper epistemology. Its scientists work steadily and productively in routine puzzle solving. Then, over time, anomalies are encountered and are often vigorously worked on to arrive at a cogent explanation. As unsolved anomalies accumulate, and as rigorous attempts to explain them continue to fail, a *paradigm crisis* may arise. At this point more scientists begin focusing on the failure of the paradigm to solve the anomalies. A scientific revolution may follow if a fresh insight or a new crucial concept then arises in the thinking of one or a few scientists that seems to solve the nagging anomalies. This new approach, if successful, becomes a new paradigm in its embryonic or heuristic form. More and more scholars begin arguing for the old paradigm to be jettisoned and the new one to be adopted. If their efforts are successful, a scientific revolution is brought to completion, a process that may take as few as ten years[10] or last as long as a generation or two. Then a new "normal science" phase resumes under the new paradigm. Obviously, rhetoric plays a major role in two phases—paradigm crisis and revolution. What does Kuhn say about the rhetoric of these phases?

For those entering the rhetoric of science, it can be astonishing to discover the frequency and power of Kuhn's comments on "persuasion." This is why his influence was so enormous in the founding of RS. This is especially true in his latter chapters on the resolution of crises. For example, in chapter 9, "The Nature and Necessity of Scientific Revolutions," Kuhn explores the messiness of revolutions in general, both political and scientific. Since fundamental institutions are at stake in the revolutionary setting, the normal political recourse fails. So the evaluative procedures of normal science no longer work, and the debate over paradigms enters an unavoidably circular mode. That is, each paradigm, both old and new, argues its case in terms of its own paradigm.

Then Kuhn points out that *such circularity is not fatal, due to certain rhetorical dynamics.* He summarizes the situation:

> The man who premises a paradigm when arguing in its defense can nonetheless provide a clear exhibit of what scientific practice will be like for those who adopt the new view of nature. That exhibit can be immensely persuasive, often compellingly so. *Yet, whatever its force, the status of the circular argument is only that of persuasion.* It cannot be made logically or even probabilistically compelling for those who refuse to step into the circle. The premises and values shared by the two parties to a debate over paradigms are not sufficiently extensive for that. As in political revolutions, so in paradigm choice—there is no standard higher than

the assent of the relevant community. *To discover how scientific revolutions are effected, we shall therefore have to examine not only the impact of nature and logic, but also the techniques of persuasive argumentation effective within the quite special groups that constitute the community of scientists.*[11] [emphasis added]

Persuasion appears in its most explicit and powerful form, Kuhn implies, during scientific revolutions. In fact, Kuhn spends nearly all of chapter 12, "The Resolution of Revolutions," probing the rhetorical processes by which a scientific revolution is brought about. I shall discuss here three of his points.

1. Kuhn seems to give some useful role to a dual "falsification-and-verification" process, whereby anomalies may discredit the old paradigm but disappear under the new rival paradigm. Nevertheless, he emphasizes that the actual process of transference to the new paradigm involves what I would characterize as a *messy cluster of profoundly human factors and indeterminate questions.* There are a host of such factors. Scientists on each side of the paradigm divide may espouse competing "standards or . . . definitions of science" (148), or different notions as to the problems that a paradigm must solve. Misunderstanding can be bred by competing vocabularies where words' meanings have subtly shifted. Over a long paradigm struggle, it may even be necessary that "opponents [of the new paradigm] eventually die, and a new generation grows up that is familiar with it." Some of the holdouts may resist stubbornly because they are the ones "whose productive careers have committed them to an older tradition of normal science" (151).

2. Related to the latter point, Kuhn states that rhetorical success can be described as a "conversion" experience, and this psychological reality may mean that some scientists are never convinced. For example, "Copernicanism made few converts for almost a century after Copernicus's death. Newton's work was not generally accepted particularly on the Continent, for more than half a century after the *Principia* appeared" (150). This conversion necessity stems from the fact that defenders of different paradigms "practice their trades in different worlds," so they "see different things when they look from the same point in the same direction." This is not to say they are making things up, but that from their radically different perspectives they see things in different relations to each other. Kuhn's point here seems compelling. It is clear that the rhetoric of a new paradigm begins with certain scientists experiencing what Kuhn describes as a conversion akin to a "gestalt switch," which is followed by their appeal to colleagues to experiment with the new way of viewing the world that is under discussion.

   The connection with the debate over the legitimacy of the intelligent design as a robust scientific explanation is obvious. In many instances of this debate, incommensurable worlds seem to be communicating.

A Darwinist holding to metaphysical naturalism tends to reject design on epistemic or procedural grounds as introducing inherently religious (or worse, irrational) factors into science's cognitive arena. Meanwhile, a theistic biologist, who has metaphysical space in his or her thinking to consider intelligent design as an explanatory notion and who may be convinced there are principled means of detecting design, argues that such a hypothesis cannot be ruled out *a priori*. The challenge is to somehow convey from the Design side to the other this new way of viewing. This is perhaps the greatest rhetorical task implicit in the exchange.[12]

3. In addition to the rhetorical challenge of inducing conversion, and the cluster of other human factors that are entailed in paradigm persuasion, rhetoric is involved in rather conventional ways in paradigm crises: the marshaling of effective arguments. In fact, Kuhn spends half of his chapter exploring the "techniques of persuasion" and "argument and counterargument in a situation in which there can be no proof" (152–58). These are the keys that generally lead either to conversion or to resistance to change. The most important sort of argument advanced by proponents of a new paradigm is that "they can solve the problems that have led the old one to a crisis. When it can legitimately be made, this claim is often the most effective one possible." This is especially a powerful line of argumentation when novel predictions can be made of phenomena outside the specific field under scrutiny. In addition to arguments that stress problem solving, another sort of argument is usually needed as well: arguments that "appeal to the individual's sense of the appropriate or the aesthetic—the new theory is said to be 'neater,' 'more suitable,' or 'simpler' than the old" (155). Often, says Kuhn, the piling on of such arguments can only lead the scientist who is nearing the point of conversion to a place of exercising "faith"—faith that the new paradigm has "future promise" and an ability to confront "many large problems that confront it" (157–58).

So in paradigm crises, rhetorical processes are deployed like troops and tanks in a ground war. Yet I would add (in line with my discussion throughout this book) that a new set of weapons has arrived on the scene. Since 1962 the battleground has been transformed by Kuhn's own conceptual revolution. Since Kuhn's analysis of science has spread and his world of paradigms has become nearly universal in academia and all branches of science, a new rhetorical dynamic has been introduced. Sensing the potential power of such historical notions when converted to projection themes, rhetors now employ Kuhn's very notions as new weapons of persuasion. The process of paradigm shift has become *acutely self-conscious,* and what was intended by Kuhn as new understanding can, and does, serve as a galvanizing battle cry. The rhetorical landscape is now a proving ground for the Kuhnian high-tech weapons—the very acts of calling for, or predicting, a paradigm shift.

Flocks of "Kuhn's children" now populate the ranks of RS, and they have extended and complexified his study of the way science really works. Perhaps no group of post-Kuhnian students of science has received as much attention as the several dozen scholars in the Social Study of Science (SSS) and its parent field, Science and Technology Studies (STS). Steve Fuller is one of these. Another implicit rhetorician of science who is exerting enormous influence in this field is Steve Woolgar. Woolgar, coauthor of the enormously influential *Laboratory Life,* summarizes key insights of his field of SSS in his lean primer, *Science: The Very Idea.*[13] Woolgar shows what is problematic about the traditional epistemic claims of science. Even though he researches science from a radically relativist starting point (reminiscent of the radicalism of Gross), his most effective discussions—about "science as rhetoric"—make points that should be just as compelling and enlightening to philosophical realists.

Several of Woolgar's points relate to the Intelligent Design debate. In his chapter on ethnography and reflexivity, he summarizes the surprisingly disorderly and indeterminate actions of scientists in their everyday laboratory work. *Local conditions,* rather than *rules of procedure,* are most influential in deciding the kinds of experiments to run and the sorts of interpretations that are most appropriate. In addition, says Woolgar, the daily

> press of affairs in the day-to-day life of the laboratory also means that decisions and activities are rarely undertaken in the manner of a dispassionate search for the truth. Scientists have little time for a reflective evaluation of the epistemological status of their actions and interpretations. "Philosophizing" of this kind is most common among the elder statesmen of the field or among the disaffected and marginal members of the community. For the majority, the main and immediate aim is to make things work. Theirs is an instrumental rather than an epistemological concern.[14]

This observation is shrewd and highly instructive to the discussions of this book. Science is pragmatically (and instrumentally) oriented along the lines of Kuhn's "normal science," and thus philosophical issues or arguments within science may be functionally foreign to "bench scientists," both cognitively and in terms of their potential impact on practice. The effectiveness of a rhetorical strategy, and the nature of its impact on either side of any scientific debate, will take this reality into consideration. The application to the Design versus Darwinism battlefront is obvious in light of the recent importance of philosophical assumptions as a key issue.

Woolgar then refers to the work of his colleague Knorr-Cetina, who argued that "scientific activity is better termed constructive than descriptive." In other words, "scientists are not engaged in the passive description of preexisting facts in the world, but are actively engaged in formulating or constructing the character of that world."[15] This much is clear, says Woolgar, from two sets of observations: first, watching the scientists construct and edit reports, memos, charts, and research papers, and second, noting that even their instrumentation is based on a previous chain of "selections" of ways of approaching and characterizing nature.

As a philosophical realist, I share few if any of the basic assumptions entailed in Woolgar's radical constructivism. I think that the critique of Latour and Woolgar's *Laboratory Life* by rhetorical theorists McGuire and Melia (1989, cf. 1991) is quite compelling. Nevertheless, I recognize the cogency of Woolgar's general argument for the "constructed" nature of scientific knowledge as theory-laden and theory-dependent. I note one other telling point as well: Sociologists have debated for years the question of "what kinds of knowledge-discourse" are subject to their analysis. Previously, under Mertonian sociology of science, only *error* was considered a topic for sociological study. Now, Woolgar points out, SSS scholars insist that scientific *knowledge* as well as *error* are equally targets for such study (40–41). This brings within the reach of social and rhetorical analysis the multitude of social and discursive ways in which knowledge-claims (whether valid or invalid) are grounded, constituted, negotiated, and adjudicated. In the claims and counterclaims of the Design debate, such an unfettered research approach is liberating.

Act 3: Steve Fuller Once Again

What are some typical major insights in Steve Fuller's own RS research and reflection that touch on the issues of Intelligent Design?

The last sociologist-rhetorician I will review is the ubiquitous Steve Fuller. Fuller is a prolific scholar,[16] and his critiques and recommendations for the "democratization" of science are now well known. His critique of "big science" and his agenda for public accountability dominate many of his books and articles, especially his *Philosophy, Rhetoric, and the End of Knowledge* (1993) and his rejoinder in *Rhetorical Hermeneutics* (1997). Both in my book and in this essay, Fuller has appeared repeatedly because his ideas intersect so often with the kinds of concerns to which Intelligent Design has drawn attention. Here I will wind up my discussion of Fuller by reviewing two of his salient ideas—his "Democratic Presumption" and his work in extending Kuhn's incomplete work in paradigm transitions.

Fuller's social epistemology contains an advocacy stance as well as a theoretical approach that lends itself to the respectful review of heretical movements in science (such as Design) that challenge pervasive and deeply held assumptions about the solidity of published claims of knowledge. This stance is clearly seen in Fuller's "Democratic Presumption," one of his foundational assumptions that inform his research strategy. This presumption states: "The fact that science can be studied scientifically by people who are themselves not credentialed in the science they study suggests that science can be scrutinized and evaluated by an appropriately informed lay public."[17] The application to our present case is obvious. Johnson, a critic not credentialed in biology, has nevertheless done his own extensive research and has submitted Darwin's macroevolution theory to his own eclectic set of tests. The results of those tests, set forth by Johnson in *Darwin on Trial* and his later books, claim to have shown reasonable doubt about Darwinian science, thereby triggering (to echo Fuller) "scrutiny and evaluation by an informed lay public" through his published criticisms.

From these premises, Fuller derived an exceedingly egalitarian reform project—the democratization of professional, government-sponsored science. Fuller depicts, as an outgrowth of the Democratic Presumption, a state of affairs in which scientists are "learning to account for their activities to larger audiences, which will thereby enable everyone to assume a stake in the outcome of research. Thus, a high-priority item for social epistemology is the design of rhetorics for channeling policy-relevant discussions in which everyone potentially can participate."[18]

Fuller's revolutionary perspective at this point in one sense virtually over-shadows that of Kuhn. Kuhn was seen in Fuller's analysis as too conservative, too trusting of the self-correcting nature of science. Thus, as mentioned above, Fuller's book *Thomas Kuhn* (2000) was far from a celebration or broad endorsement.

At the same time, Fuller does engage critically with some very specific notions of paradigm shifts, and in effect he moves these notions down the path to clearer understanding. One of Fuller's most interesting points is made as he summarizes Kuhn's *key claim* within the Kuhnian "invisibility thesis." This thesis says that "the outcome of a revolution is determined not by clashing parties coming to agreement, but by the research choices subsequently made by their students." In other words, it is the next generation that is rising, not the senior scientists who are fighting, who will *quietly choose the new paradigm* for practical research purposes, thus rendering the revolution practically invisible.

Here Fuller describes—in a novel and transformative way—this critical historical period of a paradigm clash. In my view, his description serves as a brilliant "type," a heuristic generalization that is acutely applicable to Design's rhetoric. I already quoted this in chapter 7 in connection with Michael Ruse's AAAS talk, when Ruse announced a change of stance after interacting with Phillip Johnson. Fuller's paragraph is so rich and provocative that I encourage the reader to turn back to page 147 to scan it once more.

The process of change Fuller describes is complex and subtle, yet ultimately effectual; it typically precedes, and prepares for, a scientific revolution. To paraphrase it, one might describe it as an *imperceptible softening, or quiet erosion, by rhetorical engagement.* By prolonged exposure to the challengers and the resultant dialectics, positions begin to shift, and partisans become accustomed to radical criticisms. Eventually, an intolerable point of view looms on the horizon of possible acceptance. One does not have to strain to see that this descriptive frame relates compellingly to Intelligent Design's battles of persuasion, regardless of whether the result is a scientific revolution.[19]

Fuller adds, "Thus, radical change can occur quite unradically, indeed, invisibly. However, conspicuously absent from Kuhn's account is any discussion of how argumentation may facilitate this transition."[20] Fuller not only faults Kuhn for his failure to explore the rhetorical processes at this point, he also criticizes his implicit "romantic view of history," by which practical, democratic argumentation plays a vague and only secondary role. "In Kuhn's hands," says Fuller, "the 'essential tension' between tradition [the established paradigm] and innovation in science meant that a scientist's colleagues are a purely reactive force that, through peer pressure, discourages the scientist from regarding her own contributions as decisive against their collectively held beliefs." Fuller criticizes this impoverished view of discourse as it tends to paralyze the progress of mutual understanding in potentially revolutionary situations. Showing his own revolutionary bent, Fuller adds: "However *a priori* there is no reason to take this state of affairs as incorrigible" (309–10).

After reviewing some of Stephen Toulmin's helpful notions of practical argumentation in science, Fuller ultimately bemoans the fact that "rhetoric was effectively distanced from the issues that animated historians and philosophers of science as they increasingly came under Kuhn's spell" (314). Comparing scientific rhetoric with the more familiar species of rhetoric in a legislature, Fuller observed, "Missing here [in science] is a role for the old-fashioned idea of critical argument familiar from democratic politics" (315).

Act 3 could be greatly extended because Fuller has argued and interacted vigorously on another dozen fronts of RS that touch on Intelligent Design. I lack space to go into these, but I'll mention one—his careful and respectful criticism of the approach of John Campbell's research on the rhetoric of Darwin. Fuller urges that "actual readership" of texts should predominate in rhetorical analysis, rather than an idealized, "hypothetical" reader of the *Origin*.[21]

Act 4: Charles Taylor and the Moderate Rhetoricians of Science

What is the nature of the moderate RS research, and what insights into Intelligent Design can be gathered from Charles Taylor's work?

Steve Fuller represents a somewhat pragmatist middle ground between Steve Woolgar and Alan Gross's radical vein of RS and the various moderates we have met. It would be misleading to imply that any of these approaches dominates RS. In fact, the variety of approaches is so great that when Gross's book was featured in "The Rhetorical Turn in Science Studies," a major review article,[22] it shared the spotlight with seven other significant books that had also been published around the start of the 1990s.

One of the seven was by Lawrence Prelli, whose indebtedness to Kuhn was mentioned earlier. His work, *A Rhetoric of Science*, was in one sense more ambitious than Gross in attempting a global, multidimensional synthesis of rhetorical theory as it applies to scientific texts. Prelli constructed a macrotheory of situational, inventional rhetoric suitable for any need of scientific discourse.[23] Drawing on the practical neoclassical approaches, Prelli adapted the informal logic of audience-centered discourse to fit all scientific contingencies.

Prelli's work is more in line with the moderate RS typified by Simons, McGuire and Melia, and Charles Bazerman, Peter Dear, Marcello Pera, and William Shea.[24] Another scholar in the moderate tradition is John Campbell, already introduced and noted for his prolific rhetorical studies of Charles Darwin.[25] Campbell's career ties in to Intelligent Design in an unusual way. As a theist (and member of a mainline Protestant denomination), he had assumed during the first two decades of his study of the rhetoric of Charles Darwin that some sort of macroevolution of all life forms had taken place. However, in 1991, after reading Denton *(Evolution: A Theory in Crisis)* and Johnson *(Darwin on Trial),* he witnessed an intense encounter at dinner between Johnson (who was speaking on his campus that week) and a science professor at the University of Washington. Campbell describes this moment as a cognitive epiphany, and thereafter he became sympathetic to the intellectual-rhetorical position of the Design Movement.[26] I will return to Campbell at the end of this essay when I discuss the issue of *Rhetoric and Public Affairs* (winter 1998) that reviewed Intelligent Design.

Charles Taylor's *Defining Science: A Rhetoric of Demarcation* (1998), also in the moderate RS vein, is the most important single study that sheds light on the rhetorical situation Design faces in its attempt to instigate a paradigm crisis. Demarcation—drawing the line between science and nonscience—is a foundational practice that decides whether to award the status of "science" to ideas or positions that challenge the status quo. Demarcation rhetoric excludes *on some particular principle* those feisty challengers that are judged to be "pseudo-scientific." As we have seen in the responses to Johnson and others, Design ideas are often excluded by demarcation rhetoric. Their critiques are typically dismissed as "religion, not science." We even encountered Johnson returning the favor in *Darwin on Trial,* describing Darwinism as a pseudoscience. Clearly, demarcation is the foundational issue in Design's rhetorical struggle to gain a hearing. In a sense, it is *the issue* on which the entire debate hinges.

Specifically, Taylor looks at the older genre of creation science as one of his case studies with which he concludes his book. Before this, he spends several chapters surveying the various approaches to demarcation's cluster of questions: *How is science defined? Is there a rule or set of sound principles from philosophy that can guide this demarcation? If not, on what basis is demarcation carried out?* Though he is not a radical relativist or antirealist, Taylor subscribes to a moderate constructivist position. He denies that scientific knowledge-claims spring from a dispassionate observation of phenomena. Rather, scientific knowledge-claims are shaped by human choices, cultural traditions and vocabularies, biases, philosophical preferences, and social configurations and alliances. Taylor emphasizes the "connectedness" of the culture of science to other social practices in the world. Taylor coins "the rhetorical ecology of science" as his metaphorical key to promote understanding of demarcation and concludes that "rhetoric is not everything that scientists do. Nonetheless, this story calls attention to the cultural rhetorics upon which those very 'observables' depend. At this juncture, the *sine qua non* of contemporary culture, its politics, is located at the heart of all discursive practices, including demarcation discourses."[27]

To build his case for *sociopolitical negotiation* as the key to demarcation, Taylor first assesses the classic philosophical proposals for grounding demarcation, including Popper's "falsification criteria." He concludes that falsification fails, not just because it is not deployed consistently in practice, but also because such criteria have themselves served prominently as tools of rhetoric—as "rhetorical battle weapons" in Popper's hands to rule out Marxism and Freudianism. After showing serious flaws in major philosophical attempts to solve the demarcation question, it becomes obvious that a philosophical foundation is not the key to demarcation. Rather, Taylor emphasizes, the demarcation of science is *rhetorically negotiated* between key social actors in a variety of contexts (21–57).

Philosopher Larry Laudan was blunt on philosophy's impotence to provide cogent criteria, adding that "it can be said fairly uncontroversially that there is no demarcation line . . . which would win assent from a majority" (52). However, Taylor scoffs at Laudan's advice that scientists and scholars who study science should just view demarcation as a "spurious" problem and abandon such terms as "pseudoscience," thus giving up the attempt to demarcate science. Taylor notes that such an attempt—to make the problem evaporate through scholarly exhortation—flies in the face of social reality. Obviously, *demarcation is being carried out daily through rhetoric:* Both the mainstream scientist and the marginalized dissenter who asserts his scientific legitimacy are equally vigorous in their demarcation discourses. Their defense or challenge of scientific orthodoxy seems to be predicated on the *reality and vitality of demarcation reasoning and persuasion* (53).

Thus, Taylor's goal is not to deny demarcation. On the contrary, the defining of science is the outworking of a complex set of rhetorical dynamics that involves such factors as the historical situation, paradigm commitments, social networks, value and belief systems, political relationships, and intentional goals and agendas. His survey chapters on "History and Social Studies of Science" and "Demarcating Science Rhetorically" move the reader inch by inch toward that newer, "messier," more complex rhetorical understanding. His research summaries are dense yet rich, as he combs scores of insights on demarcation from these fields. His erudite literature review is a compact "Who's Who" in RS and related fields of history, philosophy, and sociology of science.

Taylor concludes with a case study of efforts of creation scientists to demarcate themselves as valid scientifically and to portray evolutionary science as "poor science." His coverage of demarcation here is even-handed and thorough, yet (oddly) he seems unaware of creation science's new cousin—the Design Movement.[28] Despite this oversight, Taylor offers telling insights into demarcation in the rhetoric of origins:

While the claims of creationism might well seem arcane, misguided or just plain silly to many of us, it is unduly elitist to leave it at that. We cannot afford, as Burke noted, to offer a few adverse attitudinizings and call it a day. Those committed to a neo-Darwinian view of natural history do not have a monopoly on clear thinking and/or intelligence. The epistemic and rhetorical commitments of the competing camps are so radically different that both groups seem reduced to repeating threadbare and often misleading shibboleths so as to demean the other, not the least of which is the claim that creation (or evolution) is not *science*. As Marsden concluded, "Each party thinks that the members of the other are virtually crazy or irremediably perverse. Neither thinks that the other is doing science at all." [156]

Taylor does not support the epistemic claims of creation science. Indeed, he says that creationists employ a discredited Baconian view of science. He even describes the "evidentiary support for evolution" as "overwhelming."[29] Nevertheless, Taylor warns that the rhetoric of demarcation by evolutionists is just as feeble as that of creation scientists. He probes the implicit rhetorical fallout from recent evolutionary discourses within the origins debate that demarcate "scientific knowledge" from "ignorant religious myths." Quoting Geertz, Taylor observes that the religious realm is a "permanent, pervasive, and always central aspect of human life" spawned by basic human realities (173–74). Since the heart of these religious beliefs lies in *questions of value that are said to inhere in the structure of the universe* (and not, for instance, in questions of the afterlife), scientists doing scientific demarcation should be aware of the inescapable power of that perspective. Echoing Fuller's spirit of skepticism, Taylor concludes, "Blame, if it is to be assigned at all, must lie fundamentally with the discourses of expertise which claim too much for themselves and allow too little from nonexpert audiences. . . . Perhaps a deeper appreciation for those nontechnical audiences and *more critical attention to the rhetorical constructions of expertise and authority* will produce clearer understanding of inescapable symbiosis" (emphasis added).[30]

Before going to Act 5, two highly relevant insights of Taylor should be highlighted. First, demarcation is not scientifically or even philosophically established but rather is *rhetorically negotiated*. This makes it all the more important that a rhetorically sophisticated assessment be made of the claims and counterclaims on the front lines of the Design debate. This is not to deny the relevance of scientific or philosophical arguments. Rather, it is to affirm that demarcation is a rhetorical practice that uses such points in order to erode or harden existing viewpoints, in order to secure agreement in a social context. Second, just as the demarcation rhetoric used against the creation science position embodies socially constructed assumptions about the role of "expert audiences," who have the right to determine "what science is" in the first place, there rages an equally important battle over the "scientific status" of Design, pitting evolutionists' claims of expertise against critics who may or may not be credentialed scientists.

Act 5: The Rhetorical Symposium on ID

What are the major research findings that emerge from the recent work in RS focusing on the Intelligent Design Movement?

John Campbell, introduced above, has become widely known for his rhetorical research on Darwin's life and especially his revolutionary work, *The Origin of Species*. His work, published over a span of thirty years, has focused on Darwin as a rhetorician faced with a unique and challenging historical situation.[31] Campbell's work, which shared the spotlight with Gross and Prelli in a major symposium-in-print in 1997 *(Rhetorical Hermeneutics)*, was referred to by Taylor as "perhaps the standard of . . . historicocritical research." Taylor adds, "While Campbell's work spans a number of years and substantive topics, it coalesces around an implicit theme of science as reciprocally related to, yet operationally distinct from, the larger social milieu in which it is embedded."[32]

If one moves from Campbell's work on Darwin up to RS work on the recent evolution controversies, one finds that virtually all studies deal with the traditional scientific creationists. Fortunately, there is one massive, recent exception—a special issue of *Rhetoric and Public Affairs* (winter 1998) devoted entirely to a rhetorical understanding of the movement.

This issue opened with an introductory essay by Campbell and closed with a superb literature review by rhetorician Thomas Lessl. Between these essays are two RS feasts. First, there is a section of articles in which four key Design scholars (Behe, Meyer, Nelson, and Wells) summarize their main theses. These articles are critical of the neo-Darwinism paradigm and seek to persuade that their new paradigm is superior in all respects. Second, critical responses then follow by four rhetoricians of science (Steve Fuller, David Depew, John Lyne, and Celeste Condit), by Design leader Phillip Johnson, and by evolutionary biochemist Bruce Weber. This issue is a critical resource for the rhetorical understanding of Design, thanks to the scholarly interactions of the rhetoricians of science—six in all if one adds Campbell and Lessl to the four rhetorician-respondents. I will summarize key contributions from four of these six scholars.

Campbell on Intelligent Design

Campbell's essay contains a startling thesis that centers on the important and healthy role that Paley's design notions played in Darwin's own development as a scholar and in his rhetorical strategy. Campbell's argument is that Paley's arguments for design did not stunt the intellectual development of Darwin. On the contrary, his writings served as Darwin's lifelong dialectical base—a sort of intellectual sounding board from which he was able to develop his own diametrically opposed theory of origins. Campbell says Darwin was openly appreciative toward his tutoring by Paley in biological design; he viewed it as something of a stimulus or fertilizer for his own thinking.

Campbell contrasts Darwin's own positive attitude toward design arguments with that which prevails in American public schools today. There the Design-Darwinism debate is seen as dangerously confusing to students. At this point, Campbell turns a corner and highlights Martin Eger's piece, "Two Controversies: Dissonance in the Theory and Practice of Rationality" (*Zygon,* 1988)—an important but generally ignored study of public schools' inconsistent teaching strategies. Eger, a physicist at New York University with an interest in public education, observed a strange cognitive dissonance when comparing the teaching of evolution in high schools versus the teaching of ethics.

This dissonance can be boiled down to a stark contrast in methodology—the role of free questioning versus the role of indoctrination—between the biology classroom and the ethics classroom across the hall. The ethics class is dominated by pedagogical assumptions derived from Mill's *On Liberty* and elaborated by contemporary ethical theorists such as Bell and Kohlberg. The ethics teacher is bent on questioning a wide range of ethical norms and rules that have been instilled in children by parents. The goal here is not necessarily to overturn these convictions. The teacher simply aims to build into children's lives a rationally debated ethical conscience in lieu of their accepting, with an unquestioning spirit, the morality handed down at home. With this goal in mind, even the most fundamental moral notions—for instance, the evil of adultery, stealing, or murder—are brought into question by the high school ethicist.

Meanwhile, Eger observes that the biology teacher at the same school *takes the opposite pedagogical approach.* She discourages at every turn the questioning of such fundamental ideas as the naturalistic Darwinian development of all life. Any questioning of the adequacy of scientific evidence for macroevolution is out of the question. Rather, students are told that such scientific issues have been decisively settled and are not open to question, regardless of what they have heard outside the biology class.

Agreeing with Eger, Campbell notes the "agonistic symmetry" of these cases.

> The very mark of reason in one model is the criterion of unreason in the other. For the educational ethicists the giving of reasons for every belief differentiates reason from unreason. For the educational theorists of science the unquestioning mastery of a prior system is the precondition for proper understanding. Whereas in ethics the consideration of unorthodox or conventionally unacceptable alternatives—for instance that dishonesty might be the best policy—is to be met without prejudice. In science, by contrast, even permitting the bare impression that there might be some arguments in favor of creationism—or in the present case intelligent design—is a dereliction of education responsibility.[33]

Campbell and Eger advocate an overhaul of this bizarre educational incongruity.

Condit the Rhetor; Depew and Lyne the Rhetoricians

Campbell's introductory essay is followed by a cluster of four articles that outline Design's critique of neo-Darwinism and its alternative paradigm of origins. Dembski sketches the general rationale for ID, and Meyer argues for the "inference to the best explanation" in accounting for the origin of DNA. Michael Behe summarizes his familiar case for irreducible complexity of cellular machinery, and Nelson and Wells try to disprove thoroughly the popular 1973 aphorism of Dobzhansky—"nothing in biology makes sense except in the light of evolution." They close by recommending a whimsical retooling of Dobzhansky's slogan: "Nothing in biology makes sense except in the light of evidence."[34]

Rhetoric and Public Affairs editor Martin Medhurst managed to involve several leading rhetoricians in responding to the Design debate. The most caustic reply is from University of Georgia scholar Celeste Condit. One of the world's leading rhetorical critics, she sets out to refute systematically the Design arguments contained in the earlier articles. Her stilettolike reply has as much the function of a rebuttal of a rhetor as it does an analysis of a rhetorician. Her attacks on Design are fascinating in their tone and style, and her overall rhetorical strategy invites close analysis. Yet, as she rises to the defense of Darwinian notions of origin of life and of speciation, the quality of her own arguments is disappointing. More than once she betrays ignorance of basic scientific realities.[35] In the end, her essay is largely irrelevant to this study, except as an example of how hard it is to do rhetorical criticism of a position one is vigorously opposed to without sounding like a rhetor for one's own side.

Condit's response is primarily a polemic; Lyne and Depew's rejoinders are profoundly reflective and interactive. Theirs are so similar—in perspective, analytical insights, and research suggestions—that it makes sense to review them together. For example, both Lyne and Depew adhere to evolutionary explanations yet take Intelligent Design seriously as an intellectual challenge. Design's line of argumentation, alleging an overlooked "ontological fact" about organisms, is taken very seriously by both, as is Nelson and Wells's critique of Dobzhansky.[36] Both rhetoricians ask why certain arguments are now prominent and seek to find out what is "the rhetorical exigency for their new urgency."[37] Depew reminds us: "Creationist arguments, like evolutionary arguments, are developed, deployed, received, and debated under particular discursive conditions, conditions which it behooves a rhetorical critic to reconstruct" (572).

Two important discursive conditions shaping the Design debate are sketched by Lyne and Depew. I will call them (1) the "clash of pieties" factor and (2) the "functional versus historical" factor. Lyne explains the issue of pieties as a clash between the old piety (we are created for a purpose and live purposeful lives) and the new piety (we are the products of a "nature that is both blind and accidental").[38] Lyne is speaking of pieties in Kenneth Burke's sense, as "the longing to conform to the sources of one's being." Lyne then states modern man's existential dilemma: "How can we be at home in a place so apparently unlike us?"[39] Lyne notes that physics (and we could add chemistry and astronomy) has never driven a seeming wedge between science and the general culture like biology has, simply because physicists rarely touch on origins. In physics, only the big bang comes close to a tension point, but "any alert theist can with little difficulty meet that challenge by pushing the question of origins one step further into regress. . . . Evolution, by contrast, speaks unflatteringly about our ancestors and confusingly about us. It gives us a biological home only by evicting our angel half" (579).

I need to point out that this piety clash—the natural cultural aversion to the ultimacy of random processes as our "source"—obviously plays a role in the rhetorical exigency, especially when Design theorists communicate to the laypeople generally. It helps create the well-disposed listening audience.

However, the piety clash also serves in a leverage role in the persuasion process. Design theorists do emphasize the Darwinian claim to show (as Lyne puts it) "purposelessness as the ground of the cosmos and the life within it."[40] Yet, at this point, Design advocates apply leverage using the fulcrum of randomness. They raise the ante by *demanding to see the evidence* that random processes did—or even could—produce information-rich structures such as DNA and cells. Normally this challenge leads into Design's familiar two-front strategy. First comes the scientific front—reevaluating scientific evidence for the random origin of life, to show that it is weak or nonexistent. Then the attack moves to the philosophical front—since evidence for Darwinian origins is missing, then *philosophical assumptions* must be implicated as the true bedrock of Darwinists' certainty. Their conclusions are merely a "deduction from metaphysical naturalism." This is Meyer's strategy in his essay in *Rhetoric and Public Affairs* (519–56).

Besides the piety clash, a second, more important discursive factor is brought out by Depew: the "functional versus historical" factor. This refers to the intrinsic uncertainty and tentativeness that surround any attempt to establish "historical biology" (what paths of change have biological creatures taken in the passing eons) as contrasted with the certainty and empirical confidence that characterize "functional biology" (how living things function in the biosphere, over and over, as we directly observe them). Depew's comment is trenchant:

Functional biologists can figure out how an organic system works in the absence of any background theory about the origins of that system. Perhaps that is why one can find in the ranks of biologists creationists as well as evolutionists, non-Darwinians as well as Darwinians, and, among contemporary Darwinians themselves, people like Richard Dawkins, who takes a genocentric perspective on natural selection, as well as those, like myself, who dislike selfish gene theory even more than they distrust creationism. Given this gap between functional and historical biology, Jonathan Wells and Paul

Nelson are right to point out that Dobzhansky at the very least overstated his case when he proclaimed that "nothing in biology makes sense except in the light of evolution."[41]

Thus, the situation of evolutionary biology in its rhetorical challenge is made problematic in two ways. First, evolutionary biology—in its doctrinaire formulations—asserts a new piety that clashes with the ancient and intuitively powerful pieties of our culture. Second, evolutionary theories and claims are intrinsically hamstrung by the difficulty in piecing together past events from very fragmentary evidence.

Note that this second factor that evolutionists must confront can pose almost an equally difficult rhetorical challenge to Design. Difficulty in filling in historical details of the Design genesis story (or refusal to do so) is often perceived as a fatal rhetorical weakness in Design arguments. In chapter 6, I mentioned this issue in connection with the 1996 "Firing Line" debate, but let me share here part of Lyne's helpful summary:

> The idea of transmitted information, in contrast to mere order, gets us to intelligibility, but its relationship to origins is unclear. The design/information argument, as I understand it, applies only at the point of origin, because thereafter the mechanisms of replication presumably take over. Or do they? Meyer's argument suggests either that nothing has changed much since the Day of Design, or that there is intelligent intervention all along the way, perhaps at points of mutation. It seems to me the intelligent design writers have been obscure about where they stand on this. Some of the arguments go to the origin of life; others suggest that designing continues with the production of each new case of irreducible complexity. . . . It looks to me as if we have a basic issue to resolve if we are to contemplate intelligent design: are we speaking only of remote origins, or of an ongoing inflow of information through the evolutionary trajectory?[42]

Such questions spotlight the difficulty Design has in promoting a view of cosmology that must, by the current situation or strategy, remain vague or pluralistic at best.

Many more valuable insights could be noted in the essays of Lyne and Depew. Perhaps the most important concluding point to be drawn from their analyses of Design is their interest in what might be called argumentative and peda-gogical engagement—or, as I called it in my concluding chapter, *metaphysical management*. By this I mean that both Lyne and Depew, while not personally subscribing to ID perspectives, are sufficiently impressed with its rhetorical and scientific cogency that they toy with implementing its intellectual insights in an imaginary classroom.

Lyne goes further. He ponders—and experiments—on how to teach design ideas in a public school or public university context. For example, he has an extended discussion on his experience of having his undergraduate rhetoric of science class read extensively on both sides of the Darwin-Design debate. Lyne crafts perhaps the most penetrating statement in the entire collection of essays:

- For those who want to train better arguers, intelligent design theory does us the service in the pedagogical setting of taking the creationist argument out of the religious context. *It permits us to switch attention from belief (which students always feel entitled to) to reasoning (where some skill must be demonstrated).*

    Moreover, it brings into focus just what underpins the scientific discourses in question. I have found that a good many advocates of Darwinist principles have little idea what they really are. [580–81, emphasis added]

As Lyne ponders the potential awkwardness of teaching about design ideas in a public venue, he adds a comment that is both practical and shrewd:

I hope I will give no offense in aligning the intelligent design argument with theism. In the absence of any other nominee, I assume that the designer is God, although I understand that this view in no way commits the advocate to the Bible, or to any other religion, for that matter. I think the position does perforce imply a kind of theism, however, and thus it inherits the windy political problem of how it can be taught—and of giving meaningful secular content to the idea of a Designer. As this matter is normally handled, not by physics, but by metaphysics or revelation, one senses that for some of its advocates the *deus ex machina* is already waiting in the wings, textually and historically explained. It need not be so. . . .

    In trying to get some purchase on the Designer as *explanans,* one gets the sense that the theory is perhaps less an answer to the cosmic riddle than an opening to a whole new set of questions. And it is not clear which of these questions could legitimately be discussed in public schools. Does the idea of intelligent design presuppose purposiveness? Is it outside the universe or immanent within it? Do time and historical sequence apply to the operation of this intelligence? Is the designer in any sense personal or accessible? [580–81]

Lyne does not attempt to answer these questions, but the fact that he both perceives them *and chooses to raise them* shows that, in his assessment, the new Design arguments have reached a critical level of scientific cogency and sophisticated expression that forces the pedagogical questions to the forefront.

Conclusion

In five acts I have sought to fill in further a sketch of the history of RS, focusing on some of its major figures and research texts. This study has moved us steadily from the role of Kuhn himself, including his own thoughts on paradigm persuasion, through the overshadowing predominance of Steve Fuller, and finally into a focus on demarcation (by Taylor) and the explicit study of

Intelligent Design (by a cluster of RS experts). In a sense it is astonishing to see the number and richness of recent RS links with and insights into Intelligent Design. Not only is Design emerging as a major movement in American society and science, it certainly is beginning to move under the scrutiny of the world's finest rhetoricians of science. Stay tuned.

Notes

Chapter 1: Aroused from Dogmatic Slumber

1. One other such rarity of evolution arising in presidential politics was when Ronald Reagan, during the 1980 election campaign, mentioned his doubts about evolution. To many, this confession became symbolic of his ignorance of science. See Roy Gallant's essay "To Hell with Evolution," in *Science and Creationism,* ed. Ashley Montagu (Oxford: Oxford University Press, 1984): 284.

2. John Rennie, "A Total Eclipse of Reason," *Scientific American,* October 1999.

3. Ellen Goodman, "The Ever-evolving Creationists," *Boston Globe,* 19 August 1999, and *St. Petersburg Times,* 21 August 1999.

4. Phillip E. Johnson, "The Church of Darwin," *Wall Street Journal,* 16 August 1999.

5. Michael Behe, "Teach Evolution—And Ask Hard Questions," *New York Times,* 17 August 1999.

6. Michael Behe, "Darwin under the Microscope," *New York Times,* 29 October 1996; *Darwin's Black Box* (New York: Free Press, 1996). Both of Behe's *New York Times* columns were invited. The interest of the *Times* stemmed from the paper's rather positive review of Behe's book. See James Shreeve, "The Design of Life" (review of *Darwin's Black Box*), *New York Times Book Review,* 4 August 1996.

7. I now capitalize "Intelligent Design Movement," even though my doctoral dissertation, upon which this book is based, did not do so. Why the change, especially when the *New York Times* decided not to capitalize the movement in their major story? My decision to capitalize is in line with a recent trend, as seen in Dembski's foreword to Ben Wiker's *Moral Darwinism.* Also, it helps in ease of reading. Thus, "Design" (shorthand for Intelligent Design Movement) is capitalized when referring to the movement; it is in lowercase when referring to the idea of design. The same is true of "intelligent design." In lowercase, it refers to the idea of detecting design, not to the movement.

8. On 14 February 2001, by a vote of seven to three, the Kansas Board "reversed that decision, reinstating evolution with the adoption of new state science standards and essentially mandating that evolution be taught in public schools throughout the state." John Fountain, "Kansas Puts Evolution Back into Public Schools," *New York Times,* 15 February 2001, A18.

9. Jonathan Wells, *The Icons of Evolution* (Washington, D.C.: Regnery, 2000). This book, which is discussed later, is one of the most important recent works from the Design Movement and is aimed at ten of the chief visual symbols or icons of textbook evolutionary teaching, such as the peppered moth observations, embryo drawings (in comparison across species), and the "tree of life" diagrams. Wells received his doctorate in molecular biology from UC Berkeley in 1996.

10. Information was from a September 2000 e-mail from Casey Luskin, a UCSD student who attended the lecture and interviewed Scott afterward.

11. See, for example, Jerry Coyne's review in *Nature,* 13 April 2001.

250

12. James Glanz, "Darwin vs. Design: Evolutionists Face a New Opponent," *New York Times,* 8 April 2001. See also Teresa Watanabe, "Enlisting Science to Find the Fingerprints of a Creator" (inside headline: "Education: Believers in 'intelligent design' try to redirect evolution disputes along intellectual lines"), *Los Angeles Times,* 25 March 2001.

13. Glanz, "Darwin vs. Design."

14. In the rhetorician's jargon, such a "projection" is normally labeled a "fantasy theme"; more on this later. I am using here Thomas Kuhn's famous terms that have become commonplace in discussions of turning points in science. Thomas Kuhn, *The Structure of Scientific Revolutions* (Chicago: University of Chicago Press, 1962).

15. Besides law professor Phillip Johnson and Lehigh biologist Michael Behe, one could list Dean Kenyon, biologist (emeritus) at San Francisco State; Fred Sigworth, physiologist at Yale; Scott Minnich, biologist at the University of Idaho; Robert Kaita, plasma physicist at Princeton; Martin Poenie, zoologist at the University of Texas; Guillermo Gonzalez, astronomer at Iowa State University; Jed Macosco, molecular biologist at the University of California at Berkeley; Walter Bradley, professor emeritus of mechanical engineering at Texas A&M; Paul Chien, biologist at the University of San Francisco; Charles Thaxton, who taught natural sciences at Charles University in Prague; Henry F. Schaeffer, pioneer quantum chemist at the University of Georgia; George Lebo, astronomer at the University of Florida; Steven Meyer, philosopher of science who taught at Whitworth College; Alvin Plantinga, philosopher at Notre Dame; and William Dembski, mathematician at Baylor University. Their alignment is indicated by their participation in the Mere Creation conference in 1996 and by their involvement in contributing to or endorsing key Design texts, for example, P. William Davis and Dean Kenyon, *Of Pandas and People* (Dallas: Haughton Publishing, 1989, rev. 1993).

16. In the foreword to *Mere Creation* (ed. William Dembski [Downers Grove, Ill.: InterVarsity, 1998]), chemist Henry Schaeffer describes the makeup of 180 participants at the Mere Creation conference in Los Angeles, November 1996. The group was nonhomogeneous but had two things in common: "Virtually all the participants questioned the reigning paradigm of biology—namely, that natural selection and mutation can account for the origin and diversity of all living things. At the same time virtually none of the conference participants were creationists of the sort one frequently reads about in the popular press. In particular, a very large majority of the participants had no stake in treating Genesis as a scientific text."

17. The vote removed the Republican members who had supported the change, and the board has reverted to "full-evolution" teaching standards.

18. Barbara Forrest and Molleen Matsumura, "Supreme Court Rejects Evolution Disclaimer," *Reports of the National Center for Science Education,* 20, no. 1–2 (January-April 2000).

19. See, for example, Ashley Montagu, ed., *Science and Creationism* (Oxford: Oxford University Press, 1984); Niles Eldredge, *The Monkey Business* (New York: Washington Square Press, 1982); Philip Kitcher, *Abusing Science* (Cambridge, Mass.: MIT Press, 1982); Michael Ruse, *Darwinism Defended* (New York: Addison-Wesley, 1982); Ronald Numbers, *The Creationists* (New York: Alfred A. Knopf, 1992); and Michael Ruse, ed., *But Is It Science? The Philosophical Question in the Creation/Evolution Controversy* (Buffalo: Prometheus, 1988).

20. Among these are the polemics by Robert Pennock (*Tower of Babel* [Cambridge, Mass.: MIT Press, 1999]) and Kenneth Miller (*Finding Darwin's God* [New York: Cliff Street, 1999]) and a provocative "Special Issue on Intelligent Design," *Rhetoric and Public Affairs* (winter 1998). A few other studies have appeared that treat Intelligent Design: Robert Pennock, ed., *Intelligent Design Creationism and Its Critics* (Cambridge, Mass.: MIT Press, 2001), and a brief but excellent review by Donald Yerxa, "Phillip Johnson and the Origins of the Intelligent Design Movement, 1977–1991," *Perspectives on Science and Christian Faith,* March 2002.

21. In the opening chapters, many basic ideas in the rhetoric of science are woven into my discussions. I review briefly the development of the rhetoric of science in the concluding chapter, and more extensively in appendix 4.

22. There are good reasons for using the term "aroused" originally coined by Kant to describe his own "arousal from dogmatic slumber" by reading Hume. This term was adapted by Paul Maclean,

Yale brain physiologist and founder of the NIH Brain Evolution Laboratory, to describe the likely impact of *Evolution: A Theory in Crisis* by Michael Denton (Bethesda, Md.: Adler and Adler, 1986). Maclean's unpublished commentary on Denton compared Kant's arousal by Hume to that which an "entire audience" would experience after reading Denton. Such an "arousal" is seen in the scientific-awakening stories of Michael Behe, Phillip Johnson, Dean Kenyon, and many others.

23. From "So Near and Yet So Far," *New York Review of Books,* 20 October 1994, 26, quoted in Karlyn Campbell and Kathleen Hall Jamieson, "Form and Genre in Rhetorical Criticism: An Introduction," in *Readings in Rhetorical Criticism,* ed. Burgchardt (State College, Pa.: Strata, 1995), 4.

24. Walter Fisher, *Human Communication as Narration* (Columbia: University of South Carolina Press, 1987), 48. See also Fisher (1994).

25. The factual stories seem to be Fisher's emphasis, although he explicitly discusses the other level of fantasy themes and, crucially, as he evaluates them, he includes them within his narrative paradigm.

26. Fisher (*Human Communication as Narration,* 63–64) clarifies Bormann's fantasy theme and expresses my own reservations about the term: "Fantasy, Bormann holds, [means] . . . 'the creative and imaginative interpretation of events that fulfills a psychological or rhetorical need.' Fantasy themes arise 'in group interaction out of a recollection of something that happened to the group in the past or a dream of what a group might do in the future.' When woven together, they become composite dramas, which Bormann calls 'rhetorical visions.' From the narrative view, each of these concepts translates into dramatic stories constituting the fabric of social reality for those who compose them. They are, thus, 'rhetorical fictions,' constructions of fact and faith having persuasive force, rather than fantasies." My comment about fantasy themes being a mixture of "fact and faith" derives from this rich Fisher quote.

27. My research connects here with that of Ernest Bormann, "Fantasy and Rhetorical Vision: The Rhetorical Criticism of Social Reality," *Quarterly Journal of Speech* 58 (1972): 396–407; Donald Shields, "Symbolic Convergence and Special Communication Theories," *Communication Monographs* 67, no. 4 (December 2000); and many others who have worked in the subfield of rhetoric, "Symbolic Convergence Theory" (formerly "Fantasy Theme Analysis").

28. Bormann, "Fantasy and Rhetorical Vision," 398.

29. Phillip Johnson, *Darwin on Trial,* rev. ed. (Downers Grove, Ill.: InterVarsity, 1993).

30. I hesitate to cite a personal illustration of this chaining but will do so this once. Asked by faculty at Texas Tech University and the University of Minnesota to present a lecture on Intelligent Design and its empirical critiques of Darwinism, I continued the chaining process and chose the title "Is Darwin's Ship Sinking?" This lecture was videotaped and published and is available through the Apologetics.org web site.

31. Shields, "Symbolic Convergence," 397.

32. Denton, *Evolution: A Theory in Crisis,* 358–59.

33. Richard Dawkins, *The Blind Watchmaker* (New York: W. W. Norton, 1985).

34. See Johnson's "Introduction" chapter of *Reason in the Balance* (Downers Grove, Ill.: InterVarsity, 1995). His clearest and strongest statement on naturalism is found in *Darwin on Trial,* 116–18.

35. This point was made quite clearly in his Berkeley faculty seminar. The notes of that seminar are included in appendix 1, and my expanded discussion is found in chapter 4.

36. His Ph.D. in mathematics is from the University of Chicago, and his Ph.D. in the philosophy of science is from the University of Illinois at Chicago. He also has an M.Div. from Princeton Theological Seminary.

37. Charles Darwin, *The Origin of Species,* 6th ed. (London: John Murray, 1859), 182.

38. This does not mean that people in the movement see no connection between the "intelligence" manifest in biological complexity and a belief in God. Behe is a Roman Catholic, and Johnson is a Presbyterian. Both have stated their metaphysical bias—they believe in God. What they deny is that science has the power to "adjudicate" by itself the existence or nonexistence of God. Science, they say, should remain neutral or open on this question.

39. In *Nature,* Jerry Coyne invoked Duane Gish of the Institute for Creation Research (ICR), thereby striving to marginalize Behe. Marginalizing Behe is an important part of the Darwinists'

rhetorical strategy, but this has proved more difficult than expected because of his repeated coverage in the *New York Times*. In August 1996 the newspaper printed a moderately favorable book review of *Darwin's Black Box* (Shreeve, "The Design of Life"). This led to an invitation to summarize his arguments in the *Times* editorial pages in late October 1996 ("Darwin Under the Microscope"). When Behe appeared in a nationally televised PBS debate, he and Design were profiled in a positive light in the "Week in Review" section ("New Light on Creation," 29 December 1997). Behe was once again invited to write an opinion column in the wake of the Kansas uproar ("Teach Evolution," see also pages 12–13 in this chapter). Once Behe and his "irreducibly complex" cellular machines began to take their bow on the stage of the *Times,* Design became harder to stereotype as "recycled biblical creationism."

40. Johnson said, "The greatest rhetorical challenge is to gain a hearing. One must escape the 'Inherit the Wind' stereotype." Phillip Johnson, "The Rhetorical Problem of Intelligent Design," *Rhetoric and Public Affairs* 1, no. 4 (winter 1998).

41. As I mentioned earlier, a conference of 180 scholars held in Los Angeles in November 1996, entitled Mere Creation, is considered the official birth of Design as an organized scientific movement. See the book *Mere Creation,* edited by William Dembski. It contains the eighteen papers that were presented at this three-day conference, including those by Johnson and Behe. Denton attended and interacted but did not speak.

42. As Design has developed, one can observe a parallel emergence of its own *subgenre* of discourse—a "type" of literary and oral expression with a well-defined cluster of traits. I have alluded to three: *in content* the genre is minimalist (fewer claims are argued than creation science), *in terms of ethos and style* it is erudite and rhetorically sophisticated, and *in terms of fundamental purpose* it is radically critical of Darwinian claims.

43. See, for example, Eugenie Scott, "Not (Just) in Kansas Anymore," *Science,* 5 May 2000.

44. By saying "too many anomalies," I mean "too many anomalies *that have resisted all attempts to resolve, in terms of the existing paradigm's puzzle-solving procedures.*"

45. Design proponents have pointed out that this "shutting out" from publication in mainstream journals is not surprising given the radical, revolutionary nature of its explanations.

46. The words "Kuhnistically know" refer to the belief or knowledge of those in Design. While Denton did not propose intelligent design, he invited speculation along those lines by his rhetorical devastation of Darwinism and by his descriptions of the marvels of biological engineering in his chapter "The Puzzle of Perfection" in *Evolution: A Theory in Crisis.*

47. Johnson calls himself the "leading edge" of the wedge, and his colleagues such as Behe, Jonathan Wells, and Stephen Meyer enter after him. See Johnson's *Testing Darwinism* (London: Inter-Varsity, 1997) and Johnson's fifth book, *The Wedge of Truth* (Downers Grove, Ill.: InterVarsity, 2000).

48. Johnson was quoted in this vein by the *New York Times* in their front-page coverage on 12 August 1999; he used the "Vietnam" analogy to characterize the national situation symbolized by the Kansas Board vote.

49. The April 2002 special issue featured short pieces by Design theorists and rebuttals by Darwinists. The debate that month, recorded for the four participants but not available for the public, featured on the anti-Design side Kenneth Miller (biologist at Brown University) and Robert Pennock (philosopher and author of *Tower of Babel*). For news and commentary on this debate, see the front-page coverage in the *Cleveland Plain-Dealer,* 24 April 2002.

50. Perceived, that is, among audiences that are pro or neutral toward the claims of Design. Opponents would see this movement's success as the ultimate nightmare and infuriating setback for genuine science. I observe that even this gloomy viewpoint could be referred to as "dramatic supremacy"—in an inverted, negative sense.

51. Darwin's own movement, at the point *The Origin of Species* was published in 1859, also possessed "dramatic supremacy" in my sense of the phrase, and that supremacy was realized and played out in the decades that followed.

52. In the *New York Times* coverage (12 August 1999) of the Kansas decision, Phillip Johnson was quoted by name in the front-page news report, and he described the situation as the "science educators' Vietnam."

53. When AP writers were polled on the "news story of 1999," first place went to the Kansas evolutionary brouhaha.

54. Note Kuhn, *Structure of Scientific Revolutions,* especially 136–38 and chapter 11, "The Invisibility of Revolutions."

55. Kuhn had stressed (*Structure of Scientific Revolutions,* 77) the importance of a new "replacement candidate" for any existing paradigm before a revolution could occur. In this regard, Denton *(Evolution: A Theory in Crisis)* and Johnson *(Darwin on Trial)* might be seen as "pre-revolutionary." They simply *attacked Darwinism* by piling up scientific anomalies and analyzing the underlying philosophical assumptions. More recently, Behe and Dembski have made design theory revolutionary in a Kuhnian sense by proposing a new heuristic paradigm that organizes research in a different direction.

56. It is the theme of "known and tolerated falsehoods" upon which Wells has seized in *Icons of Evolution.* This is the book Eugenie Scott warned about earlier in this chapter.

57. See the videocassette *Opening Darwin's Black Box* (New Port Richey, Fla.: Trinity College, 1998). In this video, Behe recalled his anger (1) upon his reading of Denton, realizing that he had been terribly misled for years about the solidity of Darwinism, and (2) upon his reading of recent textbooks describing the "origin of life" problem as virtually solved. Behe said that this is *flagrantly false, and is widely known to be such.* The chemical evolution of life is a "massive" unsolved problem, says Behe.

58. One line of evidence that this is correct—that academia generally regards skeptics of macroevolution as absurdly irrational—is that challenges to evolution are frequently compared to arguments from the Flat Earth Society.

59. See Phillip Johnson, *Evolution as Dogma: The Establishment of Naturalism* (Richardson, Tex.: The Foundation for Thought and Ethics, 1990), and his "Notes on the Berkeley Faculty Colloquium" (see appendix 1). I will describe this event in some detail in chapter 4.

Chapter 2: Murmurs of Dissent

1. Sir Julian Huxley, *Evolution after Darwin,* ed. Sol Tax, vol. 3 (Chicago: University of Chicago Press, 1960), the Centennial Celebration of *The Origin of Species.*

2. Sir Julian Huxley, "The Emergence of Darwinism," *Evolution of Life,* ed. Sol Tax (Chicago: University of Chicago Press, 1960), 1, as quoted in Denton, *Evolution: A Theory in Crisis,* 75.

3. See the closing of Darwin's *The Origin of Species;* for a description of the beginnings of prebiotic evolutionary theory, see Robert Shapiro, *Origins: A Skeptic's Guide to the Creation of Life* (New York: Bantam Books, 1986); and Charles Thaxton, Walter Bradley, and Roger Olsen, *The Mystery of Life's Origin* (New York: Philosophical Library, 1984).

4. A definitive historical account of the modern creationist movement is Ronald Numbers's *The Creationists.*

5. George A Kerkut's *Implications of Evolution* (London: Pergamon, 1960) is one that could be viewed as casting doubt on macroevolution.

6. Denton interned at a hospital in Bath, United Kingdom, in early 1968 and was surgical intern at Bensham General Hospital in Gateshead in early 1969. Registered as Medical Practitioner in Great Britain in July 1969, he did general practice briefly before entering King's College, London. He completed his Ph.D. there in biochemistry (1974). He was lecturer at La Trobe University in Melbourne (1975), and in 1976 he entered general practice in Melbourne. In 1977–78, he was at Royal Hobart Hospital in Hobart, Australia, doing registrar work and receiving pathology training. Returning to London, he did pathology and genetics training at Hammersmith Hospital and University College, London University. In 1979–80, he was at the University of Toronto (including residencies at Sunnybrook Medical Centre and Toronto General Hospital).

7. All quotes in the paragraph come from a videotaped interview held in June 1988, during the conference "Sources of Information Content of DNA" organized by the Ad Hoc Origins Committee and held in Tacoma, Washington.

8. The immediately precipitating cause of the seminar—an encounter of several scientists on both sides of the questions at a picnic in Switzerland—is told in the opening pages of P. S. Moorehead and M. M. Kaplan, eds., *Mathematical Challenges to the Neo-Darwinian Interpretation of Evolution* (Philadelphia: Wistar, 1967).

9. Moorehead and Kaplan, *Mathematical Challenges*. Also quoted by Phillip Johnson, *Darwin on Trial,* 38–39. Clearly, it can be noted that in this conference, talk about neo-Darwinian theory had shifted from Julian Huxley's epideictic occasion of rhetoric (that is, commemorative) to a forensic-type setting or situation (legal investigation of the past).

10. Moorehead and Kaplan, *Mathematical Challenges*.

11. Denton, *Evolution: A Theory in Crisis,* 327.

12. From Koestler's *Beyond Reductionism,* as described and quoted in Denton, *Evolution: A Theory in Crisis,* 327–28.

13. A typical example is *The Mystery of Life's Origin* (Thaxton, Bradley, and Olsen), chapter 1, which has over a page devoted to the highlights of Wistar.

14. Pierre Grassé, *Evolution of Living Organisms* (New York: Academic Press, 1977).

15. Pierre Grassé, quoted in Theodosius Dobzhansky, "Darwinian or 'Oriented' Evolution?" *Evolution* 29 (June 1975): 376–78.

16. Ibid. This quote, often misunderstood, does not say that macroevolution is mythical in the sense of "an invented and untrue story." Rather, it is described (in Dobzhansky's description of Grassé's thought) as "mythical" in the sense of a story that purports to be (is spoken as if it is) *well understood* in its network of causation.

17. Quoted both in Johnson, *Darwin on Trial,* 174–75, and idem, "Darwinism's Rules of Reasoning," in *Darwinism: Science or Philosophy?* ed. Buell and Hearn (Dallas: Foundation for Thought and Ethics, 1993).

18. See Johnson, "Darwinism's Rules." Also see Johnson, *Darwin on Trial,* in which Grassé is cited and quoted a half-dozen times.

19. Stephen Jay Gould, *The Panda's Thumb* (New York: W. W. Norton, 1985).

20. At the Darwin Centennial, Harvard paleontologist George Gaylord Simpson summarized the lack of transitions at higher taxonomic levels. In *Evolution of Life* (ed. Tax), he said, "They are not, as a rule, led up to by a sequence of almost imperceptibly changing forerunners such as Darwin believed should be usual in evolution. . . . Gaps among known species are sporadic and often small. Gaps among known orders, classes, and phyla are systematic and almost always large" (135).

21. Stephen Jay Gould, "The Episodic Nature of Evolutionary Change," in *The Panda's Thumb* (New York: W. W. Norton, 1985), 182ff.

22. Gould rehabilitates Goldschmidt in "The Return of the Hopeful Monster," *Panda's Thumb,* 186–93.

23. Not surprisingly, Duane Gish of the Institute for Creation Research seized this rhetorical opportunity. See his "Evolution: The Changing Scene," an Institute for Creation Research essay published in April 1988.

24. Stephen Jay Gould, *Ever Since Darwin* (New York: W. W. Norton, 1980).

25. Sir Fred Hoyle and Chandra Wickramasinghe, *Evolution from Space* (London: J. M. Dent, 1981), 130. See also Hoyle and Wickramasinghe's other publications: *Lifecloud* (London: J. M. Dent, 1978); *Diseases from Space* (London: J. M. Dent, 1979); *Space Travelers: The Bringers of Life* (Cardiff: University of Cardiff Press, 1981); and *The Intelligent Universe* (London: J. M. Dent, 1982).

26. Hoyle and Wickramasinghe, *Evolution from Space,* 96.

27. See Dawkins, *The Blind Watchmaker,* chapter 6, where the analogy is quoted and turned on its head.

28. Hoyle and Wickramasinghe, *Evolution from Space;* see also *The Intelligent Universe.* Also discussed in the epilogue of *Mystery of Life's Origin* (Thaxton, Bradley, and Olsen).

29. See Denton, *Evolution: A Theory in Crisis,* 138–39.

30. See British journalist Thomas Bethell's article, "Agnostic Evolutionists," in *Harper's* (February 1985). The article was focused on Colin Patterson and a group of theorists called "transformed

cladists." It highlighted his famous question, as well as his doubts as to whether we really have any solid knowledge yet about evolution.

31. Kenyon's movement from a leading advocate to leading skeptic is reflected in his foreword to a skeptical review of the topic, *Mystery of Life's Origin* (Thaxton, Bradley, and Olsen).

32. Even Sidney Fox, a leading origin of life researcher, conceded in his review of *Mystery of Life's Origin* in *Quarterly Review of Biology*, June 1985, that the authors (Thaxton, Bradley, and Olsen) had made a strong case in their chapter "The Myth of the Prebiotic Soup."

33. See Kenyon's foreword to *Mystery of Life's Origin* (Thaxton, Bradley, and Olsen) for a summary of this mystery (which still remains a mystery).

34. Sir Francis Crick, "Directed Panspermia," *Icarus* 19 (1973): 341–46.

35. Sir Francis Crick, *Life Itself* (New York: Simon and Schuster, 1981).

36. "Crisis? What Crisis?" in *New Scientist*, 13 June 1985, 33.

Chapter 3: The Birth of Design

1. The book was published in Japan (1988) as *Han Shinkaron* by Dobutsu Sha and in France (1989) by Flammarion.

2. See Denton, *Evolution: A Theory in Crisis*, 345; this is one of the most often-quoted summary statements in his book.

3. In a 1985 letter to Denton, Dr. Marcel Schutzenberger, a principal figure in the Wistar conference, said, "Dr. M. Eden has communicated to me your magnificent book on evolution. I read it with great interest since it contains many facts I ignored (I am not a biologist: just a more or less stupid computer scientist with an old Ph.D. in medicine) and many extremely well thought arguments. . . . Receive again my warmest congratulations for your superb work."

4. Quoted in Thomas Bethell's review, *Washington Times Magazine*, 28 April 1986. In a letter to Denton, cited above in n. 3, Schutzenberger concurred, adding that the book was "magnificent" and "superb" (ibid.).

5. This is from a private letter (March 1986) to the publisher Adler and Adler, offering a blurb (copy in my possession).

6. Ashley Montagu's review was in both the *New Yorker* and the *Chicago Sun-Times* on 20 April 1986. His generous spirit toward Denton (with reservations) ranks Montagu's review as one of the most amazing responses ever given by an evolutionist to secular anti-Darwinist criticism. For example, "Imagine my surprise, then, when in reading Denton's critique of evolutionary theory, I found him to be a writer of the most astonishing range of knowledge in the natural sciences, and a scientist whose criticisms are, for the most part, just and telling." He "corrects" Denton in one area: the evidentiary deficiencies (which Montagu concedes!) cannot be construed at all as posing any question about the "*fact* of evolution" (his italics).

7. Lee Dembart, *Los Angeles Times*, 10 July 1986.

8. Paul Preuss, *San Francisco Chronicle*, 25 May 1986.

9. Philip Spieth, in *Zygon*, June 1987; I rank Spieth's review as the most hostile and alarmist of all reviews.

10. Comment in letter of David M. Hillis to fellow University of Texas zoologist Bassett Maguire, who had lent him the book and asked for his comments. UT biochemist Gordon Mills then wrote a letter to Maguire criticizing Hillis's dismissive rhetoric and defending Denton's handling of the biochemistry issues in chapter 12 of *Evolution: A Theory in Crisis*.

11. Niles Eldredge, review of *Evolution: A Theory in Crisis*, in *Quarterly Review of Biology* 61, no. 4 (December 1986): 541–42.

12. Eldredge's review (ibid.) and Ruse's *(New Scientist)* will be scrutinized at the end of the chapter in more detail, and Ridley's ("More Darwinian Detractors," *Nature* 318 [14 November 1985]: 124–25) will be quoted to some extent.

13. Private letter to me, 20 February 1987.

14. Denton's book, much broader in scope, overshadowed *Mystery of Life's Origin* (Thaxton, Bradley, and Olsen), which was focused on chemical evolution. Also, *Mystery of Life's Origin* was

somewhat more technical in some sections, focusing on chemical, thermodynamic, and geological issues at a moderately advanced level.

15. Two other books published at the same time that were important enough for Ridley to attack alongside Denton in *Nature* were Michael Pitman, *Adam and Evolution* (London: Century Hutchinson, 1984), and Alan Hayward, *Creation and Evolution* (London: Triangle, SPCK, 1985).

16. Chapter 3 traces the turning of Darwin's theory into a self-evident axiom, for which no proof is needed. To Denton (*Evolution: A Theory in Crisis*, 77) it "is still, as . . . in Darwin's time, a highly speculative hypothesis entirely without direct factual support and very far from that self-evident axiom some of its more aggressive advocates would have us believe."

17. Denton, *Evolution: A Theory in Crisis*, 29–35.

18. This new metanarrative, part of the founding lore of Design, can be pictured imaginatively (in my own projection) as if the scientific world is trying to find its way out of a vast labyrinth. Darwinists have found a tunnel that they are *positive* is the way out. Denton inspects this tunnel with its twists and turns, its bulges and loops, but sees that it is a dead end. He tells the Darwinists so, urging them to recognize their error, move back, and start looking for a better tunnel. The Darwinists are indignant. They point out that Denton is a "nonexpert" in tunnel morphology and insist that their research is on track—they are making "so much progress in passageway research!" Denton replies, "No, your research is just mapping, in ever finer detail, the exact contours of a cul-de-sac."

19. Kenneth Burke first describes the term "consubstantiality" as a key concept in communication in his *Permanence and Change*, 2d ed. (Berkeley: University of California Press, 1954). See the chapter-section "Identification and 'Consubstantiality'" in *A Rhetoric of Motives*, 20–23.

20. Stephen Jay Gould, *Wonderful Life* (New York: W. W. Norton, 1989).

21. Robert Jastrow, *God and the Astronomers* (New York: W. W. Norton, 1992; 2d ed., 2000); Timothy Ferris, *The Whole Shebang* (New York: Simon and Schuster, 1997).

22. Such as "empirical observation," "empirical reasons," "logically compelling evidence," "empirical basis," etc.

23. Ridley, "More Darwinian Detractors."

24. Koestler, *Beyond Reductionism.*

25. Of twenty-five reviews I read, only Eldredge's brief review *(Quarterly Review of Biology)* failed to mention Denton's credentials.

26. See Bethell's praise (in *National Review* and *Washington Times*). Berkeley geneticist Spieth says, "Denton, however, contends that molecular biology provides new evidence for a typological view of organisms. . . . [H]is conclusions are based upon an artifact produced by faulty interpretations of the data. Since Denton's professional training is . . . in molecular biology, a detailed look at the situation is in order." Spieth corrects Denton's "errors."

27. I bracketed "chance," misspelled in Ruse's *New Scientist* review as "change." See Denton, *Evolution: A Theory in Crisis*, 342.

28. Michael Ruse, *New Scientist.*

29. For example, Denton totally convinced George Lebo, an astronomer at the University of Florida, who bought copies for all his colleagues. I interviewed about fifty Design leaders, asking about Denton's relative importance to them. Almost all indicated that Denton was by far the most powerful early influence in their skepticism.

30. From Montagu's review in the *Chicago Sun-Times*, 20 April 1986.

31. "Common sense" is a strong rhetorical resource to tap into. This phrase is used repeatedly in the video interview of Denton. (See "Focus on Darwinism: An Interview with Michael Denton," available from Access Research Network at www.arn.org.)

32. See *Opening Darwin's Black Box: An Interview with Michael Behe* (Clearwater, Fla.: 1998), and also Thomas Woodward, "Meeting Darwin's Wager," *Christianity Today*, 23 April 1997. The other two texts he used were Dawkins's *The Blind Watchmaker* and Kuhn's *Structure of Scientific Revolutions.*

33. See Tim Stafford's "The Making of a Revolution," *Christianity Today*, 8 December 1997.

34. Behe and Johnson are "robust theists," by which I mean any theistic belief, whether Christian, Jewish, Muslim, or other, that holds to the existence of a transcendent creator who can act within the universe he has made.

Chapter 4: The Virus Spreads

1. Cynthia Russett, *Darwin in America* (San Francisco: W. H. Freeman, 1976), 210.

2. Denton, *Evolution: A Theory in Crisis,* 351.

3. After stressing the importance of a replacement paradigm, Denton adds: "Consequently, biologists wishing to operate within a scientific framework, even those only too well aware of the seriousness of the problems, have no alternative at present but to continue to subscribe to the Darwinian world view. It seems more than likely that, given the need for and the priority of paradigms in science, the philosophy of Darwinism will continue to dominate biology even if more by default than by merit; and that until a convincing alternative is developed the many problems and anomalies will remain unexplained and the crisis unresolved. The lack of any scientifically acceptable competitor leaves evolutionary biology in a state of crisis analogous to the crisis in medieval astronomy when, although the Ptolemaic system was admitted to be a monstrosity, the lack of any conceivable alternative imprisoned the science for centuries within the same circle of belief" (Denton, *Evolution: A Theory in Crisis,* 356–57).

4. "Suasory," a term used in communication theory (as in Bryant's definition of *rhetoric* as "the rationale of informative and suasory discourse"), means "related to the act of influencing or persuading." Here I adapt the wording used in a definition of "suasion" in *Webster's New Collegiate Dictionary* (1975 edition).

5. Evidence for Denton's influence on one example—the bird lung—is seen in noting the *total absence* (as far as I can determine) of this example of discontinuity in creationist literature before 1986, and then its frequent mention afterwards. See, for example, Gary Parker, *What Is Creation Science?* (1988); Johnson, *Darwin on Trial* (1993); Thomas Woodward, "Doubts about Darwin," *Moody Magazine,* September 1988.

6. These words appear in Eldredge's review in the *Quarterly Review of Biology,* December 1986, and the same idea ("the book belongs to the 'creation science' genre") is voiced by Spieth in *Zygon,* June 1987, 252–57.

7. Ashley Montagu, review in *Chicago Sun-Times,* 20 April 1986.

8. In March 1992, at the Darwinism Symposium in Dallas, entomologist Arthur Shapiro sketched for me his view that Denton was a significant turning point in the rise of skepticism of macroevolution—one that rightly provoked concern among evolutionists. He said, in appreciation/criticism, "What was good in Denton was not that new, and what was new [his biochemical isolation chapter] was not that good." See Spieth's review in *Zygon.*

9. All quotations in this section are found, sometimes more than once, in Spieth's review in *Zygon.*

10. Spieth's observation is probably true, but many would say that failure to pass "peer review" by biologists would not be due to inherent scientific flaws but rather to the failure of Denton to abide by the naturalistic paradigm rules. These rules imply that a deeply problematic paradigm is better than no paradigm at all.

11. The Japanese edition, entitled *Han Shinkaron,* was published by Dobutsu Sha. The French edition, originally released by Editions Londreys with a foreword by Marcel Schutzenberger, was rereleased in paperback by Flammarion. In the French popular science journal *Science et Vie* (March 1987, 40–53, 62), Sven Ortoli wrote a long review article that condensed and substantially endorsed the arguments of Denton.

12. His review argues: (1) The Darwinian mechanism is not "random" since natural selection is the opposite of random—it selects the finest genetic packages, ones that give superior aid to survival and reproduction. (2) Denton's chapter "A Biochemical Echo of Typology"—showing by protein sequences that each group is equally isolated from out-groups, unconnected by intermediates—betrays a basic confusion about how those protein sequences arose.

13. From Spieth's review in *Zygon*.

14. He met Johnson at a Berkeley campus fellowship of Christian professors (July 2000 interview with Johnson).

15. See Richard Dawkins, *The Blind Watchmaker* (New York: W. W. Norton, 1987), 4, and the listing in the bibliography, 325; and John Campbell, "Intelligent Design, Darwinism, and the Philosophy of Public Education," *Rhetoric and Public Affairs* 1, no. 4 (1998): 471.

16. In Gertrude Himmelfarb's biography of Darwin (*Darwin and the Darwinian Revolution* [Chicago: Elephant Books, 1959], 35), she notes that Darwin not only read the "required Paley reading" at Cambridge but that he "enjoyed the experience so much that he was inspired to take up another of Paley's books, *Natural Theology*, which was not required reading. For the youthful Darwin, as for [students and professors at] Cambridge, right thinking on religion began and ended with Paley." Himmelfarb adds, "And toward the end of his life he reaffirmed his early impression that the analysis of Paley was the only part of the university course which was 'of the least use to me in the education of my mind.' [From his Autobiography, found in *Life and Letters of Charles Darwin*, I, 47]." Of course, adds Himmelfarb, this comment referred not to the acknowledged truth of Paley's premises, but the educational value of interacting with Paley's thinking processes (36).

17. Dawkins, *The Blind Watchmaker*, 5.

18. This wording appeared in an early draft of *Darwin on Trial*. Sadly, these words did not survive the "final edit" and are not in the book. This is reminiscent of the tag "important if true" by Matthew Berke, in appraising Johnson's scientific critique, in his symposium *Evolution as Dogma: The Establishment of Naturalism* (booklet, Richardson, Tex.: Foundation for Thought and Ethics, 1991).

19. In Stafford, "The Making of a Revolution," Johnson said, "I read these books, and I guess almost immediately I thought, *This is it. This is where it all comes down to, the understanding of creation*" (emphasis in the original).

20. These recollections are mostly taken, almost exactly, from Russell Schoch's biographical article on Johnson, "The Evolution of a Creationist," *California Monthly*, November 1991, the University of California alumni magazine. A similar flow of recollection is in "The Making of a Revolution" (Stafford). I have injected one sentence from the Stafford version of the story here.

21. Information here is from personal interviews with Johnson, spanning from December 1988 through October 1991.

22. I would point here to some of his earliest recorded remarks, in the talk given by Johnson to the Ad Hoc Origins Committee, February 1990, videotape. Then I would cite a more recent article by Johnson, "The Rhetorical Problem of Intelligent Design." The main theme of that article is a statement of the same fundamental rhetorical obstacle.

23. This is clear from his Berkeley faculty colloquium memo and was repeated often by Johnson in later years.

24. Johnson described the reaction from law colleagues in a series of conversations during 1988–89.

25. The impression comes from Johnson's fellow professor, Michael Smith, whom I interviewed in September 2000.

26. The following account is constructed from the memo. Some descriptive sentences in my account are word-for-word parallel to Johnson's, and I did not always use quotation marks. I did not want to distract the reader and overemphasize exact wording where it was not needed. Johnson's own words are in quotation marks to emphasize it *as his own wording*.

27. Of the three alternative ways Johnson thought his critics could respond, note that points one and three could be seen as two sides of the same coin. That is, a person could conceivably say, "No, I do not assume naturalism conclusively" (option one); "rather, I conclude that Darwinism is right because I possess conclusive evidence for gradualistic macroevolution by natural selection" (option three). I simply observe here that, in a logical sense, the *three responses were not all mutually exclusive*, although option two does seem mutually exclusive with option one.

28. This comes from the faculty colloquium memo (appendix 1), and it was made clear through later conversations with Johnson.

29. When Johnson retold this moment, he mimicked the stony-faced stares of disgust on the faces of the scientists.

30. I sensed that a special rhetorical tool was being used here and groped for an appropriate term to describe it. I thought of "stigma-word," and after discussing this phenomenon and my proposed term with my dissertation chair, David Payne, I settled on this usage. I use this term from time to time through the rest of the book. One of my tasks as a historian is to note such stigma-words, trace their pattern of usage through the ongoing flow of rhetoric, and see how they functioned and how they were countered by the opposite side.

31. It was eighty-three pages single-spaced typed, including text and research notes.

32. Johnson referred often (in lectures, interviews, and conversations) to the "fun" of his Darwinian project, compared to the relative drudgery of his "day job" of teaching law at Berkeley.

33. Chapter 4, on the fossil record, is dominated by Gould and Eldredge; chapters 5 and 6 are devoted entirely to a point-by-point response to Gould's "evidences for macroevolution" in his famous essays that deal with Darwinism as fact and theory. In addition, chapters 1, 3, and 10 feature Gould quite prominently.

34. Raup is a prolific, world-renowned paleontological authority on the mass extinctions that have punctuated the earth's history. Gilkey was one of the star witnesses called by the prosecution in the Scopes II trial in Little Rock, Arkansas, 1980–81. Gilkey's testimony, as a theologian who testified that evolutionary science is fully compatible with theistic faith, helped to solidify the notion of religious neutrality of Darwinian science.

35. The Campion Summary paper, condensing his position into an outline of numbered paragraphs, is included as appendix 2.

36. Karlyn Kohrs Campbell, *The Rhetorical Act* (Belmont, Calif.: Wadsworth, 1982), 178.

37. See, for example, *The Rhetorical Act,* 119–22 and 178. These three factors were identified first by Aristotle in his own groundbreaking analysis in his treatise *Rhetoric.*

38. Phillip Johnson's letter to Jeremy Waldron, 3 October 1988, in reply to Waldron's 25 September 1988 letter to Johnson. Waldron said, "Neo-Darwinism does not make a universal claim, but invokes certain law-like propositions . . . in the explanation of . . . phenomena." He reminded Johnson of the question he was pressing at the colloquium, "[W]hat plays the role of those law-like statements in the 'explanation' that scientific creationists say they are offering?" Johnson replied with an informal essay of about two thousand words in which he clarified his own position and explained to what extent it is "scientific." His essay strategy was primarily *offense rather than defense:* He outlined the problematic role of the twin presuppositions of evolutionary biology—uniformitarianism and strict metaphysical naturalism. These are problematic, says Johnson, not because of any conflict with observation or experience, but rather because of their intrinsically assumptive character: *Science cannot prove them.* Johnson says, "Darwinists have evidence, of course, but the evidence itself has been characterized on the basis of these assumptions. This is true of supposed fossil evidence for macroevolution, just to give an example."

39. "Ethos" here refers to a *key aspect of the character* of a critic of Darwinism—*perceived bias dictated by religious assumptions.* If a speaker were introduced to a group of secular biologists as a "scientific creationist," that would produce a negative reaction before the speaker opened his mouth. Thus, Johnson here clarifies his ethos for Waldron.

40. Phillip Johnson's letter to Jeremy Waldron, 3 October 1988.

41. Montgomery Slatkin, September 1988 letter to Johnson. Earlier I outlined Spieth's "rhetorical epidemiology." Here Slatkin implies a parallel case: the need to *isolate* the Johnsonian mutant strain of Denton's doubt.

42. The comments of Johnson in his account of these rejections were made in the Portland Talk (February 1990, videocassette in my possession). It seems the book was simply espousing too radical a stance for academic press reviewers to give an endorsement. In addition, Johnson's position as a law professor, not a scientist, gave pause to publishers.

43. This publication, which originally appeared in the October and November 1990 issues of *First Things,* also contained five reactions: the strongly critical comments of geneticist Thomas Jukes (UC

Berkeley) and historian of science Will Provine (Cornell), milder or mixed reactions from paleontologist Gareth Nelson (American Museum) and Irving Kristol (editor of *Public Interest*), and sympathetic comments of *First Things* managing editor Matthew Berke. After the five responses, Johnson's "Reply to My Critics" was printed to round out the "symposium-in-print."

44. Also attending were two other conservative Christians: Michael Woodruff, president of the Christian Legal Society, and Charles Thaxton, a chemist whose book *Mystery of Life's Origin,* attacking the credibility of chemical-evolution scenarios, has been mentioned. Thaxton figures predominantly in the latter part of this chapter.

45. Johnson (in the 1990 "Portland Talk" video) reported that Gould said this.

46. The reconstruction of the events is based on (1) the videotape of the Portland meeting, two months later, in which recollections were shared by Charles Thaxton, Mike Woodruff, and Phillip Johnson from two months earlier; (2) telephone interviews with these three in November 2000, along with Johnson's e-mail in 2001; (3) David Raup's comments to me in two telephone interviews, held in October and November 2000; and (4) a phone interview with Charles Haynes in January 2003. Haynes was one of the organizers of the meeting. Raup confirmed that he views Johnson's scholarship, as seen in *Darwin on Trial,* as exemplary. Raup said that after his positive comments about Johnson's scholarship at Campion, Gould became quite angry and spoke to Raup about it afterward. Raup also confirmed that when Johnson stayed at his home for a visit later in 1990, he showed him a first draft of a foreword he had written for *Darwin on Trial.* Unfortunately, this piece never saw the light of day. Raup, who has it in his files but declined my request for a copy, said it had the analytical-rhetorical thrust of pronouncing a "pox on both of your houses" (creationist and Darwinist), or phrasing it differently, he added, "Maybe both sides are right." What does this mean? It may shed light on Raup's unique open-minded approach to reflect on what he said to Johnson at about the same time (1990–91) during a private chat: Probably, said Raup, there's a 40 percent chance that Darwinism will turn out to be the correct explanation, a 10 percent chance, at most, that some sort of creationist explanation will turn out to be right, and a 50 percent chance that the correct answer has not yet been glimpsed by anybody. In this regard (the 50 percent chance that no one has the answer), Raup appears to manifest much the same scientific spirit as Michael Denton.

47. The phrase in quotes was used by Raup in phone interviews (September and October 2000). The event was recreated with the help of memories shared by the three Campion participants, Phil Johnson, Charles Thaxton, and Mike Woodruff, who described Gould's attack as they reported on the meeting in the 1990 Portland meeting videotape. I also interviewed the latter three by phone in 2000 (and Charles Haynes in January 2003) to probe their memories on various specific points.

48. See Gould's description in "The Episodic Nature of Evolutionary Change," in *The Panda's Thumb.*

49. In Johnson's Portland Talk there was agreement about Gould's emotional body language. Thaxton (November 2000 interview) distinctly recalled an animated discussion of this phenomenon at Campion after Gould left early Saturday evening. The unusual agitation in his attack on Johnson became a subject of extended discussion. In an interview in 1990, Johnson said to me that Gould viewed him at Campion as a "menace and madman."

50. From the Portland Talk. Johnson, in a March 2001 e-mail, said: "Gould wasn't totally out of control, but he was very upset, shaking with anger—or what I call 'metaphysical panic.' It is very upsetting for a Darwinist authority to get into a debate on even terms with an outsider who doesn't let the Darwinist make the rules. He tried to bully me into submission, relying on his authority. I mainly just stayed calm and held my ground. I remember saying later that it was a draw, and others agreed. Of course a draw at that point amounted to a huge victory for me."

51. "Encouragement" in the sense of his scientific certification and encouragement to keep researching and writing. Raup even attended a private meeting of key members of Johnson's inner circle at a California resort, Pajaro Dunes, in 1993, which has now become legendary through the video documentary "Unlocking the Mystery of Life." Johnson's "frustration" reached one of its high points when Raup changed his mind and decided not to allow his foreword (of which he had shown Johnson a draft) to be used in *Darwin on Trial* (see endnote 46 above for details).

52. The word "menace" is one Johnson used to describe how Gould viewed him at Campion. See endnote 49 above.

53. Johnson is convinced this is the case—that Gould, after meeting and debating him at Campion, felt he had to try to neutralize the rhetorical impact of *Darwin on Trial.* Personal e-mail correspondence, March 2001.

54. These meetings were organized by John Wiester, a Stanford-educated geologist and rancher from southern California, and by William "Skeeter" Ellis, a Yale-educated lawyer based in Seattle. Financial support for the Ad Hoc Origins Committee meetings came from another Yale graduate, the Tacoma lumber benefactor C. Davis Weyerhaeuser.

55. Scientific and historical details of this paragraph are, in a sense, common knowledge, but a good review of them will be found in *Mystery of Life's Origin,* 18–39, 62–66, and in Shapiro's outstanding review of the topic, *Origins: A Skeptic's Guide.* For a more up-to-date glimpse of Shapiro's (slightly evolved) views on this topic, see his more recent book, *Planetary Dreams* (New York: John Wiley and Sons, 1999).

56. The field was overripe for critique. That Thaxton, Bradley, and Olsen had used an effective rhetorical strategy is borne out by many responses I got in 1985–88 as I asked scientists to review the book. For example, Paul Fuerst, chair of Molecular Genetics at Ohio State University and a strong evolutionist, was very pleased with the book's overall quality and even described the current abiogenesis scenarios as so much "hand-waving" (personal interviews).

57. For example, to give some sense of the technical "feel" of the book: Chapter 2, "Simulation of Prebiotic Monomer Synthesis," has 57 footnotes; chapter 3, "The Myth of the Prebiotic Soup," has 74 footnotes; and chapter 5, "Reassessing the Early Earth and its Atmosphere," has 97 footnotes, and most of these footnotes are from technical literature of peer-reviewed scientific journals and other similar publications.

58. Thaxton, Bradley, and Olsen, *Mystery of Life's Origin.*

59. Clifford Matthews, of the University of Illinois at Chicago, who pioneered abiogenesis research starting from cyanide compounds, praised the book as a timely, clear-headed appraisal (letter to Charles Thaxton, April 1985). Klaus Dose, a leading researcher in chemical evolution, cited *Mystery of Life's Origin* (Thaxton, Bradley, and Olsen) in a positive light in his 1988 review article, "The Origin of Life: More Questions Than Answers," *Interdisciplinary Science Reviews* 13, no. 4 (1988): 348.

60. The story of Kenyon's change is told in two video interviews, published by Access Research Network (www.arn.org). It is also retold in the 2002 videotape "Unlocking the Mystery of Life," published by Illustra Media (www.illustramedia.com).

61. See Shapiro, *Origins: A Skeptic's Guide.* Jastrow, the founder of NASA's Goddard Institute of Space Studies, has been a highly visible commentator on scientific matters for the latter third of the twentieth century. Although religiously agnostic, he showed his interest in the convergence of science and creation in the big bang event, as explained in his *God and the Astronomers* (rev. 1992, 2000).

62. Thaxton, Bradley, and Olsen received praise in reviews by University of South Florida chemist Charles Whittaker (*Florida Scientist,* June 1995), Yale Medical School epidemiologist James Jekel (*Yale Journal of Biology and Medicine,* December 1994, from which a short quote appears in the following sentence), and by public health scholar Richard Munday (*Journal of the American Scientific Affiliation,* January-March, 1985). Also, Yale biophysicist Harold Morowitz wrote a generally positive review, describing the book as an "interesting start with considerable scientific thrust."

63. The harshest review (which began with the memorable declaration, "I'm an agnostic and proud of it!") was a front-page review by Richard Lemmons in *Chemical and Engineering News,* April 1985. It sparked a dozen letters either critical of Lemmons's review or defending *Mystery of Life's Origin.* Abiogenesis researcher Sidney Fox's rather negative review in *Quarterly Review of Biology* (June 1985) nevertheless praised a key chapter, "The Myth of the Prebiotic Soup."

64. Thaxton, Bradley, and Olsen, *Mystery of Life's Origin,* 188–217.

65. Hoyle and Wickramasinghe, *Lifecloud* (1978), *Space Travelers* (1981), *Evolution from Space* (1981), and *The Intelligent Universe* (1982).

66. Thaxton's papers are "DNA, Design, and the Origin of Life" (1987), first published in *Darwinism under the Microscope* (Orlando, Fla.: Charisma, 2002), and "In Search of Intelligent Causes: Some Historical Background" (unpublished, 1988), Thaxton's paper at the Tacoma conference. Together, these papers outlined most of the main conceptual innovations of Thaxton, and they were circulated widely in the Design Movement. They became "founding documents" of Design as a scientific movement.

67. Yockey argued in the *Journal of Theoretical Biology* (vol. 67 [1977]: 377, and vol. 91 [1981]: 13) that a "mathematical identity" exists between the informational text of a human sentence and the genetic texts of DNA and proteins. His *Information Theory and Molecular Biology* (Cambridge, England: Cambridge University Press, 1992) further developed this identity.

68. Charles Thaxton, telephone interview, November 2000.

69. Michael Lemonick, "Evolution under Fire," *Newsweek*, 21 March 1993; and Erik Larson, "Darwinian Struggle: Instead of Evolution, A Textbook Proposes 'Intelligent Design,'" *Wall Street Journal*, 14 November 1994.

70. These two were Charles Thaxton and Michael Woodruff, president of the Christian Legal Society.

71. Personal letter to Thomas Woodward, 12 April 1989.

72. From interviews with Phillip Johnson (April 1990) and Charles Thaxton (October and November 2000). "Intelligent design" had made its publishing debut with *Of Pandas and People* (Davis and Kenyon), but this was a textbook supplement for teachers. There was no design proposal written for the public and released by a major publisher until *Darwin's Black Box*.

Chapter 5: Putting Darwin on Trial

1. "Pseudoscience" is the strongest stigma-word that Johnson employs in *Darwin on Trial*, and not only does it appear in the title and text of his closing chapter (in the first edition of 1991) but the concept is implied throughout the text.

2. Johnson, *Darwin on Trial*, 14. His comment on the bias of scientific creationists is one of several distancing statements that rhetorically separate Johnson from creation science.

3. Even Gould used the "fundamentalist" stigma-word on strict Darwinists Dennett and Dawkins (*New York Review of Books*, 12 June and 26 June 1997; Part 1 was titled "Darwinian Fundamentalism"). On this, see Johnson, "The Gorbachev of Darwinism," in *Objections Sustained* (Downers Grove, Ill.: InterVarsity, 1998). Language critic Kenneth Burke viewed this sort of symbolic linkage as a classic case of "perspective by incongruity." Our pious connotations of a symbol are dislodged "by violating the 'proprieties' of the word in its previous linkages." See his *Permanence and Change*, 88–92.

4. To clarify the prepositions: (1) "Between" is obvious, as in "between two extremes at the ends of the ideological range of possibilities." (2) "Above" is intended to picture his analytical stance of detachment—above and beyond the inherent constraining effect of commitments or biases of those two kinds of fundamentalism.

5. This is from Denton's letter to M. Brockey at Regnery Gateway, the publisher. This quote was used verbatim on the back cover of the 1991 hardcover edition. This blurb, with its visibility, possibly did more to spark projection themes of "devastating attack" and "embarrassment of Darwinism" than anything Johnson said in the book.

6. Stephen Jay Gould, "Impeaching a Self-Appointed Judge," *Scientific American* (July 1992): 118–22.

7. Ibid.

8. Information from a July 1992 interview with Phillip Johnson.

9. Johnson, *Darwin on Trial*, 116.

10. Norman Macbeth, *Darwin Retried* (Boston: Gambit, 1971); Grassé, *Evolution of Living Organisms;* Yockey, articles in *Journal of Theoretical Biology* (1977, 1981); Michael Pitman, *Adam and Evolution* (1984), Augos and Stanciu, *The New Story of Science* (Warner, N.H.: Principle Source, 1984) and *The New Biology* (Warner, N.H.: Principle Source, 1987); Bethell, "Agnostic Evolutionists," *Harper's,*

February 1985. Shapiro, while he wrote as an honest evolutionist, nevertheless added to skepticism of current theories of abiogenesis through his book, *Origins: A Skeptic's Guide* (1985).

11. My assertion can be contested. Some may say that Copernicus, Newton, or Einstein produced paradigms of greater import. In terms of a paradigm's effect in shaping metaphysics, I would reply that other paradigms may have indirect atheistic implications (e.g., a decentralization of the earth or the introduction of mechanism as a core notion of reality), but these paradigms imply no frontal assault on the intelligent design of nature's most complex systems. Darwinism clearly *does.* I agree with Richard Dawkins in *The Blind Watchmaker:* Objects in space, such as stars, being relatively simpler objects, do not logically suggest an intelligent-type explanation for their existence, but biological entities *do cry out* for such explanation due to their watchlike complexity. Darwin's solution to this old riddle is the *most dramatic answer* to the universe's seeming (even to Dawkins!) implicit suggestions of theism. I heartily endorse Dawkins's dictum: "Darwin made it possible to be an intellectually fulfilled atheist."

12. In most western secular universities, the naturalistic origin of the biosphere has been viewed for well over one hundred years as virtually *indisputable fact,* and this belief has had a constraining effect on the vast majority of teaching in the sciences that connects in any way to evolutionary issues. This constraining influence is roughly as powerful, though perhaps to a slightly lesser extent, in the arts and humanities. Such Darwinian assumptions marked out the "appropriate directions" of research and teaching for all persons and guilds in the sciences and humanities.

13. Furthermore, this framework of understanding the drama and the issues debated therein is also clearly part of the emotional and motivational frameworks of the actors and participants themselves.

14. Johnson's letter to Alvin Plantinga, 6 December 1990.

15. Johnson says, "We call our strategy 'the wedge.' A log is a seeming solid object, but a wedge can eventually split it by penetrating a crack and gradually widening the split. In this case the ideology of scientific materialism is the apparently solid log. The widening crack is the important but seldom-recognized difference between the facts revealed by scientific investigation and the materialist philosophy that dominates the scientific culture. What happens when the facts cast doubt on the philosophy? . . . My own books . . . represent the sharp edge of the wedge. I had two goals in writing those books and in pursuing the program of public speaking. . . . First, I wanted to make it possible to question naturalistic assumptions in the secular academic community. Second, I wanted to redefine what is at issue in the creation-evolution controversy so that Christians, and other believers in God, could find common ground in the most fundamental issue—the reality of God as our true Creator" (Johnson, *Testing Darwinism,* 92).

16. This goal is made visible in *Reason in the Balance* (Johnson), which forthrightly advocates and defends "theistic realism" as the most robust and fruitful metaphysical basis for research in all fields of the modern university.

17. Johnson writes, "[Darwinists'] scientific colleagues have allowed them to get away with pseudoscientific practices primarily because most scientists do not understand that there is a difference between the scientific method of inquiry, as articulated by Popper, and the philosophical program of scientific naturalism. *One reason that they are not inclined to recognize the difference is that they fear the growth of religious fanaticism if the power of naturalistic philosophy is weakened*" (*Darwin on Trial,* 156, emphasis added).

18. The highlights of this new phase of Design's story—rhetorical engagement with opponents in university-based conferences—would merit an entire chapter in this study. The most important conference, early on, was the "Darwinism Symposium" at SMU in March 1992 (see *Darwinism: Science or Philosophy?* ed. Buell and Hearn). Another highlight was the February 1997 conference organized at the University of Texas by UT philosopher Rob Koons: "Naturalism, Theism, and the Scientific Enterprise."

Perhaps the most important occasion of engagement was "The Nature of Nature," a conference at Baylor University, organized by Dembski in April 2000. The conference subtitle was "An Interdisciplinary Conference on the Role of Naturalism in Science." Speakers included Nobel laureates Steven Weinberg and Christian de Duve, cosmologist Alan Guth, Burgess Shale paleontologist Simon Conway

Morris, historian of science Ronald Numbers, philosophers Alvin Plantinga and John Searle, Michael Behe, Mark Ptashne, Horace Freeland Judson, and Everett Mendelsohn. The overview statement read: "Is the universe self-contained or does it require something beyond itself to explain its existence and internal function? Philosophical naturalism takes the universe to be self-contained, and it is widely presupposed throughout science. Even so, the idea that nature points beyond itself has recently been reformulated with respect to a number of issues. Consciousness, the origin of life, the unreasonable effectiveness of mathematics at modeling the physical world, and the fine-tuning of universal constants are just a few of the problems that critics have claimed are incapable of purely naturalistic explanation. Do such assertions constitute arguments from incredulity—an unwarranted appeal to ignorance? If not, is the explanation of such phenomena beyond the pale of science? Is it, perhaps, possible to offer cogent philosophical and even scientific arguments that nature does point beyond itself? The aim of this conference is to examine such questions." For a set of evaluations of this conference, see www.arn.org. I will discuss this conference briefly in chapter 9.

19. This refers to the major research universities of the United States and Europe, with their related professional associations, technical journals, and secular publishers.

20. See, in Johnson's *Reason in the Balance,* "The Subtext of Contempt" (the Philip Bishop story) and "Is God Unconstitutional?" (the story of Dean Kenyon). These cases describe what happens when a professor lets his "flag of personal faith" be shown or when he or she shares the problems with the naturalistic view on the origin of life.

21. The preceding sentences are based on Johnson's remarks as he described the situation in his Portland Talk.

22. For a helpful elaboration on Johnson's view of the secular trends in American universities and how things stand now, see his chapter "Education" in *Reason in the Balance,* the sequel to *Darwin on Trial.*

23. It may help balance things to share some observations. As a Christian theist, I experienced freedom in graduate seminars to offer criticisms from my perspective and to raise questions about philosophical assumptions I saw as dubious and vulnerable. The welcome-to-the-table I experienced in doctoral studies (and an openness to a theistic perspective) may be due to a deep and enduring (historic) resistance, within the fields of communication and rhetoric, to the impulses of scientific positivism in the modern era. For example, the works of Kenneth Burke typify the critique of two related scientific syndromes—reductionism and the prevalent epistemic hubris of "scientism."

Rarely have I observed rhetoricians marginalize the issue of the existence of God as a nonrhetorical. One rare example (with my reactions in parentheses) was in a communication course. We heard a presentation by a famous professor, who explained when a situation is rhetorical. The professor pointed out that "How many are in the room?" was *not rhetorical;* one could count and know instantly. No persuasion needed. (Agreed!) This professor then said that one moves into the truly *rhetorical realm of the uncertain*—an "exigence" or "imperfection marked by urgency" (Bitzer, "The Rhetorical Situation," 1968)—for example, the advisability of a proposed bill in Congress. (No problem.) On the far end of the scale, the professor then placed the question *"Does God exist?"* as so *radically indeterminate* that the question was inherently "nonrhetorical." (What? That's a highly rhetorical question!) I found this very revealing, and my mind raced. The professor's use of the "existence of God" as a nonrhetorical issue fit with contemporary intellectual fashion in universities, where the truth of monotheism is seen as roughly equivalent to the question "What is the best flavor of ice cream?" (Incidentally, the flavor question would have been *my chosen example* of a fairly indeterminate, hence "nonrhetorical," situation, although aesthetic questions do often enter into the realm of rhetoric as well.) I assume the professor meant that belief in God was nonrhetorical because (1) it is an *inescapably private matter,* rooted in one's personal life-narrative, or (2) it is *unconnected* with philosophical, historical, and scientific issues. I totally disagree. Note how Dawkins's comment on Darwinism as that which made it possible to be an "intellectually fulfilled atheist" makes clear how closely theism is tied in with other issues that can be rhetorically evaluated. Would not the debates in American universities over the case for/against the existence of God, by philosopher William Lane Craig and others, serve as defeaters of this classification? *Rhetoric is as rhetoric does.*

24. See Behe's and Dembski's works (listed in the bibliography). Behe's *Darwin's Black Box* is one long argument for the intelligent design of irreducibly complex biochemical systems in the cell. Dembski's works provided powerful theoretical support for Behe, especially his book *The Design Inference* (Cambridge, England: Cambridge University Press, 1998). Other leading advocates for a new paradigm, discussed in coming chapters, are Steven Meyer, Paul Nelson, and Jonathan Wells.

Johnson did not conceal his belief in design. In chapter 1 of *Darwin on Trial* he identified his theistic bias—and his acceptance of creation in a broad sense. Yet he never built a detailed case for the creation of life. He comes closest when he points out that DNA, RNA, and proteins are described with words of human communication like "messages, programmed instructions, languages, information, coding and decoding, libraries" (112). He asks why Darwinists do not consider the possibility that life is what it "so evidently seems to be—the product of creative intelligence?"

25. I should add "falsification" to the word "reassessment" (see Johnson, *Darwin on Trial,* 154–55). I see the complete dominance of Popper in chapter 12 as a key both to Johnson's overall rhetorical strategy and to his "bottom line" action that he invites the reader to initiate. In regard to this, just as a speech or sermon ideally issues a challenge to action at the conclusion, Johnson urges the reader to enter a rigorous testing of Darwinism *a la* Popper.

26. Within the circle of scientists, evolutionists and those in related fields would be pivotal, yet the hardest to influence. In the next circle would be university professors outside the natural sciences. Further out is the university-educated public, especially those in positions of leadership (e.g., publishers, educators, the media) who could exert influence as they are led into Darwinian doubt. The final circle would be laypeople with the educational background to appreciate Johnson's lean but eloquent prose in his moderately college-level vocabulary.

27. E-mail letter from Johnson, January 2001.

28. These words have been uttered often by Johnson when asked about his strategic goals in speaking on campuses.

29. Attitudes and emotions toward creationism are visible in many essays in *Scientists Confront Creationism* (ed. Godfrey). See Richard Lewontin's "Introduction": "The recent massive attack by fundamentalist Christians on the teaching of evolution in the schools has left scientists *indignant and somewhat bewildered.* Creationist arguments have seemed to them a compound of ignorance and *malevolence,* and, indeed, there has been both confusion and *dishonesty* in the creationist attack" (xxiii, emphasis added).

30. Johnson's letter to Alvin Plantinga, 6 December 1990.

31. Such attitudes and feelings were often linked to noble values. A scientist might bristle with indignation when convinced that his creationist opponent was lying or was willfully ignoring certain scientific evidence.

32. This is one side of the creationist obstacle that Michael Denton never had to face.

33. This assessment comes from my survey of a dozen published or recorded interviews with Johnson during the 1990s.

34. This is the wording used by David Raup in describing Johnson's competence. See chapter 4, 80–81, especially the footnote on 81.

35. Typical is a review of Harvard astronomer Owen Gingerich, *Perspectives on Science and Christian Faith* 44 (1992): 140–42. Gingerich was an attendee at Campion, who noted Johnson's "covetably sharp pen" and his knack at turning a clever phrase (Gingerich quoted a few of his favorites). Henry Morris, the creationist author and lecturer, wrote, "Phil Johnson has a remarkable gift of pithy expression, probably unmatched by any other writer in the creation/evolution field, and approached only by Stephen J. Gould" (personal letter, 1 July 1991). Even Ruse (*The Evolution Wars: A Guide to the Debates* [Piscataway, N.J.: Rutgers University Press, 2000]) compared Johnson's writing flair with Gould's. From my survey of sixty reviews, it seems that only Gould ("Impeaching a Self-Appointed Judge") criticized Johnson's writing (or at least his style of chapter transitions).

36. His narrative arguments work together in a way somewhat analogous to Denton's approach, and yet in significant ways Johnson's arguments are different from Denton's. One of these ways is

Johnson's narrative emphasis on Gould and the entire development of punctuated equilibrium and the other controversies that took place in the 1980s.

37. Walter Fisher, in Burgchardt, *Readings in Rhetorical Criticism* (State College, Pa.: Strata, 1989), 64–66. Fisher subjected stories to two tests: *narrative probability* ("what constitutes a coherent story") and *narrative fidelity* ("whether the stories . . . ring true with stories they know to be true in their lives"). In Johnson's attacks on Darwinism's genesis story, it is *coherence*—narrative probability—that is held up to scrutiny.

38. When I say "act" I do not imply that Johnson favors a "single act" or even a "single literal week" view of creation. I use the word "act" in the collective sense, as a "setting up," in which one may remain agnostic about the details.

39. Johnson, *Darwin on Trial,* 14.

40. Ibid., 112.

41. Johnson says he has no quarrel with the conventional time scale for the appearance of different types of animals. Creation in scattered events over eons, the implied position of Johnson, is called "progressive creation."

42. I see two levels of narrative—*factual stories* and *projection themes.* Johnson uses much factual narrative, which must be "true to the facts" to persuade. (For example, my quotations of Gould or Raup at Campion would be stripped of all power if I said that I had invented them, and had even invented the meeting.) Yet if a rhetor signals that he is extending upward from factual narrative into *quasi-imaginative* projection themes, not only are these stories enjoyed, they can catch fire and chain out. As complex rhetorical symbols, projection themes enjoy vast power. Thus, the factual and projection narratives live in vibrant symbiosis, without melding their distinct function and force or defining traits. My theory is that projection themes in paradigm-crisis discourses (e.g., Phil Johnson's Darwinian battleship that is leaking) have plausibility and rhetorical power to the degree the author can first build an underlying structure of narrative arguments, using factual narratives to convince the reader that the paradigm's problems really are *severe and fundamental,* justifying the projected story or image of the coming revolution.

43. There is a two-way relationship. Personal/incident stories *power* the three types of narrative argument, but these arguments, in turn, *make sense of* personal narratives. The vesting of meaning goes both directions. Thus, the Wistar story would appear to be an anomalous but trivial incident in the Darwinian HEB (and likely would not even be mentioned), but in the new-HOS narrative, it is important and highly meaningful; it makes perfect sense.

44. Master-narratives are either the evolutionary story or the history of Darwinism or any of *their key stories.*

45. This story could also be described as a "controversy story" that draws macroevolution into the swirl of some substantive controversy. Many of Johnson's personal or incident narratives can be described as controversy stories.

46. Gould makes exactly this point, including a reference to media coverage of Johnson ("Impeaching a Self-Appointed Judge," opening section).

47. See Timothy Ferris's *The Whole Shebang,* filled with marvelous narrative, and Robert Jastrow's *God and the Astronomers,* which uses a narrative approach to teach the discoveries leading up to the big bang theory.

48. Here are some of Johnson's key narratives: In *Reason in the Balance,* three scientific-legal stories are the Dean Kenyon case (29–30), the Phillip Bishop case (173–78, 181–82), and the Henry F. Schaeffer incident (179). He tells a scientific-theological story in the Nancey Murphy interaction (97–102) and educational stories in the Carper-Sears solution and the Provine-Johnson cooperation (185–91). In *Testing Darwinism,* Johnson devotes a chapter to the story of "Inherit the Wind" (24–34) and the Danny Phillips incident (34–36). His last chapter is a comparison of the life stories of three men: John Shelby Spong, Charles Templeton, and Billy Graham. Many essays in *Objections Sustained* are biographical or focus on incidents in the life of a key figure, such as T. H. Huxley, Paul Feyerabend, and Gould. *The Wedge of Truth* is full of anecdotes. Chapter 1 is Johnson's commentary on a true, published autobiographical piece on Phillip Wentworth's losing his faith in the transition from high

school to Harvard. The core of chapter 2, "The Information Quandary," is a debate between Paul Davies and Christian de Duve at a science conference in Italy. "The Kansas Controversy" (chapter 3) is narrative with commentary.

Chapter 6: The Matrix of Stories in *Darwin on Trial*

1. Prosecuting attorney William Jennings Bryan, who was no biblical literalist, faced a prosecution that used "scientific evidence" that was discredited a short time later.

2. Johnson, *Darwin on Trial,* 6.

3. Ibid., 8. Johnson returns to this issue when he discusses Kuhn's paradigms: "A paradigm rules until it is replaced with another paradigm, because 'To reject one paradigm without substituting another is to reject science itself'" (122). He then refers back to the NAS rule against negative argument in chapter 1.

4. Johnson's "qualifications and purpose" come up again in this chapter (13). He refers to his "specialty in analyzing the logic of arguments and identifying the assumptions that lie behind those arguments."

5. A prime example used regularly by Johnson is Richard Dawkins's and Johnson's late friend, William Provine (126–27), whose aggressive atheism is exemplified in the videotape "Darwinism: Science or Philosophy? The Stanford Debate between Phillip Johnson and William Provine." I will discuss this videotaped debate in the next chapter.

6. Johnson, *Darwin on Trial,* 116; the emphasis is Johnson's. This quote ranks as a key Design commonplace. Six years later it was quoted by William Dembski in his *Intelligent Design* (Downers Grove, Ill.: InterVarsity, 1999), 116–17.

7. A trademark Johnsonian criticism, made often in public speeches, this is also in chapter 5, pages 69–70, where the semantic manipulation involved in the word *evolution* is discussed.

8. A typical narrative is a minor kind of narrative outside my taxonomy of factual stories. The key to this type is a repetitive cycle or a typical behavior such as a life cycle, an experiment's procedure, or a biochemical cycle (e.g., the vision or blood-clotting cascades in Behe, 1996), or here, the typical verbal behavior of scientists given a problematic situation.

9. Johnson, *Darwin on Trial,* 153–54.

10. A controversy story is one in which a scientist or other scholar voices a problem with evolution, which sparks reactions. This passage by Justice Scalia can be seen as a controversy story, though I tend to use this term for passages that involve criticism by biologists or those in an academic study of science (e.g., a philosopher or historian of science).

11. The two pieces that mentioned this, outside creationist publications, were both by the Design-friendly journalist Tom Bethell, in *Harper's* ("Agnostic Evolutionists," February 1985) and *National Review* ("Deducing from Materialism," 29 August 1986).

12. Colin Patterson, November 1981 (ARN's published transcript of the American Museum talk in New York), 1–4.

13. Johnson, *Darwin on Trial,* 10.

14. Johnson says that if the "scenario of gradual adaptive change is wrong," then *evolution* is just a "label we attach to the observation that men and fish have certain common features, such as the vertebrate body plan." Later, he pans Gould's "fast-transition" to account for the Cambrian phyla: "Maybe a few . . . intermediates existed for some of the groups, although none have been conclusively identified, but otherwise just about all we have between complex multicellular animals and single cells is some words like 'fast-transition.' We can call this thoroughly un-Darwinian scenario 'evolution,' but we are just attaching a label to a mystery" (12, 56).

15. This term is from Kenneth Burke, *A Grammar of Motives* (Berkeley: University of California Press, 1945), 108–10. Burke sees money as a "god term," in "terms of which all this great complexity attains a unity transcending distinctions of climate, class, nation, cultural traditions, etc." He adds that *"the monetary motive can be a 'technical substitute for god,' in that 'God' represented the unitary substance in which all human diversity of motives was grounded."*

16. Burke, *Permanence and Change,* 179.

17. Johnson, *Darwin on Trial*, 10–11.

18. Johson, *Darwin on Trial*, 11. Here Johnson is paraphrasing Gould's comments.

19. Gould symbolizes the embarrassment of the fossil record, which yields so little in the way of transitional fossils. It was through Gould and his colleagues that this fossil mystery became known after 1980. Of course, Gould and his colleagues were convinced that somehow, through genetic "jumps" in isolated populations, macroevolution had occurred.

20. See Johnson, *Darwin on Trial*, 40, 43. Though it is not in Johnson's quote, Gould predicted that a new theory would soon take its place.

21. These questions are repeated three times on pages 10, 12, and 14. The importance of these questions is seen in their recurrence throughout the book. See, for example, Johnson's summary paragraph on page 158. The first two sentences of this crucial section state: "The argument of *Darwin on Trial* is that we know a great deal less than has been claimed. In particular, we do not know how the immensely complex organ systems of plants and animals could have been created by mindless and purposeless natural processes, as Darwinists say they must have been."

22. When Denton visited Johnson just before the release of *Darwin on Trial*, he reminded Johnson of the key point to attack—the Darwinists' lack of a mechanism for macroevolution (personal interview, October 1991). In Johnson's speeches, he asks: "What is the evidence for the Blind Watchmaker thesis? How do we know mutation-selection has creative power?"

23. That natural selection can produce diversity within a kind is affirmed by creationists, in the action of genetic evolution of the races from an original pair. Johnson questions whether production of a sibling species can be accomplished by natural selection (27), yet later in chapter 4 he seems to allow that the punctuationalism of Gould might accomplish this. Presumably, natural selection may play a part in this extended microevolution.

24. To a rhetorician, "natural selection" could be seen as a "god term" (see endnote 15 above). In the context of Kenneth Burke's "dramatistic pentad"—agent, act, scene, agency, and purpose—part of the scene (nature in action) begins to dominate the drama because it acts also as the agent of creation. See Burke, *Grammar of Motives*, xv, 3–320, where the pentad's elements are described and this perspective is then applied to major philosophies.

25. Darwin, *The Origin of Species*, 84; see also John Campbell, *Rhetoric and Public Affairs* 1, no. 4 (1998): 475.

26. I am withholding the name of the person quoted here because the material is from a private letter (dated 26 July 1991).

27. Johnson says, "We have already seen that distinguished scientists have accepted uncritically the questionable analogy between natural and artificial selection, and that they have often been undisturbed by the fallacies of the 'tautology' . . . formulations" (28–29). He adds: "It is therefore not as exceptional as it may have appeared that distinguished scientists have praised Darwin's theory as a profound tautology, or declared it to be a logically self-evident proposition requiring no empirical confirmation. A tautology or logical inevitability is precisely what the theory appears to them to be: it describes a situation that could not conceivably have been otherwise. From this perspective, 'disconfirming' evidence is profoundly uninteresting" (122).

28. These are Popper's own words, which according to Johnson's research notes are taken from Sir Karl Popper, *A Pocket Popper* (New York: HarperCollins, 1983), 242–43.

29. Johnson, *Darwin on Trial*, 17.

30. Because of space, I am only able to summarize Johnson's discussion, which lasts four pages (17–20).

31. Of course, this point could be blunted by observing that while biologists quickly came to accept common ancestry, as well as Darwin's strictly naturalistic approach, the selection mechanism remained extremely controversial among experts until well into the twentieth century. This point is made clear in Nancy Pearcey's paper, "You Guys Lost," in Dembski, *Mere Creation*. Nevertheless, Johnson's point tends to subvert the image, at the popular level at least, of the Darwinian history-of-science narrative.

32. Grassé (see chapter 2) is employed, but the "controversy story"—the shock he produced, seen in Dobzhansky's review—is not used but is told in the research notes. In the 1992 Darwinism Symposium and in the book that came from it (Buell and Hearn, eds., *Darwinism: Science or Philosophy?*), Johnson returned to Grassé and used his controversy story as the basis for a major narrative argument showing the problematic nature of Darwinism's rules of reasoning.

33. See Johnson, *Darwin on Trial*, 27–28.

34. Ibid.

35. Ibid. The parenthetical clarification (in brackets) in Grassé's quote is Johnson's. Note here almost an attitude of wonder and amazement on Johnson's part at the cognitive dullness of Darwinists on this point (i.e., the assumption that microevolutionary phenomena prove macroevolution).

36. On page 40, Johnson says Gould "tried to split the difference between Darwinism and Goldschmidtism."

37. Darwin says of saltations: "[If] I were convinced that I required such additions to the theory of natural selection, I would reject it as rubbish. . . . I would give nothing for the theory of natural selection, if it requires miraculous additions at any one stage of descent" (Johnson, *Darwin on Trial*, 33).

38. Ibid. This idea is restated no less than four times in the first few paragraphs of the chapter.

39. Ibid. The importance of Huxley's quote is seen in its being used by Gould in his chapters that explain his new theory of punctuated equilibrium ("The Episodic Nature of Evolutionary Change" and "The Return of the Hopeful Monster," both in *The Panda's Thumb*).

40. See Johnson, *Darwin on Trial*, 36–37. I added italics to emphasize this quote as Johnson's pivot of the chapter.

41. Johnson says, "If Goldschmidt really meant that all the complex interrelated parts of an animal could be reformed together in a single generation by a systemic macromutation, he was postulating a virtual miracle that had no basis either in genetic theory in experimental evidence" (ibid., 37).

42. Dawkins admits that "virtually all the mutations studied in genetics laboratories—which are pretty macro because otherwise geneticists wouldn't notice them—are deleterious to the animals possessing them." Johnson adds, "But if the necessary mutations are too small to be seen, there will have to be a great many of them (millions?) of the right type coming along when they are needed to carry on the long-term project of producing a complex organ" (ibid., 38).

43. In a minor error Johnson implies that the "acrimonious confrontation" took place in 1967. But 1967 is the date of the publication (Moorhead and Kaplan, eds., *Mathematical Challenges*). On the flyleaf of that book, the dates of the two-day conference are given—25–26 April 1966.

44. Johnson, *Darwin on Trial*, 43.

45. Ibid., 42.

46. Ibid., 45–46.

47. This is the phrase David Raup used in my interviews with him, reflecting the usage in paleontological literature.

48. For clarity, the "witness stand" metaphor is mine, not Johnson's.

49. Johnson, *Darwin on Trial*, 46–47.

50. "At this point," he suggests, "I ask the reader to stop with me for a moment and consider what an unbiased person ought to have thought about the controversy over evolution in the period immediately following the publication of *The Origin of Species*." Johnson invokes the jury's imagination, not only summing up Darwin's struggle but also considering what might have happened if Darwinian scientists had wanted to test common ancestry: "The test would not be fair to the skeptics, however, unless it was also possible for the theory to fail. Imagine, for example, that belief in Darwin's theory were to sweep through the scientific world with such irresistible power that it very quickly became an orthodoxy. . . . Suppose that paleontologists became so committed to the new way of thinking that fossil studies were published only if they supported the theory, and were discarded as failures if they showed an absence of evolutionary change. As we shall see, that is what happened. Darwinism apparently passed the fossil test, but only because it was not allowed to fail." Johnson's narrative strategy here is unusual. He retells the history-of-science narrative, but his invitation to "imagine" leads to a surprise twist: *This imagined story is what happened!*

Johnson turns to the kinds of evidence that Darwin's theory predicted, especially the "gradual pattern of extinction." (Darwin said that the extinction of a species is probably a slower process than its production.) Then Johnson shifts back to the "imagine this" mode: "Suppose, however, that it were shown that a substantial proportion of extinctions have occurred in the course of a few global catastrophes, such as might be caused by a comet hitting the earth or some sudden change in temperature." The reader who is aware of the importance of catastrophes in contemporary paleontology may realize that the "imagine this" is scientists' current best picture of what happened. If most extinctions are caused by catastrophes and not by steady competitive replacement, then Darwin's genesis story is again damaged. Johnson shows later that this is the case (Johnson, *Darwin on Trial*, 55–58).

51. In the two kinds of fossil stories, there are two questions: (1) *Cosmological*—How well does the Darwinian story fit the fossil data? (2) *History-of-science*—How did we learn of this striking "jerkiness" of the fossil record?

52. These accounts achieve a high degree of "narrative fidelity" (they are true to the life stories we experience) to which Fisher referred in his writing on the narrative paradigm. See Walter Fisher, "Narration as a Human Communication Paradigm: The Case of Public Moral Argument," in *Readings in Rhetorical Criticism*, ed. Burgchardt (State College, Pa.: Strata, 1995), 279.

53. From "The Episodic Nature of Evolutionary Change," in Gould, *The Panda's Thumb*. I quoted the points of "sudden appearance and stasis" in an earlier chapter but felt it was wise to note the rhetorical power of the quote here.

54. "Rapid" is understood as *relatively* rapid—in terms of the geological time scale, say, in ten thousand years. According to punctuationists, this rapid change may be fueled by mutations in regulatory genes that guide development.

55. Johnson has an excellent footnote on the pages introducing the Cambrian problem (54–55) that describes the Ediacaran fossils, which are complex animals found in a few locations in pre-Cambrian rock. According to Seilacher (supported by Gould) these animals are irrelevant to the Cambrian mystery—they are not precursors.

56. Johnson, *Darwin on Trial*, 57. Johnson returns to Darwin's doctrine of competitive replacement and shows how it is falsified by the discovery of the massive extinctions. Gould says that paleontologists have known about these "great dyings" all along, but their importance was minimized. Why? Gould admits that "our strong biases for gradual and continuous change force us to view mass extinctions as anomalous and threatening." This summary persuaded those readers who trusted Johnson's grasp of the evidence and his evaluations.

57. In the revised edition Johnson concludes this chapter with a review of new embryological evidence of the early divergence of developmental patterns, which appears to problematize common ancestry (70–71).

58. Gould had pointed to several groups of vertebrate fossils as his third proof of evolution, so chapter 6 tackled "The Vertebrate Sequence." The other narrative persuasion in chapter 6 centers on personal narratives, especially that of Sir Solly Zuckerman, a prestigious anthropologist, who in his own controversy story cast colorful doubt on the scientific reliability of the popular genesis story in the ape-human sequence.

59. "In print," that is, in the supplemental text for public high schools, *Of Pandas and People*, which, while written by Davis and Kenyon, was structured upon Thaxton's conceptual base in terms of its main lines of argument.

60. See Stephen Meyer's 1990 doctoral dissertation at Cambridge University, "Of Clues and Causes."

Chapter 7: The Roaring Nineties

1. In the fall of 1991 Johnson spoke on a dozen campuses, including eight major university campuses: Harvard, Yale, Princeton, Cornell, and the Universities of Pennsylvania, Delaware, Florida, and South Florida.

2. Lynn Vincent, "Science vs. Science," *World*, April 2000.

3. The others (in addition to Johnson's five later works) are J. P. Moreland, ed., *The Creation Hypothesis* (Downers Grove, Ill.: InterVarsity, 1994); Jon Buell and Virginia Hearn, eds., *Darwinism: Science or Philosophy?* (Dallas: Foundation for Thought and Ethics, 1993); Nancy R. Pearcey and Charles B. Thaxton, *The Soul of Science* (Wheaton, Ill.: Crossway, 1994); Davis and Kenyon, *Of Pandas and People;* Dembski, ed., *Mere Creation;* and two authored by William Dembski—*Intelligent Design* and *Design Inference.*

4. Johnson, *Reason in the Balance* (1995), *Defeating Darwinism by Opening Minds* (1997), *Objections Sustained* (1998), *The Wedge of Truth* (2000), and *The Right Questions* (2002). All of these were published by InterVarsity Press in Downers Grove, Illinois.

5. Johnson often recycles such revelatory or persuasive interactions. See *Reason in the Balance,* 11–12, 31, 88, 97–100, and 207. In *Defeating Darwinism by Opening Minds* (Downers Grove, Ill.: InterVarsity, 1997), Johnson tells of the initial reaction to his book, Behe's entry into the movement, and the Darwinism Symposium (90–92). Earlier he tells the story of Behe's *Darwin's Black Box,* with a summary of the book's main ideas and a survey of published responses (75–80). Johnson's struggle to publish a review of Raup's book, *Extinction: Bad Genes or Bad Luck?* is summarized in *Objections Sustained* (40–41). In the introduction, he shares Raup's personally communicated approval of what Johnson wrote, in contrast with the vehemently hostile letters to the editor of the publisher, *Atlantic Monthly.* Note that in all of these cited pages, Johnson refers to oral or written interaction with *critics in academia.*

6. At the end of the Access Research Network video "Darwinism on Trial" (lecture taped at the University of California at Irvine, 1992) is a lively question-answer period—an example of the feisty interaction with hostile questioners.

7. The word "sneering" expresses a common assessment of Provine by college students. John Wiester (at Westmont College) and I have shown the video often in our classes. Invariably students comment on Provine's attitude toward Johnson's theism, expressed both in his tone and his actual words. The videotape of the Stanford debate, "Darwinism: Science or Naturalistic Philosophy? The Provine-Johnson Debate," is available from www.arn.org.

8. Most observers on both sides agreed that the Darwinists had made a better showing, although some called it a draw. All agreed that Kenneth Miller's persuasion-by-chart was a clever and effective part of the debate, although it violated the rules that had been communicated to the Design participants.

9. See Johnson's comment in a double-asterisk footnote in *Objections Sustained,* 88–89. Some of my wording is adapted from that footnote.

10. For three years I was part of the Design e-mail discussion group and have vivid memories (and printouts) of the comments about the "Firing Line" debate setback. However, an acquaintance of Johnson's attended the debate and gave a positive evaluation. He felt that Johnson and Behe were very strong and that the Darwinist side had so cultivated openness to God's role in evolution that it undermined the classic Darwinian conception as a mindless, purposeless process. Darwinists had eroded their firm materialism by making macroevolution "religion-friendly."

11. For Johnson's comments on the Veritas Forum, see especially chapter 10, "The Beginning of Reason," in *Reason in the Balance,* 193–204. The Veritas Forum worked closely with Christian Leadership Ministries (CLM), which probably played a more important role than any other organization in the promotion of Johnson's rhetorical project. CLM used a mailing list of over ten thousand Christian university professors during the 1990s, and about two thousand of these would be considered somewhat active in the network. Organizing the network in different regions of the United States were several dozen staff members with Campus Crusade for Christ. Not only did these CLM staff and affiliated faculty serve as organizers for many of Johnson's campus lectures during this period, but CLM also gave Johnson extensive publicity in its publication, *The Real Issue,* and helped sponsor the Darwinism Symposium in Dallas in 1992. Its most important collaboration with Design was its organization and sponsorship of the Mere Creation conference in Los Angeles in November 1996. As I said in chapter 1, many view this conference as the public launch-point of Intelligent Design, although I have identified Design's earlier birth and adolescence phases, beginning with the Portland Ad Hoc Origins Committee meeting in early 1990.

12. Funds were provided by the Santa Ynez Foundation, near Santa Barbara, California. This is the same foundation that supported the sending of free copies of *Darwin on Trial* to about three hundred key professors and cultural leaders in 1991. I have recorded summaries of these projects in my research files.

13. For details on the mailing of Johnson's essay with the Ad Hoc cover letter, see Johnson's epilogue to *Darwin on Trial* and the research notes (especially page 206). I also have direct personal experience of these events, having conversed with virtually all the key participants in this mailing-project. Steven Meyer was the chief author of the letter, but a half-dozen or so of the other signatories were involved in its literary evolution. The entire list of signatories is published in the "Resources" section of www.apologetics.org as "Scientists . . . Question Darwinism." There were several well-known signatories, including Alvin Plantinga, a world-renowned philosopher, who became skeptical of Darwinism after reading Denton (personal interview, November 1990). He supplied a blurb for the earliest edition of *Darwin on Trial*. Henry Schaeffer, a friend of Johnson's at Berkeley until 1988 when he was recruited by the University of Georgia, is a pioneer in a field of chemistry called "quantum computational chemistry." After reading *Mystery of Life's Origin* (Thaxton, Bradley, and Olsen) in 1985, he became a skeptic of naturalistic macroevolution and chemical evolution. Princeton signatories include plasma physicist Robert Kaita, chemist Andrew Bocarsly, and chemical engineering professor Robert Prud'homme. All three had read *Mystery of Life's Origin* and cosponsored Thaxton's lecture at Princeton in December 1988. Of the three at Princeton, Kaita played by far the most active role, contributing a chapter to *Mere Creation,* which grew out of his 1996 talk at the Mere Creation conference. Yale physiologist Fred Sigworth was also a signatory.

14. See the partial listing in chapter 1, note 15.

15. Dembski's first Ph.D., in mathematics, is from the University of Chicago; his second Ph.D. in the philosophy of science is from the University of Illinois at Chicago.

16. Pennock, *Tower of Babel,* 29.

17. David Hull, *Darwin and His Critics* (Cambridge: Harvard University Press, 1974), 455.

18. Two other recent polemics (of somewhat lesser importance) are Nile Eldredge, *The Triumph of Evolution and the Failure of Creationism* (New York: W. H. Freeman, 2000), and Celeste Condit, "The Rhetoric of Intelligent Design: Alternatives for Science and Religion," *Rhetoric and Public Affairs* 1, no. 4 (winter 1998).

19. Jukes was referring to Johnson's words, "I am not easily intimidated," in a San Francisco newspaper article that appeared when his book was published. I scurried to my dictionary for a refresher on these two words. From the meanings in *Webster's Collegiate Dictionary* for "truculent" and "bombastic," I infer that he meant something like "fiercely belligerent and pretentious."

20. Thomas Jukes, "The Persistent Conflict," in the "Random Walking" column, *Journal of Molecular Evolution* 33 (1991): 205–6.

21. Phillip Johnson, "Response to Jukes," in the "Random Walking" column, *Journal of Molecular Evolution* 34 (1992): 93-94.

22. We saw such treatment in Ridley's review in *Nature* (1985) of Denton's *Evolution: A Theory in Crisis,* which was discussed in chapter 3. Whereas Ridley had lumped Denton's work in with several other books critical of Darwinism, Hull focuses exclusively on *Darwin on Trial.*

23. David Hull, "The God of Galapagos," *Nature* 352 (8 August 1991).

24. Ibid., 485.

25. Ibid.

26. As I consider here how Johnson might reply, my purpose is simply to expose the vulnerability (rhetorical weakness) of Hull's remark. To elaborate briefly on what I have said, Johnson could point out that biology and the law both deal with realities that (1) *function* and (2) *arose over time.* In the first case, both of these branches of academia address the function side (observable, repeatable actions) without reference to God, and this is clearly appropriate. It is when one turns to study *how these realities arose over time* that one notices that the law (leaving out possible inspiration from a deity) has a human history connected with it. In the realm of biology, the "rise over time" is not recorded in human history but is indirectly recorded in the rocks and the biological data of natural history.

Johnson asks: "Does this fragmentary data, as a whole, truly establish the inference of Darwinian evolution, or does it point in another direction?"

27. This list is actually taken from Johnson's own book. The dismissal of arguments *because they are old* is seen in the words, "He runs through the usual objections to darwinian [*sic*] version of evolutionary theory," and in the closing words (quoted above) describing the book as "yet another rehash of creationist objections to evolutionary theory" (Hull, "The God of Galapagos").

28. This type of rebuttal is similar to Gould's point that is critiqued in *Darwin on Trial:* the distinction between evolution as "fact" and as "theory." The difference is that Hull fleshes out this distinction in the individual-stories of the four experts who doubted Darwin's mechanism but who nevertheless remained true, in their own belief, to the "fact" of common ancestry.

29. Hull, "The God of Galapagos."

30. Johnson, *Darwin on Trial,* 210.

31. Hull, "The God of Galapagos."

32. Johnson, *Darwin on Trial,* 162.

33. Hull's argument has the same appeal, yet suffers from the same vulnerabilities, as Gould and Futuyma's "argument from imperfection." Johnson critiqued this argument on pages 70–74 of *Darwin on Trial* (see my discussion in the previous chapter).

34. "Johnson vs. Darwin," *Science,* 26 July 1991, 379.

35. The syllabus for Behe's freshman seminar class at Lehigh University said, "Is life on earth the inevitable result of terrestrial chemical processes? Is there evidence that life here was 'seeded' by beings from another planet, as a Nobel laureate scientist has suggested? How could bats and whales be descended from a common ancestor? To intelligently evaluate arguments on such questions requires the ability to weigh the occurrence of wildly improbable biochemical events against the availability of vast quantities of time and matter. In this course we will critically evaluate arguments that are addressed to the general public on the topic of evolution. The factual evidence used to support various positions and the reasonableness of conclusions based on extrapolation of the evidence will be examined from a scientific viewpoint."

In a letter to Johnson (18 September 1991), Behe said of this course: "The students find it astounding when they read Denton that rational, even compelling, arguments can be advanced against Darwin's theory."

36. In numerous interviews Behe has recalled his intense anger at the anti-intellectual response of *Science* to *Darwin on Trial.* See the videotaped interview *Opening Darwin's Black Box,* available from www.apologetics.org.

37. In rhetorical theory dating back to Aristotle, an "enthymeme" is an important type of argument (a kind of persuasive device) in which a conclusion is drawn jointly by the orator and the audience. This device emphasizes only one key premise and leaves unstated the other supporting premise(s). To put it in terms of a logical syllogism, an enthymeme has one stated premise—often a minor premise. Typically, the major premise is understood by the orator and the audience but is left unstated. The conclusion that flows from the enthymeme may be stated but is often inferred. The present example, in which Johnson's connections with creationism are emphasized, can be put in terms of a crude enthymeme-syllogism: The major premise (unstated): *Creation science is a dangerous pseudoscience that leads people astray with flawed, religiously biased reasoning.* The minor premise (stated): *Johnson's book employs many of the arguments of creation science and is heartily endorsed by leading creation scientists.* Conclusion (implied): *Johnson's book is a dangerous piece of creation science in disguise; the unsuspecting public may be swayed by it to doubt the factual status of evolution.* Or, to restate the conclusion in different terms: *Johnson's rhetoric is a clandestine, sophisticated version of creation science.*

38. A key point about the book notice is the use of the weasel-word "evolution"—a word with wildly differing meanings—without qualifying what evolution means. As I showed, the vague/shifting definition of "evolution" is one of Johnson's main criticisms of Darwinism.

39. "Johnson vs. Darwin," 379.

40. Ibid.

41. Johnson used this phrase in March 1999 in Tampa Bay, Florida, and I have heard him say it on several other occasions.

42. Michael Ruse, "Crisis? What Crisis?" *New Scientist,* 13 June 1985, 33.

43. "Scientific creationism" is the older genre of creationism, which seeks to use scientific evidence to support a fit between Genesis history and chronology and the scientific data. This older genre (and its movement) is similar to the term "biblical creationism" except for the emphasis on the use of scientific arguments to support a literal Genesis.

44. Ruse's many books (not to mention his scores of articles) span three decades. One of his most important earlier works is *The Darwinian Revolution: Science Red in Tooth and Claw* (Chicago: University of Chicago Press, 1979 [2d ed. 1999]). His most recent is *The Evolution Wars: A Guide to the Debates* (Piscataway, N.J.: Rutgers University Press, 2001).

45. His testimony (containing his five points) is discussed in Johnson, *Darwin on Trial,* chapter 9.

46. I have copies of this correspondence. In Ruse's rejection, he included the comments from an anonymous reviewer, who admitted he had read only the first few pages of Johnson's paper (it quickly became clear where he was headed, in questioning the foundations of Darwinism). Johnson wrote Ruse, asking if the poor academic quality of the reviewer's response was typical of the peer-review system of *Philosophy and Biology.* Ruse sent a letter of apology and explanation, responding a bit to Johnson's substantive questions and seeking to show why Johnson's paper was simply out of bounds for publication in a journal that viewed Darwinism as a factual matter.

47. On Ruse's side were Arthur Shapiro, a zoologist from the University of California at San Diego; Leslie Johnson, lecturer in biology at Princeton University; Fred Grinnel, professor of biology at the University of Texas at Arlington; and K. John Morrow, professor of biology at Texas Tech University.

48. Technically there were eleven in all: five Darwinists and six Design speakers. Johnson deferred, for his fifteen-minute reply to Ruse's paper, to Steven Meyer. Meyer did not present his own paper, so there were five Design scholars who presented papers. I represented it as five on each side to give the sense of balance that was practically intact.

49. See Johnson's introduction in Buell and Hearn, eds., *Darwinism: Science or Philosophy?* This was the first university event that showcased Behe, Johnson, Dembski, Meyer, and other Design proponents together in scholarly dialogue with mainstream evolutionists.

50. The video set may be purchased from one of the sponsors, the Foundation for Thought and Ethics. There were two other sponsors: (1) Dallas Christian Leadership, a local chapter of professors at SMU and other Dallas colleges and universities, led by Steven Sternberg, on staff with Christian Leadership Ministries, and (2) the C. S. Lewis Society, launched at Princeton University in 1975 but currently based in Tampa Bay, Florida. As the director of the Society, I was involved in inviting several of the Darwinian participants.

51. The papers and responses, published by the Foundation for Thought and Ethics, is *Darwinism: Science or Philosophy?* (Buell and Hearn, eds.).

52. Johnson indicated this in personal correspondence.

53. This and following quotes from Michael Ruse are taken from an audiotape of his speech to the American Association for the Advancement of Science (Boston, February 1993). This tape is available from the National Center for Science Education by requesting the 1993 audiocassette, "AAAS Meeting on Nonliteralist Antievolutionists."

54. In Symbolic Convergence Theory, this can be seen as a corrective chaining-out of the Ruse projection, or one could view it as merely a personal analysis—a context-sensitive assessment of a personal narrative.

55. Arthur Shapiro, "Did Ruse Give Away the Store?" *NCSE Reports* (spring 1993).

56. Ruse agreed with Scott, adding that Johnson's book was a "slippery" piece of advocacy. Nevertheless, said Ruse, the path of progress is to acknowledge the points he had just made in his talk on Johnson and to "move on."

57. The last question of the paragraph is of course inspired by the explanatory filter concept of William Dembski.

58. Steve Fuller, *Thomas Kuhn: A Philosophical History for Our Times* (Chicago: University of Chicago Press, 2000).

59. Ibid., 306.

60. "Qualified admission" refers to Ruse's qualification—that Darwinism's metaphysical base in naturalism, while true and somewhat constraining, is nevertheless vindicated by the tests of pragmatism.

61. Richard Dawkins, Will Provine, and even Stephen Jay Gould have frequently spoken out on the atheistic implications of Darwinism, and Design advocates have used such published statements as weapons.

62. Michael Behe's works include *Darwin's Black Box* (1996); "Darwin Under the Microscope" (1996); *Opening Darwin's Black Box* (1997); *Irreducible Complexity: A Lecture by Michael Behe at Princeton University* (Colorado Springs: Access Research Network, 1998), videotape; "Intelligent Design Theory as a Tool for Analyzing Biochemical Systems," in *Mere Creation* (ed. Dembski, 1998); "Teach Evolution" (1999).

63. William Dembski, *Design Inference, Mere Creation,* and *Intelligent Design.*

64. The ground rules for these special theories can be readily observed and systematized by reading letters or published communications written between or among members of the same community or by listening to audiotapes or videotapes of their meetings.

65. I have a copy of videotape footage from the August 1993 Ad Hoc Origins Committee meeting at Seattle Pacific University, where Behe presented this embryonic research outline in an informal talk.

Chapter 8: The Dam Breaks

1. Fifty-six letters were published, and many more were received, according to David Berlinski.

2. The "give us more time" responses are evaluated later. Shreeve's review was actually quite complimentary.

3. Behe, *Darwin's Black Box,* 5.

4. Ibid., 233–53. Behe showed how the big bang was not rejected even though it had powerful theistic implications.

5. Ibid., 196. This quote is imbedded in the final three-chapter section (187–253) that sets forth a compact manifesto—a scientific/philosophical rationale—for design theory.

6. This was Johnson's *ethos* problem. He had to deal with his status as a law professor, not a scientist. The charge of "not understanding how science works" applied also to Berlinski in the rhetorical counterattack on his article.

7. As of April 2002, Michael Behe had counted over 110 reviews or articles that focused primarily on his book, not to mention other published articles or columns that mentioned it or discussed it in other contexts.

8. Based on an interview with Bruce Nichols at Free Press in April 1998 and an interview with Behe in March 2001.

9. See the headline and descriptive paragraph on the back of the videotaped interview with Behe, *Opening Darwin's Black Box,* published by Apologetics.org. The headline reads, "Brace Yourself for a Cultural Earthquake."

10. For the context and details, see Woodward, "Meeting Darwin's Wager."

11. In "Darwin Under the Microscope," Behe says: "In parochial school I was taught that He could use natural processes to produce life. Contrary to conventional wisdom, religion has made room for science for a long time. But as biology uncovers startling complexity in life, the question becomes, can science make room for religion?"

12. Behe received his promotion to full professor a year after publishing *Darwin's Black Box.*

13. Yet the critique of Behe was leveled not so much at his own field (other than to state that biochemistry books were weak on explaining how irreducible complexity evolved); rather, his critique was aimed at evolutionary biology. Behe has limited contact with this group. My comments here are based on personal interviews.

14. In *Opening Darwin's Black Box,* a 1998 video interview mentioned earlier, Behe says twice that he became "angry" about the blatantly false statements that had recently appeared in published biology texts.

15. Behe's evolution into a Design theorist is gleaned from *Opening Darwin's Black Box* (Behe) and personal interviews with Behe.

16. Jerry Coyne, "God in the Details," *Nature,* 19 September 1996, 227.

17. See Bitzer, "The Rhetorical Situation."

18. Denton, *Evolution: A Theory in Crisis,* discusses "cilium" (107–9) and "flagellum" (223–25).

19. The words in quotes come from Ruse's review of Denton in *New Scientist.*

20. "Factual-conceptual" refers to Behe's dual rhetoric: The discoveries were based on a review of *empirical data* (protein parts, tallies of papers in journals), but they were equally *conceptual descriptions* of the data.

21. Behe says that no paper "in *JME* over the entire course of its life as a journal has ever proposed a detailed model by which a complex biochemical system might have been produced in a gradual, step-by-step Darwinian fashion. . . . The very fact that none of these problems is even addressed, let alone solved, is a very strong indication that Darwinism is an inadequate framework for understanding the origin of complex biochemical systems" (*Darwin's Black Box,* 176).

22. Karlyn Kohrs Campbell in *The Rhetorical Act* wrote of the triad: "As Aristotle also noted, however, proof comes in different forms. He wrote about three modes of proof arising out of the discursive or rational linkages *(logos),* out of the demographics and the feelings, attitudes, or state of mind of the audience *(pathos),* and out of the audience's perceptions of the rhetor *(ethos)"* (178).

23. Coyne, "God in the Details," 227.

24. Lee Cullum, "Evolution's Foes Take New Tack," *Dallas Morning News,* 22 April 2001, 5J.

25. Behe, *Darwin's Black Box,* 5.

26. Ibid.

27. Ibid., 97.

28. Darwin, *The Origin of Species,* 154.

29. Behe, *Darwin's Black Box,* 39.

30. In the debate (see chapter 7) Miller attacked the mousetrap argument by showing a cruder (yet functional) mousetrap with four parts. Behe said that Miller still had all five parts, but the spring was acting as two parts.

31. Andrew Pomiankowski, "A Tiny Mystery," *New Scientist,* October 1996: 45; Coyne, "God in the Details," 227; Shreeve, "The Design of Life," 8.

32. Behe, *Darwin's Black Box,* 20.

33. My interview with Behe, March 2001, confirmed this rhetorical purpose in the extensive scientific detail.

34. Behe, *Darwin's Black Box,* 6. He gives as an example the inner workings of a computer.

35. Ibid., 24. The first quote is Behe's quote of Haeckel; the second is Behe's paraphrase.

36. Such quotes (microsnippets of individual narrative) are from many of the world's most renowned biologists, such as Lynn Margulis, Richard Goldschmidt, Niles Eldredge, Stephen Jay Gould, George Miklos, Jerry Coyne, Hubert Yockey, the Wistar meeting participants, Stuart Kauffman, and St. George Mivart.

37. That is, based on its *actual* dramatic supremacy.

38. See Behe, *Darwin's Black Box,* 192.

39. "Scientific chauvinism" is illustrated by Shapiro *(Origins: A Skeptics Guide),* who would cling to the "best material solution" if the origin of life remained unsolved after many more decades of research. Behe responds, "Maybe the origin of life just didn't happen by undirected chemical reactions, as Shapiro hopes. To an active participant in the search, however, a conclusion of design can be deeply unsatisfying. The thought that knowledge of the mechanisms used to produce life might be forever beyond their reach is admittedly frustrating to many scientists. Nonetheless, we must be careful not to allow distaste for a theory to prejudice us against a fair reading of the data" (*Darwin's Black Box,* 235).

40. Richard Dickerson's article can be found in *Perspectives on Science and the Christian Faith,* July 1994.

41. This was clarified through interviews with Michael Behe in June 2001 and July 2002.

Chapter 9: Mere Creation and Beyond

1. On Behe's scale, one is a total bashing; ten is pure praise. He repeats this in *Opening Darwin's Black Box.*

2. "Darwin Under the Microscope" was published 29 October 1996. The story of this column began when Behe received a call from a *Times* editor in early fall. The editor had read *Darwin's Black Box,* was impressed with its arguments, and wondered if Behe would consider writing an opinion piece for the *Times.* Behe immediately wrote and e-mailed a draft, but it languished for over a month during the editorial overload of the Clinton-Dole election. However, in late October, the Pope released a statement through the Pontifical Academy of Science supporting the cogency of several "theories of evolution," while criticizing attempts to posit the naturalistic evolution of the human spirit. On Friday, 26 October, the *Times* editor contacted Behe about rewriting his opinion piece in a way that would connect it to these events. Behe did so immediately, and the piece appeared on 29 October, the following Monday.

3. See Dembski, *Mere Creation,* 9–10. When I say "the Design community," I note that a few attendees (e.g., cosmologist John Leslie and Princeton geologist John Suppe) were friends of members of the movement and were interested in the presentations but were not necessarily in harmony with all of the movement's beliefs, goals, and assumptions.

4. Ibid.

5. Nelson's dissertation is in press as of 2003 as volume 16 of the prestigious "Evolutionary Monographs" series at the University of Chicago. The evidence that Wells has used against common descent comes from his biochemical training at Berkeley in embryonic development.

6. Dembski, *Mere Creation,* 104.

7. In this description, I am aware that some wording will echo the phrasing of Dembski's descriptions. This is inevitable when one is quite familiar with not only a concept but also the phraseology employed frequently by the creator of this concept. Rather than cite specific quotes from Dembski, I have decided to employ my own wording in order to describe his ideas, as absorbed from my reading of Dembski and hearing him on several occasions.

8. Dembski, *Mere Creation,* 108.

9. Lynn Vincent, "Science vs. Science," *World,* April 2000.

10. See the end of Johnson's chapter, "The Intelligent Design Movement: Challenging the Modernist Monopoly on Science," in *Signs of Intelligence,* ed. William Dembski (Grand Rapids: Brazos, 2001), 38–39.

11. Dembski, *Design Inference,* chapter 3. I have tried to reproduce, as accurately as possible, the appearance of these two paragraphs, right down to the lack of indent on the opening paragraph and a very short indent on the second.

12. Ibid., 222.

13. Ibid., dust jacket.

14. Dembski, *Intelligent Design,* 9–10.

15. Caputo, the most famous example, is found in nearly all of Dembski's discourses on the filter. See Dembski, *Design Inference,* 9–19, 226–27; *Mere Creation,* 94–96. I suggest that this story is both a commonplace and possibly also a projection theme (fantasy theme) because of its *frequent repetition in various contexts.* Though Bormann tended to portray fantasy chaining as a social act that projects imaginatively into the past or the future, he says that the essence of such activity is not the invented nature of the plot but the *energizing or motivating response it produces, leading to social repetition* (personal phone interview, June 2001).

16. See Moreland, ed., *The Creation Hypothesis,* 113–35. For a critic's response, see Pennock, *Tower of Babel,* 230–33.

17. See Dembski, "Signs of Intelligence," especially pages 76–79, in the special issue of *Touchstone* devoted to Intelligent Design (1999). This special issue was republished as a book by Brazos Press in 2001, *Signs of Intelligence.*

18. These two examples (including both paragraph excerpts) are from Dembski's own chapter "Redesigning Science" in *Mere Creation,* ed. Dembski, 108.

19. Both quotes come from Larry Brumley's Public Relations Press Release, "Baylor Releases Polanyi Center Committee Report," 17 October 2000 (www.pr.baylor.edu). The first quote appears to be a word-for-word quote from the committee, while the second is in Brumley's own words.

20. E-mail from Dembski, originally circulated to the Design e-mail discussion group in October 2000 and resent to me in June 2001.

21. An e-mail to me from Dembski in June 2001 conveyed this as his main interpersonal task.

22. Christian de Duve and Steven Weinberg were the Nobel laureates.

23. In April 2002 I chatted about this conference with William Lane Craig, one of the philosophers who spoke on the "fine-tuned universe" from the Design side. His recollection was quite positive, and he said he sensed no serious "tilt" in terms of numbers or strength of presentation toward the Darwinist side.

24. This is excerpted from Paul Nelson's e-mail message to Johnson's e-mail discussion group, which read in part: "Todd Moody wrote: 'As conferences go, this one was delightful, from gavel to gavel.' [Right.] Not only was the conference tremendous fun, it demonstrated that ID approaches are every bit the match for 'the standard model' (naturalism)." (Nelson's comment about de Duve's toast fits in at this point.) "Later, Dr. Beatrice Anner, a Swiss biomedical researcher new to ID circles, gave very much the same sort of toast. She had traveled a long way, she said, and the conference was more than worth the trip."

25. When I speak of the "consciousness of the media," I refer primarily to the authors, but implied also is the approval of editors and the publisher who agreed in the final analysis to publish the articles, with its design-friendly wording.

26. Of course, the plot can be construed both at the level of an empirical generalization (for which several kinds of evidence must be marshaled) and also at the level of a powerful projection theme ready for chaining.

27. See, for example, the rhetorical commonplace of the Klaus Dose quote, used in Thaxton and Bradley (1994) and in Johnson, *Darwin on Trial:* "At present all discussions on principal theories and experiments in the field either end in stalemate or in a confession of ignorance" (109). For examples of critiques of the origin of life field, see Charles Thaxton and Walter Bradley's chapter in *The Creation Hypothesis,* ed. Moreland; Thaxton and Pearcey's *The Soul of Science* (1994); and Wells, *Icons of Evolution,* chapter 2.

28. Such descriptions and analyses are now almost ubiquitous in Design texts. Dembski's *Design Inference* and *Intelligent Design* are devoted to such critiques of prevailing biological evolution, and large portions of Johnson's newest book, *The Wedge of Truth,* treat this subject as well. For a recent article-length treatment, see Steven Meyer's article in *First Things* (2000).

29. See Wells, *Icons of Evolution,* 40–41. The Chengjiang fossils (along with the reinterpretation of the Burgess Shale fossils from Canada) were described in a cover story in *Time,* "Biology's Big Bang," 16 December 1995. See Wells, ibid., all of chapter 3. One place where Design emphasized the Chinese fossils and Chen's humorous remark that has become famous is the opening of Johnson's *Wall Street Journal* piece after the Kansas decision ("The Church of Darwin").

30. In Wells's appendix *(Icons of Evolution)*, all ten textbooks are graded, A through F, according to the accuracy with which they discuss seven of the ten icons. A detailed set of grading criteria for each of the icons is included. On the grading chart, with seventy potential grade slots, there is one B, one C, and two Xs (meaning "contains no image, but uncritically repeats the standard story in the text"), and the rest of the sixty-six grades are Ds and Fs. Half of the texts received a few Ds, and the rest Fs. The rhetorical effect of the chart is, in my view, quite devastating.

31. In Wells (ibid.) the ten textbook examples of misinformation are major "evidences of macroevolution" (or chemical evolution) or are supposed summaries of discoveries. Wells insists that each of these presentations conceals numerous problems that have arisen in the professional-level analysis and discussion of such examples. Included (besides the embryos) are the peppered moth observations in England, the Miller-Urey experiments, the "tree of life," the horse series, the ape-to-man fossil series, the four-winged fruit fly, and the Galapagos finches.

32. Ibid.

33. Michael Behe told Johnson's e-mail discussion group his wife became angry when she found out that the embryo drawings used in her high school biology class were fraudulent—faked to show a "similarity" that in fact was not there. Behe has often included an exposé of this fraud in his own talks on irreducible complexity.

34. Jonathan Wells and Paul Nelson, "Some Things in Biology Don't Make Sense in Light of Evidence," *Rhetoric and Public Affairs* 1, no. 4 (1998): 560–61, give examples of explicit "paradigm talk" by Design authors. See also Johnson, "The Intelligent Design Movement."

35. Thomas Lessl, "Intelligent Design: A Look at Some of the Relevant Literature," *Rhetoric and Public Affairs* 1, no. 4 (winter 1998): 630. Lessl conveys the criticism found in Design texts with regard to the failure of Darwinian rhetoric: "But these generally accepted premises do not prove the blind-watchmaker theory of evolution as it is typically constructed. In order for the logical sufficiency of evolutionary theory to be established, scientists would have to be able to demonstrate with some reasonable degree of probability that these known processes of evolution are really capable of creating the kind of complexity that pervades every corner of the biosphere. Design theorists argue that evidence of this kind is nonexistent for the most part and that the case for the neo-Darwinian paradigm remains circumstantial at best—despite a century and a half of effort to substantiate it."

36. Miller, *Finding Darwin's God,* 173, 187–90.

37. This distortion would have become very much less likely if those who foster it had participated in academic forums like that at Baylor, where a clash of ideas (in person-to-person engagement) cleared heads, clarified story lines, and cleansed projection themes. Assumptions about motivation—that which truly undergirds acts of persuasion—tend to be exposed and put to the test in such intense, interactive venues.

38. I asked Johnson's e-mail discussion group about the impact of reading Denton or Johnson as a memorable experience leading to persuasion. I asked for detailed recollections and received over a dozen that were quite revealing. To a large extent, these accounts paralleled the experiences of Johnson and Behe, which I have described. Most fit my metanarrative capsule summary, "aroused from dogmatic slumber."

39. David Berlinski used this image in his blurb on Dembski's *Design Inference.* The "resurrections" are true and historic, in a *symbolic sense of entry into public discourse,* and they are "real" in a scientific-consensus sense or a generally recognized sense, within the fantasy themes of the Design Movement.

Chapter 10: A Revolution Built on Recalcitrance

1. My essay (appendix 4), traces (1) the recent growth of the "rhetoric of science," (2) its connection (via Kuhn) with Design, (3) how my study fits into its research on the evolution debate, and (4) what its leading scholars say about Design.

2. One building, the Institute for Defense Analysis, was only partially seized; the lobby and courtyard became the scene of an unscheduled teach-in. The protest led, I am told, to changes in Princeton's administrative policies. The other building seized, New South Administration Building, had nothing to do with Vietnam but was taken over by ABC (Association of Black Collegians) to protest Princeton's alleged immoral investment in corporations working in South Africa, who thus were (or were seen as) complicit with that country's apartheid racial policy.

3. A brilliant and sensitive analysis of the counterculture, from a C. S. Lewis–type perspective, is Os Guinness, *The Dust of Death* (Downers Grove, Ill.: InterVarsity, 1975). It is out of print but is a must-read gem.

4. Although Kuhn taught most recently at MIT, he was a philosopher and historian of science at Princeton in 1964–79. In Kuhn's 19 June 1996 *New York Times* obituary, Lawrence Van Gelder listed the prestigious universities where Kuhn taught: "He received master's and doctoral degrees in physics from Harvard in 1946 and 1949. From 1948 to 1956, he held various posts at Harvard, rising to an assistant professorship in general education and the history of science. He then joined the faculty of the University of California at Berkeley, where he was named a professor of history of science in 1961.

In 1964, he joined the faculty at Princeton, where he was the M. Taylor Pyne Professor of Philosophy and History of Science until 1979, when he joined the faculty of MIT."

5. Deirdre McCloskey in *Rhetorical Hermeneutics: Invention and Interpretation in the Age of Science,* ed. Alan Gross and William Keith (Albany, N.Y.: State University Press, 1997), 102. Contra this notion, to maintain balance, Gaonkar says, "But it is not uncommon to find those who are celebrated as masters of 'implicit rhetorical analysis' react indifferently, if not with hostility, to such interpretations of their work. To the best of my knowledge, none of those masters (and the list is formidable: Kuhn, Feyerabend, Gadamer, Habermas; Toulmin is the possible exception) so far has either conceded that what they have been doing all along is a form of rhetorical reading, or gone on to incorporate rhetorical vocabulary in their subsequent work." Gross and Keith, *Rhetorical Hermeneutics,* 74.

6. By the 1960s, departments of rhetoric or speech communication (focusing on oral discourse) were being renamed departments of communication. The University of South Florida's Department of Communication, through which I received my Ph.D., has four fields: rhetoric, interpersonal communication, organizational communication, and performance studies. I specialized in rhetoric, with a subspecialty in the rhetoric of science.

7. John Campbell's work is extensive. In my bibliography (far from exhaustive) I list ten articles or book chapters.

8. Simons adds charitably, "In the final analysis, that is what defenders of science *mean* by 'scientific objectivity.'" He acknowledges that scientists can lapse into deceptive "sophistic rhetoric" but cautions his fellow rhetoricians not to judge this as the norm: "There is merit in a skeptical posture . . . but rather than carping at science in general, rhetorical critics would be more persuasive were they to reserve their slings and arrows for more limited and vulnerable targets such as individual practitioners or communities in the time bound sense. Of special concern should be willful violations of scientific canons, and here critics can stand not as enemies of science, but as defenders of its time-honored norms." Herbert Simons, "Are Scientists Rhetors in Disguise?" in *Rhetoric in Transition: Studies in the Nature and Uses of Rhetoric,* ed. Eugene White (University Park, Pa.: Penn State University Press, 1980), 127–29.

For all of Simons's wisdom and goodwill here, there are lurking problems that relate to the debate over design. For example, *who* is to decide what the "scientific canons" or "time-honored norms" are, especially when a canon is in hot dispute? *How* is that evaluation to be made? Also, what if violations are not "willful" but inadvertent? Johnson perceives a blindness to such violations when two key norms—naturalism and primacy of empirical data—come into conflict. Should not such cases be prime targets for rhetoricians? Both sides would subscribe to Simons's caution, yet they would disagree on the validity of methodological naturalism as a mandatory norm.

9. For example, Gross adapts, in terms of scientific rhetoric, the Aristotelian triad of logos, ethos, pathos, as well as the various situational "stases," and even the three types of discourse—epideictic, forensic, and deliberative.

10. Alan Gross, *The Rhetoric of Science* (Cambridge, Mass.: Harvard University Press, 1990), 4.

11. Ibid., 4–16.

12. J. E. McGuire and Trevor Melia, "Some Cautionary Strictures on the Writing of the Rhetoric of Science," *Rhetorica* 7, no. 1 (winter 1989); "The Rhetoric of the Radical Rhetoric of Science," *Rhetorica* 9 no. 4 (autumn 1991). In the first article they were targeting Bruno Latour and Steve Woolgar's radically constructivist and increasingly influential *Laboratory Life* (Princeton, N.J.: Princeton University Press, 1986).

13. *Webster's New Collegiate Dictionary,* 10th ed., s.v. "recalcitrance" and "recalcitrant."

14. Quoted by McGuire and Melia, "Some Cautionary Strictures," 89. Their footnote refers to Burke, *Permanence and Change,* 257.

15. McGuire and Melia, "Some Cautionary Strictures," 97.

16. Gross "does not foreclose the possibility that McGuire and Melia are correct that there is a limit to rhetorical analysis." This quote is taken from McGuire and Melia, "Rhetoric of the Radical Rhetoric of Science."

17. I tackled an ambitious goal: a rhetorical history of an entire social movement in science. Although movement studies have become common in rhetoric generally since theoretical groundwork was laid by Leland Griffin in 1952, this kind of study is rare in the rhetoric of science, where typical case studies focus on smaller targets: individual scientists, single publications or experiments, special literary genres, or key episodes in the history of science. One reason for the rarity of such studies in the rhetoric of science is, of course, that it is uncommon to have self-identified *rhetorical* movements among scientists.

18. I benefited from all of Walter Fisher's work on the narrative paradigm (especially his 1987 work, *Human Communication as Narration*) and key articles by Ernest Bormann (e.g., "Fantasy and Rhetorical Vision") and Donald Shields ("Symbolic Convergence and Special Communication Theories," 2000) on "fantasy theme analysis." I decided that both analytical approaches were relevant to the rise of Design. I discussed this with Bormann in a telephone conversation in June 2000 and in e-mail correspondence with Shields in August 2000. Both said they were unaware of any use of fantasy theme (SCT) analysis in a rhetorical study of science. I also had encouragement from Walter Fisher (via e-mail 2001–02). I am indebted to the three in many ways; they made helpful suggestions and encouraged me along the way.

19. See Glanz, "Darwin vs. Design."

20. By "speaking charitably," I am not raising the real possibility that some Darwinian rhetors know the Design story and yet purposefully distort it to their advantage.

21. I use the broader term "excitement" to capture the sense of the breakthrough both within Design and the similar descriptions outside. It conveys the projection theme, noted already, of Behe's "cultural earthquake." This term embraces the nature of the discussions of Behe on radio shows and television and in articles, books, and lectures. My term even embraces the negative and hostile reviews. Two chief indicators of the excitement is the list of over one hundred published reviews of Behe in the first two years, coupled with the torrid sales pace recorded by the Free Press.

22. See Shreeve, "The Design of Life"; Behe, "Darwin Under the Microscope"; Behe, "Teach Evolution"; Goodwin, "New Light on Creation" (29 December 1997); Glanz, "Design vs. Darwin."

23. Of course, such allegations have other problems, related to empirical plausibility, in relation to these descriptions being accurate concerning the social reality of Design rhetors' motives, understanding, dreams, and plans.

24. Here are two reordering scenarios: (1) The publication of Design-oriented articles in a major journal that was controlled by Darwinist editors and peer-reviewers. (2) The publication of books or articles by members of the National Academy of Science embracing the legitimacy of the concerns, critiques, or methodology of Design.

25. See Dennett, *Darwin's Dangerous Ideas*, 263; Johnson, *Objections Sustained*, 22; and Johnson, *The Wedge of Truth*, 65 and 78. I described the Gallup "recent creation" view as vague because this option only says that humankind was brought into its present form by a divine act of creation in the last ten thousand years. It does not mention previous evolution or creation of other species. Conceivably, a progressive creationist, who holds that the earth is four billion years old and that God created the Cambrian marine species in a direct creative act 540 million years ago, could answer yes to the "recent creation" option if that is where he or she places the time of God's creation of humankind on the timeline.

26. I am trying to convey the wording that Johnson has used, in person and in his writings, since 1992.

27. Underneath this process would be a "permissive connection" with religious or metaphysical assumptions. By "permissive connection," I mean that such conversions *will take place much more easily when there is theistic metaphysical space in the worldview of the reader.*

28. In chapter 1, I described that Bormann, Shields, and others had toiled since the 1970s developing an approach of rhetorical analysis that pivots on this notion of fantasy themes (which I have been calling projection themes) that build a rhetorical vision. See Shields (2000) for a superb overview and application of this theoretical system, now called "Symbolic Convergence Theory."

29. Obviously, Behe's effectiveness in degrading the story of the macroevolution of irreducibly complex systems is what counts. His acceptance of a significant part of the cosmology of Darwin is overlooked as an odd personal perspective.

30. There are many such factors—worldview and biases of author and audience, type of artistic creativity, strength of argumentative construction, and more.

31. Here I could have added a fifth point of recalcitrance: those seemingly unbridgeable morphological gaps, such as the appearance of the bird lung and the other novel structures in Denton's "Bridging the Gaps" chapter in *Evolution: A Theory in Crisis.*

32. Johnson, "How to Sink a Battleship," in *Mere Creation,* ed. Dembski.

33. One could argue that since 1996 or so this has been the leading argumentative theme of Johnson's rhetoric. A leading example is found in his essay "The Intelligent Design Movement."

34. I am referring, of course, to David Raup's endorsement of the quality of Johnson's critique. See chapter 4 for the details.

35. Denton and Berlinski are two who are counted as Design proponents but do not argue for a creator. Both are open to a preexisting intelligence but are not on the record as being convinced that the intervention of such a being is how the creation of complexity occurred.

36. There are basically two sides to this new HOS—the *controversies* that impinge upon evolutionary science and the Design story itself with all the rich variety of narrative that it embraces.

37. The vindication of Paley poses three questions: (1) Why, in the history of science, were Paley's arguments thrown out? (2) What were the scientific, philosophical, and even theological factors involved in Paley's sudden intellectual demise? (3) How does modern Design theory address and solve each of the problems or weaknesses of Paley's early argument for design? This trend in updating the story of Paley is visible in Dembski's *Intelligent Design* and the end of Behe's *Darwin's Black Box.* See also Nancy Pearcey's "You Guys Lost" in *Mere Creation,* ed. Dembski. This narrative argument is necessitated by the consciousness of the scientifically literate audience that Design is trying to influence.

38. These stories are *directly recalcitrant,* by which I mean that if they are not recorded on paper or on tape (as was Patterson's speech), they are at least recorded on human memories, such as those of four Campion witnesses whom I probed in order to research and then retell the amazing, recalcitrant story of the verbal duel between Johnson and Gould. In addition, all kinds of interaction stories between the two sides of the debate, including book reviews, are also *highly recalcitrant* and can be used in the same way.

39. Thus, due to the inherent nature of certain Design stories as anomalous and recalcitrant, I conclude that such *individual narratives are the most powerful type of story among the types of factual narrative.*

40. Any close analysis of Design looking for such themes makes good sense according to the thought of Symbolic Convergence Theory (SCT), which has pioneered this approach. That is because such themes are known to arise and powerfully chain out—be repeated with variations and embellishments—especially in a tight-knit social context oriented to some goal.

41. It appears, from my own servey of the field, that rhetoricians of science had not attempted previously to integrate the study of factual narrative reasoning with projection themes and related symbolic systems (Symbolic Convergence Theory's "cues" and "special communication theories"). In experimenting with this dual approach, I found that they work together quite well in close textual analysis and the broader context of a movement study.

42. I sense a proportionality between the *strength or perceived effectiveness* of Design's creation-or-destruction of factual stories and the power of its projection themes *that spring from the narrative creation/destruction.*

43. This is the thrust of Wells's *Icons of Evolution,* and it is the original (and enduring) motivation of Behe as he read Denton and became angry that no one had ever hinted at the existence of such an array of evidential problems for Darwinism.

44. The quote from Lynn Vincent in chapter 6 stresses Johnson's goal in making friends of his strongest opponents.

45. To naturalists, God may exist (as source of meaning and morality), but there is no room for a deity who can intervene in nature and *who might choose to do so.* Most naturalists are atheists or agnostics who see no compelling reason to believe in a creator of the cosmos who can (and might) act in the cosmos. This issue became apparent in the 1990s. Compare Johnson's *Reason in the Balance* with

Dembski's *Intelligent Design*. Both critique the deadening tyranny of naturalism, with its *underlying assumption of the irrelevance of the theistic possibility.*

46. There are plenty of dissenters to this statement in academia. On the conservative Christian side is physicist Howard Van Till, who separates *evolution* (making no reference to God) from *evolutionism* (which extrapolates from Darwinism to a godless universe). Gould, on the atheist side, would argue for "nonoverlapping magisteria"—a pair of separate but equal realms for both science and religion.

47. This philosophy is allegedly supported by Darwinian evolution, and yet in turn macroevolution's credibility, says Design, is dependent upon the truth of naturalism. This is cited as circular reasoning, and scientists who embrace such thinking (after its flaws have been pointed out) are said to have become accomplices with naturalism.

48. William Dembski, *No Free Lunch* (New York: Rowman and Littlefield, 2001).

49. See Edward Larson and Larry Witham, "Scientists and Religion in America," *Scientific American* (1999), 90, an attempted replication of the survey of religious beliefs of scientists in 1914 and 1933 by Bryn Mawr psychologist James Leuba. Theistic belief was described as belief in a personal, all-powerful God who answers prayers and gives some sort of afterlife.

50. Ibid.

51. Ruse's actions are retold in chapter 7. Another example is David Raup's sympathy toward Johnson's critique and his endorsement of the accuracy of *Darwin on Trial.* Raup's story is in the text and footnotes of chapter 4.

52. See Phillip Johnson, "Do You Sincerely Want to Be Radical?" monograph reprint from *Stanford Law Review* 36, nos. 1 and 2 (January 1984): 278. "Value relativism supports liberalism at one level, because it undercuts any claim to absolute or divine preference for any one value system. The corresponding problem is that it undercuts any such claim for 'liberal' values as well. That is why the value choices have to be hidden, why liberal rationalism must always seem to be invoking only reason, or values like 'liberty,' 'equality,' or 'tolerance,' that seem to be uncontroversial because they are so vague. The struggle to establish which values will predominate is to liberal rationalism what sex was to the Victorians: We know that it has to go on, but decent people don't talk about it in public. Fortunately, ideological conflict is not as bitter in our society as in many others, a circumstance I would credit mainly to widespread prosperity and our common religious heritage."

53. By "intelligent agent" the Design advocates refer to any being who conforms to the defining traits of the genus of "intelligent beings," whether humans, extraterrestrial intelligent races, or even a transcendent being such as God.

54. These three types of global narratives are capitalized just to bring out their sense of *ultimate* or *grand narratives.*

55. I have referred on several occasions to SCT's special communication theories, which are unique systems of communication that guide certain human social activities (particular games like bridge, academic-communicative acts like producing a journal article, social events like roasts, and more). The Design discourse heuristic I refer to here qualifies as an important new *special theory* in this perspective.

56. For example, of the response forms, letters, and e-mail reviews of the 1992 Darwinism Symposium and also of the April 2000 "Nature of Nature" conference, almost all Darwinist participants in these conferences report a positive experience. (This is true for *all* in the Darwinism Symposium.)

57. From Gould's column in the March 1997 issue of *Natural History.* This distortion is discussed in "The Unraveling of Scientific Materialism," in Johnson, *Objections Sustained,* 75.

Appendix 4: The Rhetoric of Science and Intelligent Design

1. I use the word *assault* here as a metaphor of "analysis yielding understanding," not of opposition, destruction, or subjugation. It is akin to the "assault" of climbers on a mountain peak of the unknown, as pictured by Robert Jastrow at the end of *God and the Astronomers.*

2. McCloskey in *Rhetorical Hermeneutics,* ed. Gross and Keith, 102.

3. Gaonkar, in *Rhetorical Hermeneutics,* ed. Gross and Keith, 42. See also McCloskey, ibid., 102. The tag of "implicit rhetoric of science" refers to a kind of analytical writing rather than a field desig-

nation. Gaonkar notes, "Implicit analysis occurs when the analyst neither understands nor describes what she is doing as a form of rhetorical analysis, but it would be recognized as such by someone familiar with rhetoric" (74). This *redescription* of existing research occurs when any analysis of science emphasizes "argumentation, figurative language, generic constraints, modes of constructing authority, and addressivity." Some, especially the sociologists of scientific knowledge, seem happy to wear the implicit (or even explicit) RS badge. Others, such as the late philosopher of science Feyerabend, have not been so eager to have their work classed as "implicit rhetoric of science research" (Gaonkar, 39, 74–75; Gross and Keith, 7).

4. Steve Fuller, *Philosophy, Rhetoric, and the End of Knowledge* (Madison: University of Wisconsin Press, 1993), 9.

5. Fuller, *Thomas Kuhn: A Philosophical History.*

6. In his 1998 article Fuller closes with the following statement: "I heartily support the development of IDT [Intelligent Design Theory] as part of this general strategy of converting the dominant political image of knowledge from that of tributaries to that of a delta. In that spirit, IDT's defenders should expect my support in the future." Steve Fuller, "An Intelligent Person's Guide to Intelligent Design Theory," *Rhetoric and Public Affairs* 1, no. 4 (winter 1998).

7. Fuller, *Philosophy, Rhetoric, and the End of Knowledge,* 9.

8. Ibid., xi–17.

9. Ibid., 8.

10. Ten years would be at the rapid end of the spectrum of paradigm shifts. Two examples of such rapid shifts would be Darwin's own revolution (which essentially was complete in the rejection of special creation by the time of his death) and the lightning-fast takeover of plate tectonics over the old geosynclinal paradigm in the 1960s and 1970s.

11. Kuhn, *Structure of Scientific Revolutions,* 94.

12. See Dembski and Richards, *Unapologetic Apologetics,* in which Dembski speaks of "epistemic perspectives" and how humans enjoy an important intellectual ability in this area: "I need to stress . . . that we can and do move quite freely among distinct perspectives. The reason I cannot stress this point strongly enough is because all too frequently in our day we are led to imagine ourselves inescapably imprisoned within our perspectives. Thus we are led to believe that our perspectives are unalterably fixed and that whatever we see is solely a function of the perspective where fate has stuck us." Dembski contends that we are blessed with "mobile minds with which to change our epistemic perspectives. To change our physical perspective we simply have to move our bodies, an activity to which all normal bodies are ideally suited. So too, to change our epistemic perspective, we have to move our minds, an activity to which all normal minds are ideally suited. Of course, we don't typically speak of 'moving our minds.' We have another word for that, and it is called *inquiry.*"

13. Latour and Woolgar, *Laboratory Life;* Steve Woolgar, *Science: The Very Idea* (New York: Routledge, 1988).

14. Woolgar, *Science: The Very Idea,* 87.

15. Ibid., drawing on Knorr-Cetina, *The Manufacture of Knowledge* (Oxford: Pergamon, 1981).

16. In his references section of *Thomas Kuhn: A Philosophical History,* Fuller lists his own works as including five books and thirty-seven scholarly articles (not counting book reviews). All this from a scholar who entered Columbia University as a freshman in 1976, which means in 2003 he was only forty-four years old!

17. Fuller, *Philosophy, Rhetoric, and the End of Knowledge,* 25.

18. Ibid., 26.

19. One can imagine such a process resulting, say, in a sort of "synthetic revision" of Darwinism. The synthetic revision might have the appearance of a "reformed" or "supplemented" Darwinism. Perhaps one scenario would be along the lines of Denton's vision of microevolution as well-confirmed, with macroevolution relegated to a more mysterious, "unsolved" status. I consider this an extremely remote likelihood, but it still is conceivable.

20. Fuller, *Thomas Kuhn: A Philosophical History,* 306.

21. Appreciative of Campbell's labors, Fuller nevertheless posed some caveats. First, Campbell in general has not pursued the "empirical question of whether Darwin's strategy actually worked." In other words, how successful was Darwin? Were Victorian readers of his text really persuaded by Darwin? Did the members of his audience uniformly (or to a sizable degree) even hold the inductivist belief system that Campbell seems to think they did, and which was crucial to his approach? On all three of these questions, Fuller would say an emphatic "no."

Fuller makes a cogent point here. See Nancy Pearcey's "You Guys Lost" in *Mere Creation* (ed. Dembski). Pearcey, following the work of Peter Bowler, is convinced that most scientists imbibed Darwin's evolutionary notions in a vague way (common ancestry, descent-with-modification) while hesitating to embrace natural selection as the key mechanism of that gradual change and adaptation.

Fuller argues that *The Origin of Species,* or any other scientific text, should be evaluated not on its performative virtues but on the actual reader response in the lives of those who read it. What was actually going on in the typical reader's interaction? How often was there a "leisured and learned reader genuinely interested in what the author has to say?" Or "How much time and effort did people give to reading, and with what other social practices was it associated?" Fuller sums up his partial dissent to Campbell's approach: "While I follow rhetoricians of science in believing that the best way to understand texts is in terms of what happens once they get into the hands of readers, I deny, on empirical grounds, that the best way to understand what happens is in terms of the theory of reading most closely associated with classical rhetoric and the humanistic tradition." Steve Fuller, "'Rhetoric of Science': Double the Trouble?" in *Rhetorical Hermeneutics,* ed. Gross and Keith, 290, 292.

Fuller concludes that accurate, empirical rhetoric of science is that which studies closely the real readership of texts. From this procedural stance, key questions are asked: Who actually has read the texts, and in what way and in what context? What purpose, response, or results can be seen [295]? This point is a key contribution of Fuller to rhetorical methodology, and it is one that I vigorously adopted in my approach to rhetorical history of science.

22. John Angus Campbell and Keith Benson, "The Rhetorical Turn in Science Studies," *Quarterly Journal of Speech* 82 (1996).

23. Prelli views rhetoric as "effective expression," which takes into consideration (1) the role of language, (2) the nature of audiences, (3) the nature of situations, (4) criteria for evaluating material for expression, and (5) methods for finding the materials. These points are explicitly or implicitly woven into his theory of RS practice, both in the creation and critical evaluation of scientific rhetoric.

24. Herbert Simons, "Requirements, Problems, and Strategies: A Theory of Persuasion for Social Movements," in *Readings in Rhetorical Criticism,* ed. Carl Burgchardt (State College, Pa.: Strata, 1995); McGuire and Melia, "Some Cautionary Strictures"; Charles Bazerman. *Shaping Written Knowledge: The Genre and Activity of the Experimental Article in Science* (Madison: University of Wisconsin Press, 1988); Peter Dear, *The Literary Structure of Scientific Argument* (Philadelphia: University of Pennsylvania Press, 1991), Marcello Pera and William Shea, eds., *Persuading Science: The Art of Scientific Rhetoric* (Canton, Mass.: Science History Publications/Watson Publishing, 1991).

25. See the entries for John Campbell in the bibliography.

26. Pennock even listed Campbell as a member of Design. See Pennock, *Tower of Babel,* 29, where Campbell described the members of "our movement" in an article in *Origins and Design* on the Mere Creation Conference. Pennock also says that Campbell considered Darwin's *The Origin of Species* as "little more than clever rhetoric." (This phrase is how Pennock erroneously characterized Campbell's paper at a conference at the University of Texas.)

27. Charles Taylor, *Defining Science: A Rhetoric of Demarcation* (Madison: University of Wisconsin Press, 1996), 17–18.

28. I suggest that Taylor is unaware of Design rhetoric in light of his description of the evidence for evolution as "overwhelming." I realize that evolutionary scientists still maintain this estimation of the evidentiary situation, but one would assume that Taylor, as a nonspecialist, could hardly make this firm claim if he were conversant with the arguments of Denton and Johnson, which argue that the *evidence against macroevolution is overwhelming.* Oddly, even though the publication date is 1996, there is no mention of any of the major new Design critiques or their forerunners, such as Denton

(Evolution: A Theory in Crisis), Thaxton, Bradley, and Olsen *(Mystery of Life's Origin)*, or Johnson *(Darwin on Trial)*.

29. Taylor, *Defining Science,* 155–56, 163. Taylor sides with the standard evolutionary defense when creationists expose problems in the fossil record. He says here also that there is a "clear evidential superiority of evolution."

30. Ibid., 174. Taylor's affinity with the strong critique of Fuller against the rhetoric of "big science" is quite clear here.

31. Darwin had to persuade his Victorian English audience—an audience educated to see biological creatures as the marvels of God's handiwork—to consider the radical notion of the creative power of natural selection. Campbell has not only shown how Darwin, a religious agnostic, persuaded theists to view natural selection as part of the creator's handiwork; he has also emphasized the value of "intelligent design" notions (from Paley) as useful to Darwin both in the dialectical development of his evolutionary ideas and in their presentation to the public. This is especially seen in his lead article, "Intelligent Design, Darwinism, and the Philosophy of Public Education," in *Rhetoric and Public Affairs* 1, no. 4 (winter 1998).

32. Ibid., 108. See my brief discussion of Campbell and his work in chapter 10.

33. Campbell, 1988, 488–89. I have broken the long sentence quote into two, for ease of understanding.

34. Nelson and Wells, "Some Things in Biology Don't Make Sense."

35. Condit, "The Rhetoric of Intelligent Design," *Rhetoric and Public Affairs* 1, no. 4 (winter 1998). I was surprised at the problems with scientific understanding in this piece. Two examples will suffice:

First, she says (597) that gaps between species are not clear because "scientists cannot even draw clean boundaries between species." Here Design rhetors "up the ante to the level of phyla." She recognizes that the "size of the gaps at the level of phyla are, of course, apparently larger than those at the species level." Then, she says these gaps are no threat to Darwinism. Why? Condit explains that classification of phyla changes over time as new phyla are discovered, and this classification rests in part on "differences of philosophy among the classifiers themselves." She adds that "even distinctions among phyla require partially artificial judgments of sameness and difference. Therefore, we ought not take the 'differences' among phyla too seriously. If we cannot even agree definitively . . . where the boundaries between these categories lie, it seems rather bizarre to argue that there are 'gaps' between these groupings that are so great as to make evolutionary relationships among them impossible."

Condit invites withering criticism here. She overlooks the fact that the sudden appearance of disparate phyla is not a creationist problem at root but a dogged, frustrating, recalcitrant mystery that has confounded evolutionists since the time of Darwin. Astonishingly, she implies that *all biologists* (oriented toward Design *or* Darwinism) should not worry about the gaps between phyla because they rest on rather subjective human constructions, and the seriousness of such problems may be more apparent than real—generated perhaps by the inherent foibles of language. Here she fails to understand the intrinsic quandary that crossing large morphological gaps poses for *any naturalistic process*, whether it be a classic Darwinian "step-by-tiny-step" scenario or one that proposes punctuated equilibrium. As I have argued, a cogent scenario is needed for a satisfactory and plausible genesis story of any kind. The problem is not primarily with language or human classifiers but rather with the tension between naturalistic scenarios and the recalcitrant patterns in the fossil record.

Second, Condit dismisses Behe's argument, saying that in our experience, apparently designed structures like a natural rock bridge, carved by erosion, were naturally produced. This reply is weak. A cogent natural explanation exists for such a simple structure, with no challenge to nature's probabilistic resources. Also, a natural bridge does not even qualify as irreducibly complex using Behe's "multiple parts" definition. Further confusion is also seen in her assertion: "Similar re-imagining works equally well with all of Behe's so-called irreducibly complex biological machines." She makes this bold assertion, but unfortunately, she fails to explain specifically how this is so. Not one example is given. We are given only hand-waving descriptions. In this vein she also asserts that "we do have evolutionary histories that explain many apparently irreducible complexities." (She refers vaguely to "organs and features" that have been explained.) She seems not to grasp two crucial points. First, only at the molecular level do we reach the weird recalcitrance of complexity that is compared with current high technology. Second, no such structure that has been identified as irreducibly complex has ever had a published testable explanation.

36. David Depew, "Intelligent Design and Irreducible Complexity: A Rejoinder," *Rhetoric and Public Affairs* 1, no. 4 (winter 1998): 571. Depew calls Design scholars "a new breed of scientific creationists." Their arguments rest not only on supposed failures of Darwinian evidences but "more deeply on an overlooked ontological fact about organisms," referring to "irreducible complexity" and Meyer's DNA arguments. Also, Depew and Lyne admit that Nelson and Wells make strong points in refuting Dobzhansky. In the end, Depew joins biochemist Weber in arguing that the "self-organization" of living systems is a cogent alternative to design.

37. Depew, "Intelligent Design and Irreducible Complexity," 571.

38. John Lyne, "Intelligent *Dasein*," *Rhetoric and Public Affairs* 1, no. 4 (winter 1998): 579.

39. Ibid.

40. Ibid. Johnson often quotes prestigious Darwinists who made this point. Why does he emphasize this point so strongly? One important purpose is to make it clear that "theistic evolution" does not fit well with the strong statements by the preeminent leaders of the scientific establishment.

41. Depew, "Intelligent Design and Irreducible Complexity," 573. This difficulty in studying the past, entailing careful reasoning to a cogent explanation, is where Meyer uses "abductive reasoning." An abductive methodology of "inference to the best explanation" would work well even in functional biology, but is very well suited to the difficult terrain of historical biology.

42. Lyne, "Intelligent *Dasein*," 583.

Select Bibliography

Bazerman, Charles. *Shaping Written Knowledge: The Genre and Activity of the Experimental Article in Science.* Madison: University of Wisconsin Press, 1988.

Behe, Michael. "Darwin Under the Microscope." *New York Times* (29 October 1996).

——. *Darwin's Black Box.* New York: Free Press, 1996.

——. *Irreducible Complexity: A Lecture by Michael Behe at Princeton University.* Videotape. Colorado Springs: Access Research Network, 1997.

——. "Intelligent Design Theory as a Tool for Analyzing Biochemical Systems." In *Mere Creation,* edited by William Dembski. Downers Grove, Ill.: InterVarsity, 1998.

——. *Opening Darwin's Black Box: An Interview with Michael Behe.* Videotape. New Port Richey, Fla.: Trinity College and the C. S. Lewis Society, 1998.

——. "Teach Evolution—And Ask Hard Questions." *New York Times,* 17 August 1999.

Berlinski, David. "The Deniable Darwin." *Commentary,* June 1996.

——. "David Berlinski Replies." *Commentary,* September 1996.

Bitzer, Lloyd. "The Rhetorical Situation." In *Readings in Rhetorical Criticism,* edited by Carl Burgchardt. State College, Pa.: Strata, 1995. Originally published in 1968.

Bormann, Ernest. "Fantasy and Rhetorical Vision: The Rhetorical Criticism of Social Reality." In *Readings in Rhetorical Criticism,* edited by Carl Burgchardt. State College, Pa.: Strata, 1995. Originally published in *Quarterly Journal of Speech* 58 (1972): 396–407.

Buell, Jon, and Virginia Hearn, eds. *Darwinism: Science or Philosophy?* Dallas: Foundation for Thought and Ethics, 1993.

Burgchardt, Carl, ed. *Readings in Rhetorical Criticism.* State College, Pa.: Strata, 1995.

Burke, Kenneth. *A Grammar of Motives.* Berkeley: University of California Press, 1945.

——. *A Rhetoric of Motives.* Berkeley: University of California Press, 1950.

————. *Permanence and Change.* 2nd rev. ed. Berkeley: University of California Press, 1954.

————. *The Philosophy of Literary Form.* 3d ed. Berkeley: University of California Press, 1973.

Campbell, John A. "Darwin and *The Origin of Species:* The Rhetorical Ancestry of an Idea." *Speech Monographs* 37 (1970): 1–14.

————. "Charles Darwin and the Crisis of Ecology." *Quarterly Journal of Speech* 60 (1974): 442–49.

————. "Science, Religion, and Emotional Response: A Consideration of Darwin's Affective Decline." *Victorian Studies* 18 (1974): 159–74.

————. "The Polemical Mr. Darwin." *Quarterly Journal of Speech* 61 (1975): 375–90.

————. "Scientific Revolution and the Grammar of Culture: The Case of Darwin's *Origin.*" *Quarterly Journal of Speech* 72 (1986): 351–76.

————. "Charles Darwin: Rhetorician of Science." In *The Rhetoric of the Human Sciences,* edited by J. S. Nelson, A. Megill, and D. N. McCloskey, 69–86. Madison: University of Wisconsin Press, 1987.

————. "On the Way to the *Origin*: Darwin's Evolutionary Insight and Its Rhetorical Transformation." *The Van Zelst Lecture in Communication.* Evanston, Ill.: Northwestern University School of Speech, 1990.

————. "Scientific Discovery and Rhetorical Invention: The Path to Darwin's *Origin.*" In *The Rhetorical Turn,* edited by Herbert Simons. Chicago: University of Chicago Press, 1990.

————. "Of Orchids, Insects, and Natural Theology: Timing, Tactics, and Cultural Critique in Darwin's Post-'*Origin*' Strategy." *Argumentation* 8 (1994): 63–80.

————. "The Cosmic Frame and the Rhetoric of Science: Epistemology and Ethics in Darwin's *Origin.*" *RSQ* 24, nos. 1 and 2 (1995): 27–50.

Campbell, Karlyn Kohrs, and Kathleen Hall Jamieson. "Form and Genre in Rhetorical Criticism: An Introduction." In *Readings in Rhetorical Criticism,* edited by Carl Burgchardt. State College, Pa.: Strata, 1995.

Campbell, Paul. "The *Personae* of Scientific Discourse." *Quarterly Journal of Speech* 61 (December 1975).

Condit, Celeste. "The Rhetoric of Intelligent Design: Alternatives for Science and Religion." *Rhetoric and Public Affairs* 1, no. 4 (winter 1998).

Coyne, Jerry. "God in the Details." Review of *Darwin's Black Box. Nature,* 19 September 1996.

Darwin, Charles. *The Origin of Species.* 6th ed. London: John Murray, 1859.

Davis, P. William, and Dean Kenyon. *Of Pandas and People.* Dallas: Haughton Publishing, 1989, rev. 1993.

Dawkins, Richard. *The Blind Watchmaker.* New York: W. W. Norton, 1985.

————. *Climbing Mount Improbable.* New York: W. W. Norton, 1996.

Dear, Peter. *The Literary Structure of Scientific Argument.* Philadelphia: University of Pennsylvania Press, 1991.

Dembski, William. *The Design Inference*. Cambridge: Cambridge University Press, 1998.

———. *Intelligent Design*. Downers Grove, Ill.: InterVarsity, 1999.

Dembski, William, ed. *Mere Creation*. Downers Grove, Ill.: InterVarsity, 1998.

Dembski, William, and Jay Richards, eds. *Unapologetic Apologetics*. Downers Grove, Ill.: InterVarsity, 2001.

Denton, Michael. *Evolution: A Theory in Crisis*. Bethesda, Md.: Adler and Adler, 1986.

———. *Nature's Destiny: How the Laws of Biology Reveal Purpose in the Universe*. New York: Free Press, 1998.

Depew, David. "Intelligent Design and Irreducible Complexity: A Rejoinder." *Rhetoric and Public Affairs* 1, no. 4 (winter 1998).

Eldredge, Niles. Review of *Evolution: A Theory in Crisis*. *Quarterly Review of Biology* 61, no. 4 (December 1986): 541–42.

Ferris, Timothy. *The Whole Shebang*. New York: Simon and Schuster, 1997.

Fisher, Walter R. *Human Communication as Narration*. Columbia: University of South Carolina Press, 1987.

———. "Narrative Rationality and the Logic of Scientific Discourse." *Argumentation* 8 (1994): 21–32.

———. "Narration as a Human Communication Paradigm: The Case of Public Moral Argument." In *Readings in Rhetorical Criticism,* edited by Carl Burgchardt. State College, Pa.: Strata, 1995.

Fuller, Steve. *Philosophy, Rhetoric, and the End of Knowledge*. Madison: University of Wisconsin Press, 1993.

———. "'Rhetoric of Science': Double the Trouble?" In *Rhetorical Hermeneutics: Invention and Interpretation in the Age of Science,* edited by A. G. Gross and W. M. Keith. Albany, N.Y.: State University of New York Press, 1997.

———. *Science*. Minneapolis: University of Minnesota Press, 1997.

———. "An Intelligent Person's Guide to Intelligent Design Theory." *Rhetoric and Public Affairs* 1, no. 4 (winter 1998).

———. *Thomas Kuhn: A Philosophical History for Our Times*. Chicago: University of Chicago Press, 2000.

Gills, James, and Thomas Woodward, eds. *Darwinism under the Microscope*. Orlando, Fla.: Charisma, 2002.

Glanz, James. "Darwin vs. Design: Evolutionists face a New Opponent." *New York Times,* 8 April 2001.

Gould, Stephen Jay. *Ever Since Darwin*. New York: W. W. Norton, 1980.

———. *Hens' Teeth and Horses' Toes*. New York: W. W. Norton, 1983.

———. *The Panda's Thumb*. New York: W. W. Norton, 1985.

———. *The Flamingo's Smile*. New York: W. W. Norton, 1987.

———. *Wonderful Life*. New York: W. W. Norton, 1989.

———. *Bully for Brontosaurus*. New York: W. W. Norton, 1992.

———. "Impeaching a Self-Appointed Judge." Review of *Darwin on Trial*. *Scientific American* (July 1992).

Gross, Alan. *The Rhetoric of Science*. Cambridge, Mass: Harvard University Press, 1990.

Gross, Alan, and William Keith, eds. *Rhetorical Hermeneutics: Invention and Interpretation in the Age of Science*. Albany, N.Y.: State University Press, 1997.

Gross, Paul. "The Dissent of Man." Review of *Darwin's Black Box*. *Wall Street Journal*, 20 July 1996.

Himmelfarb, Gertrude. *Darwin and the Darwinian Revolution*. Chicago: Elephant Books, 1959.

Hoyle, Fred, and N. Chandra Wickramasinghe. *Lifecloud*. London: J. M. Dent, 1978.

———. *Diseases from Space*. London: J. M. Dent, 1979.

———. *Evolution from Space*. London: J. M. Dent, 1981.

———. *Space Travelers: The Bringers of Life*. Cardiff, England: University of Cardiff Press, 1981.

———. *The Intelligent Universe*. London: J. M. Dent, 1982.

Hull, David. *Darwin and His Critics*. Cambridge: Harvard University Press, 1973.

———. *Philosophy of Biological Science*. Englewood Cliffs, N.J.: Prentice-Hall, 1974.

———. *The Metaphysics of Evolution*. Albany, N.Y.: State University of New York Press, 1989.

———. "The God of Galapagos." Review of *Darwin on Trial*. *Nature* 352 (8 August 1991).

Johnson, Phillip. *Evolution as Dogma: The Establishment of Naturalism*. Richardson, Tex.: Foundation for Thought and Ethics, 1990.

———. "The Religion of the Blind Watchmaker." *Perspectives on Science and Christian Faith* (March 1993): 46–48. Originally published and distributed by Trinity College of Florida, 1992.

———. "Response to Jukes." In the "Random Walking" column. *Journal of Molecular Evolution* 34 (1992): 93–94.

———. *Darwin on Trial*. Rev. ed. Downers Grove, Ill.: InterVarsity, 1993.

———. *Reason in the Balance*. Downers Grove, Ill.: InterVarsity, 1995.

———. "What (If Anything) Hath God Wrought? Academic Freedom and the Religious Professor." *Academe*, September/October 1995.

———. *Testing Darwinism*. London: Inter-Varsity, 1997. Published in the United States as *Defeating Darwinism by Opening Minds*. Downers Grove, Ill.: InterVarsity, 1997.

———. *Objections Sustained*. Downers Grove, Ill.: InterVarsity, 1998.

———. "The Church of Darwin." *Wall Street Journal*, 16 August 1999.

———. *The Wedge of Truth*. Downers Grove, Ill.: InterVarsity, 2000.

———. *The Right Questions*. Downers Grove, Ill.: InterVarsity, 2002.

Jukes, Thomas. "The Persistent Conflict." In the "Random Walking" column. *Journal of Molecular Evolution* 33: 205–6.

Kitcher, Philip. *Abusing Science.* Cambridge, Mass.: MIT Press, 1982.

Kuhn, Thomas. *The Structure of Scientific Revolutions.* 3d ed. Chicago: University of Chicago Press, 1996. Originally published in 1962.

Larson, Edward, and Larry Witham. "Scientists and Religion in America." *Scientific American* (September 1999): 88–93.

Lessl, Thomas. "Intelligent Design: A Look at Some of the Relevant Literature." *Rhetoric and Public Affairs* 1, no. 4 (winter 1998).

Lyne, John. "Intelligent *Dasein.*" *Rhetoric and Public Affairs* 1, no. 4 (winter 1998).

McGuire, J. E., and Trevor Melia. "Some Cautionary Strictures on the Writing of the Rhetoric of Science." *Rhetorica* 7, no. 1 (winter 1989).

———. "The Rhetoric of the Radical Rhetoric of Science." *Rhetorica* 9, no. 4 (autumn 1991).

Miller, Kenneth R. *Finding Darwin's God.* New York: Cliff Street Books, 1999.

Montagu, Ashley, ed. *Science and Creationism.* Oxford: Oxford University Press, 1984.

———. Review of *Evolution: A Theory in Crisis. Chicago Sun-Times,* 20 April 1986.

Moorehead, P. S., and M. M. Kaplan, eds. *Mathematical Challenges to the Neo-Darwinian Interpretation of Evolution.* Philadelphia: Wistar Press, 1967.

Moreland, J. P., ed. *The Creation Hypothesis.* Downers Grove, Ill.: InterVarsity, 1994.

Numbers, Ronald. *The Creationists.* New York: Alfred A. Knopf, 1992.

Payne, David. "Adaptation, Mortification, and Social Reform." *Southern Speech Communication Journal* 51, no. 3 (spring 1986).

Pennock, Robert. *Tower of Babel.* Cambridge, Mass.: MIT Press, 1999.

Pera, Marcello, and William Shea, eds. *Persuading Science: The Art of Scientific Rhetoric.* Canton, Mass.: Science History Publications/Watson Publishing, 1991.

Pitman, Michael. *Adam and Evolution.* London: Century Hutchinson, 1984. Reprint, Grand Rapids: Baker, 1984.

Pomiankowski, Andrew. "A Tiny Mystery." Review of *Darwin's Black Box. New Scientist,* October 1996.

Prelli, Lawrence. *A Rhetoric of Science: Inventing Scientific Discourse.* Columbia: University of South Carolina Press, 1989.

Rhetoric and Public Affairs 1, no. 4 (winter 1998). "A Special Issue on the Intelligent Design Argument." East Lansing: Michigan State University Press.

Ridley, Mark. "More Darwinian Detractors." *Nature* 318 (14 November 1985): 124–25.

Ruse, Michael. *The Darwinian Revolution: Science Red in Tooth and Claw.* Chicago: University of Chicago Press, 1979.

———. "Transcript of Speech by Professor Michael Ruse, Saturday, February 13, 1993." (Available from the National Center for Science Education, Berkeley, Calif.)

———. *The Evolution Wars: A Guide to the Debates.* Piscataway, N.J.: Rutgers University Press, 2001.

Ruse, Michael, ed. *But Is It Science? The Philosophical Question in the Creation/Evolution Controversy.* Buffalo: Prometheus, 1988.

Schoch, Russell. "The Evolution of a Creationist." *California Monthly,* November 1991.

Shapiro, James. Review of *Darwin's Black Box. National Review,* 10 September 1996.

Shapiro, Robert. *Origins: A Skeptic's Guide to the Creation of Life on Earth.* New York: Bantam Books, 1986.

Shields, Donald. "Symbolic Convergence and Special Communication Theories: Sensing and Examining Dis/Enchantment with the Theoretical Robustness of Critical Auto-ethnography." *Communication Monographs* 67, no. 4 (December 2000): 392–421.

Shreeve, James. "The Design of Life." Review of *Darwin's Black Box. New York Times Book Review,* 4 August 1996.

Simons, Herbert. "Are Scientists Rhetors in Disguise?" In *Rhetoric in Transition: Studies in the Nature and Uses of Rhetoric,* edited by Eugene White. University Park, Pa.: Penn State University Press, 1980.

Stafford, Tim. "The Making of a Revolution." *Christianity Today,* 8 December 1997.

Taylor, Charles Alan. *Defining Science: A Rhetoric of Demarcation.* Madison: University of Wisconsin Press, 1996.

Tax, Sol, ed. *Evolution after Darwin.* Three volumes. Chicago: University of Chicago Press, 1960.

Thaxton, Charles. "In Pursuit of Intelligent Causes: Some Historical Background." Unpublished paper, 1988.

———. "DNA, Design, and the Origin of Life." In *Darwinism under the Microscope,* edited by James Gills and Tom Woodward. Orlando, Fla.: Charisma, 2002.

Thaxton, Charles, Walter Bradley, and Roger Olsen. *The Mystery of Life's Origin.* New York: Philosophical Library, 1984.

Toulmin, Stephen. *The Return to Cosmology: Postmodern Science and the Theology of Nature.* Berkeley: University of California Press, 1980.

Vincent, Lynn. "Science vs. Science." *World* 15, no. 8 (26 February 2000).

Watanabe, Teresa. "Enlisting Science to Find the Fingerprints of a Creator." *Los Angeles Times,* 25 March 2001.

Wells, Jonathan. *The Icons of Evolution.* Washington, D.C.: Regnery, 2000.

Wheeler, David. "A Biochemist Urges Darwinists to Acknowledge the Role Played by an 'Intelligent Designer.'" *Chronicle of Higher Education,* 1 November 1996.

Woolgar, Steve. *Science: The Very Idea.* New York: Routledge, 1988. Republished in 1993.

Woodward, Thomas. "Doubts about Darwin." *Moody Magazine,* September 1988.

———. "A Professor Takes Darwin to Court." *Christianity Today,* 15 September 1991.

———. "Meeting Darwin's Wager." *Christianity Today,* 23 April 1997.

Yerxa, Donald. "Phillip Johnson and the Origins of the Intellectual Design Movement, 1977–1991." *Perspectives in Science and Christian Faith* 54, no. 1 (March 2002): 47–52.

Index

Thomas Woodward (Ph.D., University of South Florida) is a professor at Trinity College of Florida, where he teaches the history of science, communication, and systematic theology. He is founder and director of the C. S. Lewis Society and lectures in universities on scientific, apologetic, and religious topics. He has been published in *Moody* magazine as well as *Christianity Today*.

576.82
W9123

108150

LINCOLN CHRISTIAN COLLEGE AND SEMINARY